Experimental Designs: Exercises and Solutions

D. G. Kabe
A. K. Gupta

Experimental Designs: Exercises and Solutions

Springer

D. G. Kabe
Department of Mathematics and Computer Science
Saint Mary's University
Halifax, Nova Scotia
Canada
pkabe@rogers.com

A. K. Gupta
Department of Mathematics and Statistics
Bowling Green State University
Bowling Green, OH 43403
USA
gupta@bgsu.edu

Library of Congress Control Number: 2006925853

ISBN-10: 0-387-33892-6 e-ISBN-10: 0-387-33893-4
ISBN-13: 978-0-387-33892-7 e-ISBN-13: 978-0-387-33893-4

Printed on acid-free paper.

9 8 7 6 5 4 3 2 1

springer.com

Dedicated to the memory of my dear wife Ansuya
DGK

Dedicated to:
Alka, Mita, Nisha, Sharad and Patrick
AKG

CONTENTS

PREFACE

This volume is a collection of exercises with their solutions in Design and Analysis of Experiments. At present there is not a single book which collects such exercises. These exercises have been collected by the authors during the last four decades during their student and teaching years. They should prove useful to graduate students and research workers in Statistics.

In Chapter 1, theoretical results that are needed for understanding the material in this book, are given. Chapter 2 lists the exercises which have been collected by the authors. The solutions of these problems are given in Chapter 3. Finally an index is provided for quick reference.

Grateful appreciation for financial support for Dr. Kabe's research at St. Mary's University is extended to National Research Council of Canada and St. May's University Senate Research Committee. For his visit to the Department of Mathematics and Statistics the authors are thankful to the Bowling Green State University.

The authors are thankful to all the graduate students who contributed to this collection. They would like to thank Keshav Jagannathan for his technical assistance with the computer graphics included here. They are also indebted to Cynthia Patterson for the excellent word processing for the final version of the book. She did a wonderful job not only in this but also in correcting errors and patiently going through the book over and over again, carrying out the changes. We are sure there are still many errors left in this book and of course they are our sole responsibility. Finally, we would like to convey our gratitude to the Springer editor John Kimmel, for his counsel on matters of design, form, and style.

March, 2006

D. G. Kabe
A. K. Gupta

CHAPTER 1

THEORETICAL RESULTS

This chapter is devoted to an exposition of the basic results which are used throughout the book. The material from matrix theory, estimation, testing of hypothesis and design of experiments has been summarized. The proofs have been omitted since these topics have been well studied elsewhere e.g. see John (1998), Anderson and McLean (1974), Montgomery (2005), and Winer (1971), and other references given in the bibliography.

1.1 Linear Estimation and Testing of Hypothesis

1. **Definition.** A *generalized inverse (g-inverse)* of any $n \times m$ matrix A is an $m \times n$ matrix A^- which satisfies the relation $AA^-A = A$.

2. Let A be any $n \times m$ matrix of rank r, and A^- be its any *g-inverse*. Let $H = A^-A$ and $T = AA^-$. Then

   (i) H and T are idempotent matrices
   (ii) rank (H) = rank (T) = rank (A)
   (iii) rank $(I - H) = m - r$ and rank $(I - T) = n - r$.

3. **Definition.** If A is any $n \times m$ matrix, then its *Moore-Penrose inverse* will be defined to be the $m \times n$ matrix A^+ which satisfies the following four conditions:

   (i) $AA^+A = A$
   (ii) $A^+AA^+ = A^+$
   (iii) $(AA^+)' = AA^+$
   (iv) $(A^+A)' = A^+A$

4. Let A be any $n \times n$ symmetric matrix of rank r ($<n$). Suppose $\lambda_1, \lambda_2, \ldots, \lambda_r$ are the non-zero characteristic roots of A and $\xi_1, \xi_2, \ldots, \xi_r$ are the corresponding unit and mutually orthogonal characteristic vectors. Then

   $$A = P \Lambda P',$$
   $$A^+ = P\Lambda^{-1}P',$$

   where $P = [\boldsymbol{\xi}_1, \boldsymbol{\xi}_2, \ldots, \boldsymbol{\xi}_r]$ and $\Lambda = \mathrm{diag}(\lambda_1, \lambda_2, \ldots, \lambda_r)$.

5. Let A be any $n \times m$ matrix. Then

   $$A^+ = (A'A)^+A'.$$

6. The Moore-Penrose inverse of $aI_n + bE_{nn}$, for $a \neq 0$ and $a + nb = 0$ is given by $(\frac{1}{a})I_n + (\frac{b}{a^2})E_{nn}$, where I_n is the $n \times n$ identity matrix and E_{nn} is the $n \times n$ matrix with unit elements.

7. If the equations $A\mathbf{x} = \mathbf{b}$ are consistent, then a general solution of $A\mathbf{x} = \mathbf{b}$ is given by $\mathbf{x} = A^{-}\mathbf{b} + (I - A^{-}A)\mathbf{z}$, where A^{-} is any g-inverse of A and $\mathbf{z}$ any arbitrary vector.

8. Consider the linear model

$$\mathbf{y} = A\theta + \mathbf{e}, \quad \mathcal{E}(\mathbf{e}) = \mathbf{0}, \quad \mathcal{E}(\mathbf{e}\mathbf{e}') = \sigma^2 I_n$$

where $\mathbf{y}$ is an $n \times 1$ vector, A is an $n \times m$ matrix of known numbers, $\boldsymbol{\theta} = (\theta_1, \ldots, \theta_m)'$ is a $m \times 1$ vector of unknown parameters, $\mathbf{e}$ is an $n \times 1$ vector of random errors, and rank $(A) = \text{rank}(A'A) = m$. Then the BLUE of $\boldsymbol{\theta}$ is $\hat{\boldsymbol{\theta}} = (A'A)^{-1}A'\mathbf{y}$ and $\text{var}(\hat{\theta}) = \sigma^2(A'A)^{-1}$. Further, the BLUE of $\mathbf{b}'\boldsymbol{\theta}$ is $\mathbf{b}'\hat{\boldsymbol{\theta}}$ and $\text{var}(\mathbf{b}'\hat{\boldsymbol{\theta}}) = \sigma^2\mathbf{b}(A'A)^{-1}\mathbf{b}$.

9. Consider the linear model

$$\mathbf{y} = A\boldsymbol{\theta} + \mathbf{e}, \quad \mathcal{E}(\mathbf{e}) = \mathbf{0}, \quad \mathcal{E}(\mathbf{e}\mathbf{e}') = \sigma^2 I_n,$$

where $\mathbf{y}$ is an $n \times 1$ vector, A is an $n \times m$ matrix of known numbers, $\boldsymbol{\theta}$ is an $m \times 1$ vector of unknown parameters, and $\mathbf{e}$ is an $n \times 1$ vector of random errors. Further it is assumed that $A'A$ is singular.

(i) A necessary and sufficient condition that $\mathbf{b}'\boldsymbol{\theta}$ is *estimable* is that

$$\text{rank}\,(A') = \text{rank}\,(A', \mathbf{b}),$$
$$\text{or} \quad \text{rank}\,(A'A) = \text{rank}\,(A'A, \mathbf{b}).$$

(ii) If $\mathbf{b}'\boldsymbol{\theta}$ is estimable, then its BLUE is given by $\mathbf{b}'\hat{\boldsymbol{\theta}}$, where $\hat{\boldsymbol{\theta}} = GA'\mathbf{y}$ is any solution of $A'\mathbf{y} = A'A\boldsymbol{\theta}$, and G is any g-inverse of $A'A$. Further, $\text{var}(\mathbf{b}'\hat{\boldsymbol{\theta}}) = \sigma^2\mathbf{b}'G\mathbf{b}$.

(iii) Let $\mathbf{b}'\boldsymbol{\theta}$ and $\mathbf{d}'\boldsymbol{\theta}$ be two estimable parametric functions. Let $\hat{\boldsymbol{\theta}} = GA'\mathbf{y}$ be any solution of $A'\mathbf{y} = A'A\boldsymbol{\theta}$, where G is any g-inverse of $A'A$. Then,

$$\text{cov}(\mathbf{b}'\hat{\boldsymbol{\theta}}, \mathbf{d}'\hat{\boldsymbol{\theta}}) = \sigma^2\mathbf{b}'G\mathbf{d}.$$

(iv) $A'\mathbf{y} = A'A\hat{\boldsymbol{\theta}}$ are called *normal equations*.

(v) $\mathcal{E}(A'\mathbf{y}) = A'A\boldsymbol{\theta}$.

(vi) $\text{var}(A'\mathbf{y}) = \sigma^2 A'A$.

(vii) The *sum of squares due to regression* when $\theta_1, \theta_2, \ldots, \theta_m$ are fitted is given by

$$\text{SSR}(\boldsymbol{\theta}) = \hat{\boldsymbol{\theta}}'A'\mathbf{y}, \quad \text{with r d.f.}$$

where $\hat{\boldsymbol{\theta}}$ is any solution of the normal equations, $A'\mathbf{y} = A'A\hat{\boldsymbol{\theta}}$ and $r = \text{rank}\,(A'A) =$ number of independent normal equations.

(viii) The *error sum of squares* is given by

$$\text{SSE} = \mathbf{y}'\mathbf{y} - \hat{\boldsymbol{\theta}}'A'\mathbf{y} \text{ with } (n - r) \text{ d.f.}$$

(ix) The unbiased estimator of σ^2 is given by

$$\hat{\sigma}^2 = \frac{\mathrm{SSE}}{n - r}.$$

(x) $\mathcal{E}(\mathrm{SSR}(\boldsymbol{\theta})) = \mathcal{E}(\hat{\boldsymbol{\theta}}'A'\mathbf{y}) = r\sigma^2 + \boldsymbol{\theta}'A'A\boldsymbol{\theta}$.

(xi) $\hat{\boldsymbol{\theta}}$ is distributed as $N(GA'A\boldsymbol{\theta}, \sigma^2 GA'AG')$, where $\hat{\boldsymbol{\theta}} = GA'\mathbf{y}$ is any solution of $A'\mathbf{y} = A'A\hat{\boldsymbol{\theta}}$, and G is any g-inverse of A′A.

(xii) $\hat{\boldsymbol{\theta}}$ and SSE are independent.

(xiii) $\frac{\mathrm{SSE}}{\sigma^2}$ is distributed as $\chi^2\,(n - r)$.

(xiv) $\frac{\mathrm{SSR}(\boldsymbol{\theta})}{\sigma^2}$ is distributed independently of SSE and has a non-central chi-square distribution $\chi'^2\left(r, \frac{\boldsymbol{\theta}'A'A\boldsymbol{\theta}}{2\sigma^2}\right)$, where $r = \mathrm{rank}\,(A'A)$.

(xv) $\frac{\mathrm{SSR}(\boldsymbol{\theta})/r}{\mathrm{SSE}/(n - r)}$ follows a non-central F distribution with r and (n − r) d.f. and non-centrality parameter $\frac{\boldsymbol{\theta}'A'A\boldsymbol{\theta}}{2\sigma^2}$.

(xvi) For testing $\mathbf{b}'\boldsymbol{\theta} = \mathbf{b}_0$, where $\mathbf{b}'\boldsymbol{\theta}$ is an estimable function, the t statistic is

$$t = \frac{(\mathbf{b}'\hat{\boldsymbol{\theta}} - b_0)/\sqrt{\mathrm{var}(\mathbf{b}'\hat{\boldsymbol{\theta}})/\sigma^2}}{\sqrt{\mathrm{SSE}/(n - r)}}$$

with (n − r) d.f. or equivalently, the F statistic is

$$F = \frac{(\mathbf{b}'\hat{\boldsymbol{\theta}} - b_0)^2/\mathrm{var}(\mathbf{b}'\hat{\boldsymbol{\theta}})/\sigma^2}{\mathrm{SSE}/(n - r)}$$

with 1 and (n − r) d.f.

(xvii) The $100\,(1 - \alpha)\%$ confidence limits for $\mathbf{b}'\boldsymbol{\theta}$ are given by

$$\mathbf{b}'\hat{\boldsymbol{\theta}} \pm t_{\alpha/2}(n - r)\sqrt{\frac{\mathrm{var}(\mathbf{b}'\hat{\boldsymbol{\theta}})}{\sigma^2} \cdot \frac{\mathrm{SSE}}{(n - r)}}$$

where $t_{\alpha/2}\,(n - r)$ is the value of t distribution with (n − r) d.f. such that the probability of a t value exceeding it is $\frac{\alpha}{2}$.

(xviii) To test $\boldsymbol{\theta} = \mathbf{0}$, the appropriate F statistic is

$$F = \frac{\mathrm{SSR}(\boldsymbol{\theta})/r}{\mathrm{SSE}/(n - r)}$$

with r and (n − r) d.f.

(xix) To test $\boldsymbol{\theta}_2 = \mathbf{0}$, where $\boldsymbol{\theta}' = (\boldsymbol{\theta}_1', \boldsymbol{\theta}_2')$, the appropriate F statistic is given by

$$F = \frac{[\mathrm{SSR}(\boldsymbol{\theta}) - \mathrm{SSR}(\boldsymbol{\theta}_1)]/(r - s)}{\mathrm{SSE}/(n - r)}$$

with $(r - s)$ and $(n - r)$ d.f., where $r = \text{rank}\,(A'A)$ and $s = \text{rank}\,(A_1'A_1)$, $A = (A_1, A_2)$, and A_1 consists of columns corresponding to $\boldsymbol{\theta}_1$.

10. Consider the linear model

$$\mathbf{y} = A\boldsymbol{\theta} + \mathbf{e},$$

where, $\mathbf{y}$ is an $n \times 1$ vector of observations, A is an $n \times m$ matrix of known numbers, $\boldsymbol{\theta}$ is an $m \times 1$ vector of unknown constants, $\mathbf{e}$ is an $n \times 1$ vector of random errors distributed as $N(\mathbf{0}, V)$, with V a positive definite matrix.

(i) A necessary and sufficient condition for the estimability of $\mathbf{b}'\boldsymbol{\theta}$ is that $\text{rank}\,(A'V^{-1}A) = \text{rank}\,(A'V^{-1}A, \mathbf{b})$.

(ii) The BLUE of an estimable parametric function $\mathbf{b}'\boldsymbol{\theta}$ is given by $\mathbf{b}'\hat{\boldsymbol{\theta}}$, where $\hat{\boldsymbol{\theta}}$ is any solution of the normal equations $A'V^{-1}\mathbf{y} = A'V^{-1}A\hat{\boldsymbol{\theta}}$.

(iii) $\text{var}(\mathbf{b}'\hat{\boldsymbol{\theta}}) = \mathbf{b}'G\mathbf{b}$, where $\hat{\boldsymbol{\theta}} = GA'V^{-1}\mathbf{y}$, G being any g-inverse of $A'V^{-1}A$.

(iv) Suppose $(A'A)$ is non-singular. Then

$$\hat{\boldsymbol{\theta}} = (A'V^{-1}A)^{-1}A'V^{-1}\mathbf{y}$$
$$\text{var}(\hat{\boldsymbol{\theta}}) = (A'V^{-1}A)^{-1},$$

and θ is estimable.

(v) $A'V^{-1}\mathbf{y} = A'V^{-1}A\hat{\boldsymbol{\theta}}$ are called *normal equations*.

(vi) $\mathcal{E}(A'V^{-1}\mathbf{y}) = A'V^{-1}A\boldsymbol{\theta}$.

(vii) $\text{var}(A'V^{-1}y) = A'V^{-1}A$.

(viii) $SSR(\boldsymbol{\theta}) = \hat{\boldsymbol{\theta}}'A'V^{-1}\mathbf{y}$, with r d.f.
$SSE = \mathbf{y}'V^{-1}\mathbf{y} - \hat{\boldsymbol{\theta}}'A'V^{-1}\mathbf{y}$, with $(n - r)$ d.f. where $r = \text{rank}\,(A'A)$.

(ix) $\mathcal{E}(SSE) = (n - r)$.

(x) $\mathcal{E}(SSR(\boldsymbol{\theta})) = r + \boldsymbol{\theta}'A'V^{-1}A\boldsymbol{\theta}$.

(xi) $\hat{\boldsymbol{\theta}}$ is distributed as $N(GA'V^{-1}A\boldsymbol{\theta}, GA'V^{-1}AG')$, where $\hat{\boldsymbol{\theta}} = GA'V^{-1}\mathbf{y}$, G being any g-inverse of $A'V^{-1}A$.

(xii) SSE is distributed as $\chi^2\,(n - r)$.

(xiii) $\hat{\boldsymbol{\theta}}$ and SSE are independent.

(xiv) $SSR(\boldsymbol{\theta})$ is distributed independently of SSE and has non-central chi-square distribution $\chi'^2(r, \frac{\boldsymbol{\theta}'A'V^{-1}A\boldsymbol{\theta}}{2})$.

(xv) To test $\mathbf{b}'\boldsymbol{\theta} = b_0$, the t statistic is given by

$$t = \frac{(\mathbf{b}'\hat{\boldsymbol{\theta}} - b_0)/\sqrt{(\text{var}\,\mathbf{b}'\hat{\boldsymbol{\theta}})}}{\sqrt{SSE/(n - r)}}$$

with $(n - r)$ d.f.

(xvi) To test $\boldsymbol{\theta}_2 = \mathbf{0}$, where $\boldsymbol{\theta}' = (\boldsymbol{\theta}_1', \boldsymbol{\theta}_2')$, the F statistic is given by

$$F = \frac{[\hat{\boldsymbol{\theta}}'A'V^{-1}\mathbf{y} - \hat{\boldsymbol{\theta}}_1'A_1'V^{-1}\mathbf{y}]/(r - s)}{SSE/(n - r)}$$

where $A = (A_1, A_2)$, $r = \text{rank}\,(A'A)$, $s = \text{rank}\,(A_1'A_1)$, and A_1 consists of columns corresponding to $\boldsymbol{\theta}_1$.

(xvii) To test $\boldsymbol{\theta} = \mathbf{0}$, we use F statistic, where

$$F = \frac{\hat{\boldsymbol{\theta}}'A'V^{-1}\mathbf{y}/r}{[\mathbf{y}'V^{-1}\mathbf{y} - \hat{\boldsymbol{\theta}}'A'V^{-1}\mathbf{y}]/(n-r)}$$

with r and (n − r) d.f.

1.2 General Properties of Block Designs

11. **Notations**

n = number of plots in a block design
b = number of blocks
v = number of treatments
n_{ij} = number of times the i-th treatment occurs in the j-th block, $i = 1, 2, \ldots, v, j = 1, 2, \ldots, b$
$\mathbf{k}' = (k_1, k_2, \ldots, k_b)$ = the vector of block sizes
$\mathbf{r}' = (r_1, r_2, \ldots, r_v)$ = the vector of numbers of replications of treatments
$K = \text{diag}(k_1, k_2, \ldots, k_b)$
$R = \text{diag}(r_1, r_2, \ldots, r_v)$
$N = (n_{ij})$ = a $v \times b$ matrix, called the *incidence matrix* of the design
E_{pq} = a $p \times q$ matrix with all elements unity
I_t = an identity matrix of order t
$\lambda_{ii}{}'$ = number of times a pair of treatments i' and i' occur together in a block, $i, i' = 1, 2, \ldots, v$
$\lambda_{ii} = r_i, i = 1, 2, \ldots, v$
$\ell_{jj'}$ = number of common treatments between the j-th and j′-th block, $j, j' = 1, 2, \ldots, b$
$\ell_{jj} = k_j, j = 1, 2, \ldots, b$
$\mathbf{B}' = (B_1, B_2, \ldots, B_b)$ = the vector of block totals
$\mathbf{T}' = (T_1, T_2, \ldots, T_v)$ = the vector of treatment totals
G = total yield of all plots
$\mathbf{B}'K^{-1}\mathbf{B} - (G^2/n)$ = unadjusted block sum of squares
$\mathbf{T}'R^{-1}\mathbf{T} - (G^2/n)$ = unadjusted treatment sum of squares
$C = R - NK^{-1}N'$ = the C-matrix of the design
$D = K - N'R^{-1}N$
$\mathbf{Q} = \mathbf{T} - NK^{-1}\mathbf{B}$ = the vector of adjusted treatment totals
$\mathbf{P} = \mathbf{B} - N'R^{-1}\mathbf{T}$ = the vector of adjusted block totals
$|S|$ = determinant of a square matrix S and also the cardinality of a set S.

12. **Relations**

$\sum_{j=1}^{b} n_{ij} = r_i, \sum_{i=1}^{v} n_{ij} = k_j, E_{1b}\mathbf{k} = E_{1v}\mathbf{r} = n$
$NE_{b1} = \mathbf{r}, E_{1v}N = \mathbf{k}', E_{1v}NE_{b1} = n$
$KE_{b1} = \mathbf{k}, RE_{v1} = \mathbf{r}$
$KE_{bb}K = \mathbf{kk}', RE_{vv}R = \mathbf{rr}'$
$\mathbf{k}'K^{-1} = E_{1b}, \mathbf{k}'K^{-1}\mathbf{k} = n$
$\mathbf{r}'R^{-1} = E_{1v}, \mathbf{r}'R^{-1}\mathbf{r} = n$

$E_{1b}\mathbf{B} = E_{1v}I = G$
$E_{1v}\mathbf{Q} = 0, E_{1b}\mathbf{P} = 0$
$E_{1v}C = \mathbf{0}, E_{1b}D = \mathbf{0}$

13. **Definitions**

(i) *Complete block design*: A block design will be called a complete block design if $k_j = v$ for all $j = 1, 2, \ldots, b$.

(ii) *Incomplete block design*: A block design will be called an incomplete block design if $k_j < v$ for $j = 1, 2, \ldots, b$.

(iii) *Regular design*: A block design will be called proper or regular if $k_j = k$ for all $j = 1, 2, \ldots, b$.

(iv) *Equi-replicate design*: A block design will be called equi-replicate if $r_i = r$ for all $i = 1, 2, \ldots, v$.

(v) *Binary design*: A block design will be called binary if $n_{ij} = 0$ or $n_{ij} = 1$ for all $i = 1, 2, \ldots, v$ and $j = 1, 2, \ldots, b$.

14. For binary designs,

$$\sum_{j=1}^{b} n_{ij}n_{i'j} = \lambda_{ii'},\ i, i' = 1, 2, \ldots, v$$

$$\sum_{i=1}^{v} n_{ij}n_{ij'} = \ell_{jj'}, j, j' = 1, 2, \ldots, b$$

$$NN' = (\lambda_{ii'}), N'N = (\ell_{jj'}).$$

15. For any block design,

$$v + \text{rank}\,(D) = b + \text{rank}\,(C)$$

16. **Definitions**

(i) *Treatment contrast*: A linear function $\boldsymbol{\ell}'\mathbf{t} = \ell_1 t_1 + \ell_2 t_2 + \cdots + \ell_v t_v$ is said to be a treatment contrast if $E_{1v}\boldsymbol{\ell} = 0$. It will be said to be an *elementary treatment contrast* if ℓ contains only two non-zero elements $+1$ and -1.

(ii) *Normalized treatment contrast*: A treatment contrast $\boldsymbol{\ell}'\mathbf{t}$, $E_{1v}\boldsymbol{\ell} = 0$, where $\boldsymbol{\ell}'\boldsymbol{\ell} = 1$, will be called a normalized treatment contrast.

(iii) *Connected design*: A design is said to be connected if every elementary treatment contrast is estimable. Otherwise it is called a disconnected design.

Another definition: A design will be called connected if for any two given treatments θ and ϕ, it is possible to construct a chain of treatments $\theta = \theta_0, \theta_1, \ldots, \theta_n = \phi$ such that every consecutive pair of treatments in the chain occurs together in a block.

(iv) *Pairwise balanced design*: A design is said to be a pairwise balanced design of index λ if every pair of treatments occurs together in exactly λ blocks.

(v) *Variance balanced or simply balanced design*: A design will be said to be balanced if the BLUE of every normalized estimable treatment contrast has the same variance.

(vi) *A connected balanced design*: A connected design is balanced if the BLUE of every elementary contrast has the same variance.

(vii) *Orthogonal design*: A design is said to be orthogonal if the BLUE of every estimable treatment contrast is uncorrelated with the BLUE of every estimable block contrast.

(viii) *Efficiencey of a design*: Let $\overline{V}$ denote the average variance of the intrablock BLUEs of estimable elementary treatment contrasts in a given design and $\overline{V}_R$ denote the corresponding quantity in a randomized block design with the same number of treatments and experimental blocks, $\overline{V}$ and $\overline{V}_R$ being computed on the assumption that the intrablock error variance σ^2 per plot remains the same in both designs. Then the efficiency E of the given design is defined as

$$E = \frac{\overline{V}_R}{\overline{V}} = \frac{2\sigma^2}{\bar{r}\overline{V}},$$

where $\bar{r} = \sum_{1}^{v} r_i/v$ = average number of replications of treatments in the given design.

17. **Results about Connectedness, Balancedness and Orthogonality**

(i) A necessary and sufficient condition for $\boldsymbol{\ell}'\mathbf{t}$ to be estimable is that

$$\text{rank}\,(C) = \text{rank}\,(C, \boldsymbol{\ell}).$$

(ii) A necessary conditon for $\boldsymbol{\ell}'\mathbf{t}$ to be estimable is that

$$E_{1v}\boldsymbol{\ell} = 0.$$

(iii) A necessary and sufficient condition for a design to be connected is that

$$\text{rank}\,(C) = v - 1.$$

(iv) In a block design, the number of estimable linearly independent treatment contrasts is equal to the rank of the matrix C.

(v) The average variance of the BLUEs of elementary treatment contrasts in a connected design is

$$\frac{2\sigma^2}{(v-1)} \sum_{i=1}^{v-1} (1/\theta_i) = \frac{2\sigma^2}{H},$$

where $\theta_1, \theta_2, \ldots, \theta_{v-1}$ are the $(v-1)$ non-zero characteristic roots of the matrix C and H is their harmonic mean.

(vi) A necessary and sufficient condition for a connected design to be balanced is that all the $(v-1)$ non-zero characteristic roots of the matrix C are equal.

(vii) A necessary and sufficient condition for a design to be a connected balanced design is that its C matrix is given by

$$C = a[I_v - (1/v)E_{vv}],$$

for some positive constant a.

(viii) The C matrix of a connected balanced design is given by

$$C = \theta[I_v - (1/v)E_{vv}],$$

where θ is the non-zero characteristic root of the matrix C.

(ix) A necessary and sufficient condition for a disconnected design to be balanced is that the non-zero characteristic roots of the matrix C are all equal.

(x) For a binary design, tr $C = n - b$.

(xi) The efficiency E of any binary connected design satisfies the inequality

$$E \leq \frac{v(n-b)}{n(v-1)}.$$

(xii) The efficiency E of a binary connected balanced design is given by

$$E = \frac{v(n-b)}{n(v-1)}.$$

(xiii) For a balanced equi-replicated incomplete block design, $b \geq v$.

(xiv) A necessary and sufficient condition for a design to be orthogonal is that $CR^{-1}N = 0$ or equivalently $DK^{-1}N' = 0$.

(xv) A necessary and sufficient condition for a connected design to be orthogonal is that $n_{ij} = r_i k_j/n, i = 1, 2, \ldots, v, j = 1, 2, \ldots, b$.

18. **Results about Intrablock Analysis**

(i) *Assumptions*: Let y_{xij} be the yield of the x-th plot among the n_{ij} plots of the j-th block to which i-th treatment is applied, $x = 0, 1, \ldots, n_{ij}$, $i = 1, 2, \ldots, v$ and $j = 1, 2, \ldots, b$. We assume,

$$y_{xij} = \mu + \alpha_j + t_i + e_{xij}, \quad x = 0, 1, \ldots, n_{ij}$$
$$i = 1, 2, \ldots, v, \quad j = 1, 2, \ldots, b$$

where μ, α_j and t_i represent respectively the general mean effect, the effect of the j-th block and the effect of the i-th treatment and e's are independent random errors normally distributed with mean 0 and variance σ^2. The effects μ, α's and t's are assumed to be fixed effects.

(ii) *Normal Equations*:

$$\begin{bmatrix} G \\ \mathbf{B} \\ \mathbf{T} \end{bmatrix} = \begin{bmatrix} nI_1 & k' & \mathbf{r}' \\ \mathbf{k} & K & N' \\ \mathbf{r} & N & R \end{bmatrix} \begin{bmatrix} \hat{\mu} \\ \hat{\alpha} \\ \hat{\mathbf{t}} \end{bmatrix}$$

(iii) *Reduced Normal Equations*:

$$\mathbf{Q} = \mathrm{C}\hat{\mathbf{t}} \qquad \mathbf{P} = \mathrm{D}\hat{\boldsymbol{\alpha}}.$$

(iv) *A set of solutions*:

$$\hat{\mu} = G/n,$$
$$\hat{\boldsymbol{\alpha}} = K^{-1}\mathbf{B} - (G/n)E_{b1} - K^{-1}N'\hat{\mathbf{t}},$$
$$\mathbf{Q} = C\hat{\mathbf{t}}.$$

(v) *Error Sum of Squares*:

$$SSE = (\mathbf{y}'\mathbf{y} - G^2/n) - (\mathbf{B}'K^{-1}\mathbf{B} - G^2/n) - \hat{\mathbf{t}}'\mathbf{Q},$$

with d.f. $(n - b - v + g)$, where $\text{rank}(C) = v - g$.

(vi) *Adjusted Treatment Sum of Squares*:

$$SST(ad_j) = \hat{\mathbf{t}}'\mathbf{Q}, \qquad \text{with d.f.}(v - g).$$

(vii) *Adjusted Block Sum of Squares*:

$$SSB(ad_j.) = (\mathbf{y}'\mathbf{y} - C^2/n) - (\mathbf{T}R^{-1}\mathbf{T} - G^2/n) - SSE$$

with d.f. $(b - g)$.

(viii) *F-tests*:

For testing the significance of treatment differences, the F statistic is given by

$$F = \frac{SST(adj.)/(v - g)}{SSE/(n - b - v + g)}$$

with d.f. $(v - g)$ and $(n - b - v + g)$. For testing significance of block differences, the F statistic is given by

$$F = \frac{SSB(adj.)/(b - g)}{SSE/(n - b - v + g)}$$

with d.f. $(b - g)$ and $(n - b - v + g)$.

(ix) *Variance and Covariances*:

$$\mathcal{E}\begin{bmatrix} G \\ \mathbf{B} \\ \mathbf{T} \end{bmatrix} = \begin{bmatrix} nI_1 & \mathbf{k}' & \mathbf{r}' \\ \mathbf{k} & K & N' \\ \mathbf{r} & N & R \end{bmatrix} \begin{bmatrix} \mu \\ \boldsymbol{\alpha} \\ \mathbf{t} \end{bmatrix}$$

$$\text{var}\begin{bmatrix} G \\ \mathbf{B} \\ \mathbf{T} \end{bmatrix} = \sigma^2 \begin{bmatrix} nI_1 & \mathbf{k}' & \mathbf{r}' \\ \mathbf{k} & K & N' \\ \mathbf{r} & N & R \end{bmatrix}$$

$$\text{var}\,(\mathbf{Q}) = \sigma^2 C$$
$$\text{var}\,(\mathbf{P}) = \sigma^2 D$$
$$\text{cov}\,(\mathbf{Q}, \mathbf{P}) = -\sigma^2 CR^{-1}N = -\sigma^2 NK^{-1}D.$$

(x) *Method For Solving* $\mathbf{Q} = C\hat{t}$:

Suppose rank $C = v - g$. Take a $g \times v$ matrix H such that $H\mathbf{t} = 0$ and rank $[C, H] = v$, and rank $(H) = g$. Then, there exists a $v \times g$ matrix M such that $CM = 0$, and rank $(M) = g$. Then

(a) $(C + H'H)^{-1}$ is a g-inverse of C, and
(b) $(C + H'H)^{-1} - M(M'H'HM)^{-1}M'$ is a g-inverse of C

(xi) *The BLUE of an Estimable Treatment Contrast*:

The BLUE of an estimable treatment contrast $\boldsymbol{\ell}'\mathbf{t}$ is given by

$$\boldsymbol{\ell}'\hat{\mathbf{t}} = \boldsymbol{\ell}'(C + H'H)^{-1}\mathbf{Q}$$

and

$$\text{var}(\boldsymbol{\ell}'\hat{\mathbf{t}}) = \sigma^2\boldsymbol{\ell}'[(C + H'H)^{-1} - M(M'H'HM)^{-1}M']\boldsymbol{\ell}$$

(xii) *Expected Values of Sum of Squares In Intrablock Analysis of Variance:*

(a) Expected Value of Total SS:

$$\mathcal{E}[\mathbf{y}'\mathbf{y} - G^2/n] = (n-1)\sigma^2 + \boldsymbol{\alpha}'K\boldsymbol{\alpha} + \mathbf{t}'\mathbf{R}\mathbf{t} + 2\mathbf{t}'N\boldsymbol{\alpha} - n^{-1}(\mathbf{k}'\boldsymbol{\alpha} + \mathbf{r}'\mathbf{t})^2$$

(b) Expected Value of Unadjusted Block SS:

$$\mathcal{E}[\mathbf{B}'K^{-1}\mathbf{B} - G^2/n] = (b-1)\sigma^2 + \boldsymbol{\alpha}'K\boldsymbol{\alpha} + 2\mathbf{t}'N\boldsymbol{\alpha} + \mathbf{t}'NK^{-1}N'\mathbf{t} - n^{-1}(\mathbf{k}'\boldsymbol{\alpha} + \mathbf{r}'\mathbf{t})^2.$$

(c) Expected Value of Adjusted Block SS:

$$\mathcal{E}[\hat{\boldsymbol{\alpha}}'\mathbf{P}] = (b-g)\sigma^2 + \boldsymbol{\alpha}'D\boldsymbol{\alpha},$$

where $g = v - \text{rank}(C)$.

(d) Expected Value of Unadjusted Treatment SS:

$$\mathcal{E}[\hat{\mathbf{t}}'\mathbf{Q}] = (v-g)\sigma^2 + \mathbf{t}'C\boldsymbol{\alpha},$$

(e) Expected Value of Intrablock Error SS:

$$\mathcal{E}[\text{Intrablock Error SS}] = (n-b-v+g)\sigma^2.$$

19. **Results In the Analysis with Recovery of Interblock Information**

(i) *Interblock Treatment Estimates*: We assume that blocks are of the same size k, and the design is connected

$$\mathcal{E}(B_j) = k\mu + \sum_{i=1}^{v} n_{ij}t_i$$

$$\text{var}(B_j) = k(\sigma_e^2 + k\sigma_b^2),\ j = 1, 2, \ldots, b$$

$$\text{cov}(B_j, B_j') = 0,\ j \neq j' = 1, 2, \ldots, b.$$

The normal equations are given by

$$\begin{bmatrix} kG \\ \mathbf{NB} \end{bmatrix} = \begin{bmatrix} bk^2 & k\mathbf{r}' \\ k\mathbf{r} & NN' \end{bmatrix} \begin{bmatrix} \tilde{\mu} \\ \tilde{\mathbf{t}} \end{bmatrix}.$$

The reduced normal equations for $\mathbf{t}$ are $\mathbf{Q}_1 = C_1\tilde{\mathbf{t}}$, where $\mathbf{Q}_1 = (1/k)\mathbf{NB} - (G/bk)\mathbf{r}$ and $C_1 = (1/k)NN' - (1/bk)\mathbf{rr}'$.

The solutions of $\mathbf{t}$ obtained by solving the above normal equations are called *interblock treatment estimates.*

Let $w_2{}' = \sigma_{e'}^2 + k\sigma_b^2$, $w_2 = 1/w_2{}'$, $w_1{}' = \sigma_e^2$, $w_1 = 1/w_1'$. Then from the above normal equations, we get

$$\mathcal{E}\begin{bmatrix} kG \\ \mathbf{NB} \end{bmatrix} = \begin{bmatrix} bk^2 & k\mathbf{r}' \\ k\mathbf{r} & NN' \end{bmatrix} \begin{bmatrix} \mu \\ \mathbf{t} \end{bmatrix}$$

$$\text{var}\begin{bmatrix} kG \\ \mathbf{NB} \end{bmatrix} = w_2' \begin{bmatrix} bk^2 & k\mathbf{r}' \\ k\mathbf{r} & NN' \end{bmatrix}.$$

If $\boldsymbol{\ell}'\mathbf{t}$ is any estimable treatment contrast, then $\boldsymbol{\ell}'\tilde{\mathbf{t}}$ is called its *interblock estimator*, where $\tilde{\mathbf{t}}$ is any solution of the above normal equations.

(ii) *With Recovery of Interblock Information*:

Here, the normal equations are

$$\begin{bmatrix} w_2G \\ w_1\mathbf{Q} + (w_2/k)\mathbf{NB} \end{bmatrix} = \begin{bmatrix} w_2bk & w_2\mathbf{r}' \\ w_2\mathbf{r} & w_1C + (w_2/k)NN' \end{bmatrix} \begin{bmatrix} \overset{*}{\mu} \\ \overset{*}{t} \end{bmatrix}.$$

The reduced normal equations for $\mathbf{t}$ are

$$w_1\mathbf{Q} + w_2\mathbf{Q}_1 = (w_1C + w_2C_1)\overset{*}{\mathbf{t}}.$$

The solutions of $\mathbf{t}$ obtained by solving the above normal equations are known as the combined intra and interblock treatment estimates. From the above normal equations, we obtain

$$\mathcal{E}\begin{bmatrix} w_2G \\ w_1\mathbf{Q} + (w_2/k)\mathbf{NB} \end{bmatrix} = \begin{bmatrix} w_2bk & w_2\mathbf{r}' \\ w_2\mathbf{r} & w_1C + (w_2/k)NN' \end{bmatrix} \begin{bmatrix} \mu \\ t \end{bmatrix}$$

$$\text{var}\begin{bmatrix} w_2G \\ w_1\mathbf{Q} + (w_2/k)\mathbf{NB} \end{bmatrix} = \begin{bmatrix} w_2bk & w_2\mathbf{r}' \\ w_2\mathbf{r} & w_1C + (w_2/k)NN' \end{bmatrix}.$$

(iii) *The F test*:

The F statistic for testing $\mathbf{t} = \mathbf{0}$ is given by

$$F = \frac{\overset{*}{\mathbf{t}}'(w_1\mathbf{Q} + w_2k^{-1}\mathbf{NB})/(v-1)}{w_1E_e}$$

with $(v - 1)$ and $(bk - b - v + 1)$ d.f. where E_e = mean intrablock error SS. However to apply the above F test, we must know w_1 and w_2. Since w_1 and w_2 are unknown, we use their estimates to calculate the above F statistic. The estimates of w_1 and w_2 are obtained as

$$\hat{w}_1 = 1/E_e,\ \hat{w}_2 = (bk - v)/[k(b-1)E_b - (v-k)E_e]$$

where E_b = mean adjusted intrablock SS. For a connected resolvable design, the weights w_1 and w_2 are estimated by

$$\hat{w}_1 = 1/E_e,\ \hat{w}_2 = (r-1)/(rE_b - E_e),$$

where rE_b = mean adjusted intrablock with replications block SS.

(iv) *Variance and Covariances*

$$\text{var}(\mathbf{Q}) = \sigma_e^2 C = C/w_1$$
$$\text{var}(\mathbf{Q}_1) = C_1/w_2$$
$$\text{cov}(G, K_1) = 0$$
$$\text{var}(w_1\mathbf{Q} + w_2\mathbf{Q}_1) = w_1 C + w_2 C_1$$
$$\text{cov}(\mathbf{Q}, \mathbf{Q}_1) = \mathbf{0}$$
$$\text{cov}(G, w_1\mathbf{Q} + w_2\mathbf{Q}_1) = 0.$$

(v) *Expected Values of Sums of Squares in the Analysis with Recovery of Interblock Information*:

$$\mathcal{E}(\text{Total SS}) = (bk-1)\sigma_e^2 + k(b-1)\sigma_b^2 + \mathbf{t}'\mathbf{R}\mathbf{t} - (bk)^{-1}(\mathbf{r}'\mathbf{t})^2$$
$$\mathcal{E}(\text{unadj. Treatment SS}) = (v-1)\sigma_e^2 + (v-k)\sigma_b^2 + \mathbf{t}'\mathbf{R}\mathbf{t} - (bk)^{-1}(\mathbf{r}'\mathbf{t})^2$$
$$\mathcal{E}(\text{unadj. BlockSS}) = (b-1)(\sigma_e^2 + k\sigma_b^2) - \mathbf{t}'\mathbf{C}\mathbf{t} + \mathbf{t}'\mathbf{R}\mathbf{t} - (bk)^{-1}(\mathbf{r}'\mathbf{t})^2$$
$$\mathcal{E}(\text{adj. Treatment SS}) = (v-1)\sigma_e^2 + \mathbf{t}'\mathbf{C}\mathbf{t}$$
$$\mathcal{E}(\text{Intrablock Error SS}) = \sigma_e^2(bk - b - v + 1)$$
$$\mathcal{E}(\text{adj. Block SS}) = (b-1)\sigma_e^2 + (bk-v)\sigma_b^2.$$

1.3 Standard Designs

20. **For a RBD with** b **Blocks and** v **Treatments**

(i) $N = E_{vb},\ C = b[I_v - \frac{1}{v}E_{vv}]$

(ii) $D = v[I_b - \frac{1}{b}E_{bb}]$

(iii) $\mathbf{Q} = I - \frac{G}{v}E_{v1}$

(iv) A solution of $\mathbf{Q} = C\hat{\mathbf{t}}$ is taken as $\hat{\mathbf{t}} = \mathbf{Q}/b$

(v) rank(C) = v − 1. The characteristic roots of C are o with multiplicity 1, and b with multiplicity (v − 1).

(vi) The BLUE of $t_i - t_j$ is given by $(Q_i - Q_j)/b = (T_i - T_j)/b$ with variance $2\sigma^2/b$.

(vii) The design is connected, balanced and orthogonal.

21. **BIBD**

An incomplete block design with v treatments in b blocks of k plots each is called a BIBD if (i) each treatment occurs at most once in a block, (ii) each treatment occurs in exactly r blocks and (iii) every pair of treatments occurs in exactly λ blocks. For a BIBD (c, b, r, k, λ), we have the following results:

(i) $NN' = (r - \lambda)I_v + \lambda E_{vv}$

(ii) $|NN'| = rk(r - \lambda)^{v-1}$

(iii) rank (N) = rank (NN′) = rank (N′N) = v

(iv) bk = vr, r(k − 1) = λ(v − 1), b ≥ v (Fisher's inequality)

(v) $(NN')^{-1} = \frac{1}{(r-\lambda)}I_v - \frac{\lambda}{(r-\lambda)rk}E_{vv}$

(vi) $C = \frac{\lambda v}{k}I_v - \frac{\lambda}{k}E_{vv}$, rank (C) = v − 1

(vii) The characteristic roots of C are o with multiplicity 1 and λv/k with multiplicity (v − 1).

(viii) A solution of $\mathbf{Q} = C\hat{\mathbf{t}}$ is given by $\hat{\mathbf{t}} = k\mathbf{Q}/\lambda v$. The adj. treatment $SS = k\mathbf{Q}'\mathbf{Q}/\lambda v$.

(ix) $\mathbf{Q} = \mathbf{T} - \frac{1}{k}N\mathbf{B}$, $Q_i = T_i - \frac{1}{k}B_{(i)}$, where $B_{(i)}$ = sum of block totals in which the i-th treatment occurs.

(x) The BLUE of $(t_i - t_j)$ is $k(Q_i - Q_j)/\lambda v$, with variance $2\sigma^2 k/\lambda v$.

(xi) The efficiency factor E is given by $E = \frac{\lambda v}{kr}$.

(xii) For a SBIBD (b = v), N′N = NN′. Hence, there are λ treatments common between any two blocks of a SBIBD.

(xiii) For a SBIBD with even number of treatments (r − λ) is a perfect square.

(xiv) A BIBD is called *resolvable* if the blocks can be arranged in r sets such that every treatment occurs exactly once in each set. For a resolvable BIBD, b ≥ v + r − 1. If b = v + r − 1, then the resolvable BIBD is called *affine*. In an *affine resolvable* BIBD, k^2/v is an integer, and two blocks in different replications contain k^2/v treatments in common.

(xv) The intrablock analysis of variance table is as follows:

SOURCE	SS	d.f.	SS	SOURCE
Blocks (unadj.)	$\frac{1}{k}\mathbf{B}'\mathbf{B} - \frac{G^2}{bk}$	b − 1	†	Blocks (adj.)
Treatments (adj.)	$\frac{k}{\lambda v}\mathbf{Q}'\mathbf{Q}$	v − 1	$\frac{1}{r}\mathbf{T}'\mathbf{T} - \frac{G^2}{bk}$	Treatments (unadj.)
Error (intrablock)	†	bk − b − v + 1	→	Error (intrablock)
Total	$\mathbf{y}'\mathbf{y} - \frac{G^2}{bk}$	bk − 1	→	Total

Note: † means obtained by subtraction; → means carried forward.

(xvi) The expected values of sums of squares of the intrablock analysis of variance table is given below:

SOURCE	SS	$\mathcal{E}$(SS)	$\mathcal{E}$(SS)	SS	SOURCE
Blocks (unadj.)	$\frac{1}{k}\mathbf{B}'\mathbf{B} - \frac{G^2}{bk}$	$(b-1)\sigma_b^2 + k\boldsymbol{\alpha}'\boldsymbol{\alpha} + 2\mathbf{t}'N\boldsymbol{\alpha} + K^{-1}\mathbf{t}'NN'\mathbf{t} - (bk)^{-1}(k + \mathbf{r}\Sigma\mathbf{t})^2$	$(b-1)\sigma_b^2 + \boldsymbol{\alpha}'D\boldsymbol{\alpha}$	$(b-1)E_b$	Blocks (adj.)
Treatments (adj)	$\frac{k}{\lambda v}\mathbf{Q}'\mathbf{Q}$	$(v-1)\sigma^2 + \mathbf{t}'C\mathbf{t}$	$(v-1)\sigma^2 + r\mathbf{t}'\mathbf{t} + 2\mathbf{t}'N\boldsymbol{\alpha} + r^{-1}\boldsymbol{\alpha}'N'N\boldsymbol{\alpha} - (bk)^{-1}(k\Sigma\boldsymbol{\alpha} + \mathbf{r}\Sigma\mathbf{t})^2$	$\frac{1}{r}T'T - \frac{G^2}{bk}$	Treatments (adj.)
Error	$(bk-b-v+1)E_e$	$(bk-b-v+1)\sigma^2$	$(bk-b-v+1)\sigma^2$	$(bk-b-v+1)E_e$	Error (interblock)

Note: E_b and E_e are respectively the mean intrablock adjusted block SS and the mean intrablock error SS.

(xvii) *Analysis with Recovery of Interblock Information*:

(a) The equations for combined intra and interblock treatment estimates are given by

$$w_1\mathbf{Q} + w_2\mathbf{Q}_1 = (w_1C + w_2C_1)\overset{*}{\mathbf{t}}$$

where $\mathbf{Q} = \mathbf{T} - \frac{1}{k}N\mathbf{B}$, $\mathbf{Q}_1 = \frac{1}{k}N\mathbf{B} - \frac{G}{v}E_{v1}$, $C = \frac{\lambda v}{k}[I_v - \frac{1}{v}E_{vv}]$, $C_1 = \frac{r-\lambda}{vk}[I_v - \frac{1}{v}E_{vv}]$, $w_1 = 1/\sigma_e^2$, and $w_2 = 1/\sigma_b^2$.

The combined intra and interblock treatment estimates are given by

$$\overset{*}{\mathbf{t}} = \frac{k[w_1Q + w_2Q_1]}{w_1\lambda v + (r-\lambda)w_2}$$

and its variance by var $(\mathbf{t}^*) = \dfrac{k[I_v - \frac{1}{v}E_{vv}]}{[w_1\lambda v + (r-\lambda)w_2]}$.

The combined intra and interblock BLUE of a treatment contrast $\boldsymbol{\ell}'\mathbf{t}$ is given by

$$\boldsymbol{\ell}'\mathbf{t}^* = \frac{\boldsymbol{\ell}'[kw_1\mathbf{T} - (w_1 - w_2)N\mathbf{B}]}{w_1\lambda v + (r-\lambda)w_2}$$

and its variance is var $(\boldsymbol{\ell}'\mathbf{t}^*) = \dfrac{k\boldsymbol{\ell}'\boldsymbol{\ell}}{w_1\lambda v + (r-\lambda)w_2}$.

(b) The approximate F statistic for testing the significance of treatment differences is given by

$$F = \frac{\frac{1}{r}\sum_i (T_i + \theta w_i)^2 - \frac{G^2}{bk}/(v-1)}{E'_e},$$

with (v − 1) and (bk − b − v + 1) d.f. and where

$$\theta = \frac{(b-1)(E_b - E_e)}{v(k-1)(b-1)E_b + (b-v)(v-k)E_b}$$

$$w_i = (v-k)T_i - (v-1)B_{(i)} + (k-1)G,\ i = 1, 2, \ldots, v$$

$$E'_e = [1 + (v-k)\theta]E_e$$

E_e = mean intrablock error SS,
E_b = mean intrablock adjusted block SS,
T_i = total yield for the i-th treatment,
$B_{(i)}$ = Sum of blocks in which the i-th treatment occurs.
If $E_b < E_e$, then θ is taken to be zero.
The standard error of $\hat{t}_i - \hat{t}_j$ is given by S.E. $(\hat{t}_i - \hat{t}_j) = 2E'_e/r$.

(c) If the design is resolvable BIBD, then the approximate F statistic for testing the significance of treatment differences is given by

$$F = \frac{\left[\frac{1}{r}\sum_i (T_i + \theta W_i)^2 - \frac{G^2}{bk}\right] \Big/ (v-1)}{E'_e}$$

with (v − 1) and (bk − b − v + 1) d.f. and where

$$\theta = \frac{r(E_b - E_e)}{vr(k-1)E_b + k(b-v-k+1)E_e}$$

$E'_e = [1 + (v-k)\theta]E_e$
E_e = mean intrablock error SS
E_b = mean intrablock adjusted block SS within replications.

22. PBIBD

(i) **Definition.** Given v treatments 1, 2, ..., v, a relation satisfying the following conditions is said to be an *m*-class *association scheme*:

(a) Any two treatments are either 1st, 2nd, ..., or m-th associates, the relation of association being symmetrical, i.e., if the treatment α is the i-th association of the treatment β, then the treatment β is the i-th associate of the treatment α.

(b) Each treatment α has n_i i-th associates, the number n_i being independent of the treatment α, i = 1, 2, ..., m.

(c) If the treatments α and β are i-th associates, then the number of treatments which are both the j-th associates of α and the k-th associates of β is $p_{jk}{}^{i}$ and is independent of the pair of the i-th associates α and β, $i, j, k = 1, 2, \ldots, m$.

The numbers $v, n_i, p_{jk}{}^{i}, i, j, k = 1, 2, \ldots, m$ are called the parameters of the m-class association scheme. The parameters $p_{jk}{}^{i}$ are written in the form of m matrices $P_i, i = 1, 2, \ldots, m$, which has $p_{jk}{}^{i}$ as the element in the j-th row and the k-th column.

Definition. An arrangement of v treatments in b blocks of $k\ (< v)$ plots each will be called a *partially balanced incomplete block design* (PBIBD) with m-associate classes if

(a) the v treatments satisfy an m-class association scheme;
(b) each block contains k different treatments,
(c) each treatment occurs in exactly r blocks;
(d) if two treatments α and β are i-th associates, then they occur together in exactly λ_i blocks, the number λ_i being independent of the pair of the i-th associates α and β, $i = 1, 2, \ldots, m$.

The parameters $v, n_i, p_{jk}{}^{i}, i, j, k = 1, 2, \ldots, m$ are called the *parameters of the first kind*, while the parameters b, r, k, λ_i, $i = 1, 2, \ldots, m$ are called the *parameters of the second kind.*

Definition. The m $v \times v$ matrices $B_i = (b^{i}_{\alpha\beta}), i = 1, 2, \ldots, m$, where

$$
\begin{aligned}
b^{i}_{\alpha\beta} &= (\alpha, \beta)\text{-th element of } B_i \\
&= 1, \text{ if } \alpha \text{ and } \beta \text{ are i-th associates,} \\
&= 0, \text{ otherwise}
\end{aligned}
$$

are called *association matrices*. Further, we define

$$
\begin{aligned}
B_0 &= I_v, n_o = 1, \lambda_o = r \\
p^{o}_{ij} &= n_i, \text{ if } j = i \\
&= 0, \text{ if } j \neq i \\
pok^{i} &= 1, \text{ if } i = k \\
&= 0, \text{ if } i \neq k.
\end{aligned}
$$

(i) *Properties*:

(a) $\sum_{i=0}^{m} B_i = E_{vv}$

$$B_jB_k = \sum_{i=0}^{m} p_{jk}{}^{i}B_i, \; j, k = 0, 1, 2, \ldots, m$$

$$\sum_{i=0}^{m} c_iB_i = o_{vv}, \text{ iff } c_i = 0, i = 0, 1, 2, \ldots, m, \ldots, m$$

where c_i are scalar numbers, and $B_jB_k = B_kB_j, j, k = 0, 1, 2, \ldots, m$.

(b) If N is the incidence matrix of a PBIBD, then

$$NN' = \sum_{i=0}^{m} \lambda_i B_i.$$

(c) If $v, b, r, k, n_i, \lambda_i, i = 0, 1, 2, \ldots, m$ are the parameters of a PBIBD, then

$$vr = bk, \sum_{i=0}^{m} n_i = v, \sum_{i=0}^{m} n_i \lambda_i = rk.$$

(d) The parameters $p_{jk}{}^{i}$, $i, j, k = 0, 1, 2, \ldots m$ of a PBIBD satisfy the following relations:

$$p_{jk}{}^{i} = p_{kj}{}^{i}, \sum_{k=0}^{m} p_{jk}{}^{i} = n_j, \text{ and } n_i p_{jk}{}^{i} = n_j p_{ik}{}^{j} = n_k p_{ji}{}^{k}.$$

(e) Let $S_i(t_\alpha)$ denote the sum of treatments which are i-th associates of the treatment $t_\alpha, i = 0, 1, 2, \ldots, m$, that is, $S_i(t_\alpha) = \sum_{u=1}^{v} b_{\alpha u} t_u$. Further, let $S_j S_i(t_\alpha)$ denotes the sum of treatments which are j-th associates of the i-th associates of t_α, that is, $S_j S_i(t_\alpha) = \sum_{x=1}^{V} b_{\alpha}{}^{i}{}_{\chi} S_i(t_\chi)$. Then, we have

$$S_j S_i(t_\alpha) = n_i t_\alpha + \sum_{u=1}^{m} p_{ii}{}^{u} S_u(t_\alpha), \text{ if } j = 1$$

$$= \sum_{u=1}^{m} p_{ji}{}^{u} S_u(t_\alpha), \text{ if } j \neq i.$$

(iii) *Intrablock Analysis of a PBIBD*: Let $A = (a_{ju})$ be an $m \times m$ matrix, whose (j, u)-th element a_{ju} is defined as

$$a_{jj} = r(k-1) + n_j \lambda_j - \sum_{i=1}^{m} \lambda_i p_{ji}{}^{j}$$

$$a_{ju} = n_j \lambda_j - \sum_{i=1}^{m} \lambda_i p_{ji}{}^{u}, j \neq u, j, u = 1, 2, \ldots, m$$

Let $A^- = (a^{ju})$ be a g-inverse of A. Further let $S_j(Q_s)$ denote the sum of Q's over treatments which are j-th associates of t_s. Then, the solutions of the normal equations $\mathbf{Q} = \hat{C}\mathbf{t}$ are given by

$$\hat{t}'_s = \frac{k}{r(k-1)} \left[Q_s + \sum_{j=1}^{m} \sum_{i=1}^{m} \lambda_i a^{ij} S_j(Q_s) \right], \quad s = 1, 2, \ldots, v.$$

The adjusted treatment sum of squares is given by

$$\hat{\mathbf{t}}'\mathbf{Q} = \frac{k}{r(k-1)} \left[\sum_{s=1}^{v} Q_s^2 + \sum_{s=1}^{v} \left\{ \sum_{j=1}^{m} \sum_{i=1}^{m} \lambda_i a^{ij} S_j(Q_s) \right\} Q_s \right]$$

with (v − 1) d.f. Also,

$$\mathrm{var}\,(\hat{t}_s) = \sigma^2 k/r(k-1),\ s = 1, 2, \ldots, v.$$

$$\mathrm{cov}\,(\hat{t}_s, \hat{t}_{s'}) = \sigma^2 k \sum_{\ell=1}^{m} \lambda_\ell a^{\ell i}/r(k-1),$$

if t_s and $t_{s'}$ are i-th associates, $s \neq s' = 1, 2, \ldots, v$.

Further, if t_s and $t_{s'}$ are i-th associates, then

$$\mathrm{var}\,(\hat{t}_s - \hat{t}_{s'}) = \frac{2\sigma^2 k}{r(k-1)}\left[1 - \sum_{\ell=1}^{m} \lambda_\ell a^{\ell i}\right].$$

(iv) *Intrablock Analysis of a 2-Associate Class PBIBD:*

The intrablock treatment estimates are given by

$$\hat{t}_s = k[DQ_s - BS_i(Q_S)]/\Delta,\ S = 1, 2, \ldots, v$$

where $\Delta = AD - BC$ and

$$A = r(k-1) + \lambda_2,\ B = \lambda_2 - \lambda_1,$$
$$C = (\lambda_2 - \lambda_1)p_{12}{}^2,\ D = A + B(p_{11}{}^1 - p_{11}{}^2).$$

The adjusted treatment sum of squares is given by

$$\sum_{s=1}^{v} \hat{t}_s Q_s = \left[D\sum_{s=1}^{v}(kQ_s)^2 - B\sum_{s=1}^{v}(kQ_s)S_1(kQ_s)\right]/k\Delta.$$

Further,

$$\mathrm{var}(\hat{t}_s) = \sigma^2 kD/\Delta,\ s = 1, 2, \ldots, v$$

$$\mathrm{cov}(\hat{t}_s, \hat{t}_{s'}) = -\sigma^2 kB/\Delta, \quad \text{if } t_s \text{ and } t_{s'} \text{ are first associates}$$
$$= 0, \text{ it } t_s \text{ and } t_{s'} \text{ are second associates}$$

$$\mathrm{var}(\hat{t}_s - \hat{t}_{s'}) = \frac{2k\sigma^2(B+D)}{\Delta}, \text{ if } t_s \text{ and } t_{s'} \text{ are first associates}$$
$$= \frac{2kD\sigma^2}{\Delta}, \text{ if } t_s \text{ and } t_{s'} \text{ are second associates.}$$

(v) *Analysis of PBIBD with Recovery of Interblock Information*:

Let

$$P_s = w_1 T_s - \frac{1}{k}(w_1 - w_2)B_{(s)} - \frac{1}{k}w_2 G,\ s = 1, 2, \ldots, v$$

where T_s = total yield for the s-th treatment, B_s = Sum of blocks in which the s-th treatment occurs, and G = Grand total.

Let $H = (h_{ju})$ be an $m \times m$ matrix with elements h_{ju} defined by

$$h_{jj} = rw + (w_1 - w_2)n_{jj} - (w_1 - w_2)\sum_{i=1}^{m} \lambda_i p_{ji}{}^{j}$$

$$h_{ju} = (w_1 - w_2)n_{jj} - (w_1 - w_2)\sum_{i=1}^{u} \lambda_i p_{ji}{}^{u}, j \neq u$$

$$j, u = 1, 2, \ldots, m.$$

Let $H^{-} = (h^{ju})$ be a generalized inverse of H. Then the combined intra and interblock treatment estimates are given by

$$t_s^* = \frac{k}{wr}[P_s + (w_1 - w_2)\sum_{j=1}^{m}\sum_{i=1}^{m} \lambda_i h^{ij} S_j(P_s)],$$

$$s = 1, 2, \ldots, v$$

where $w = (k-1)w_1 + w_2$.

The adjusted treatment sum of squares is given by $\Sigma t_s^* P_s$, with $(v-1)$ d.f., assuming that the design is a connected one. For testing the significance of treatment differences, the F statistic is given by

$$F = \frac{\Sigma t_s^* P_s/(v-1)}{w_1 E_e}$$

with $(v-1)$ and $(bk - b - b + 1)$ d.f.

Since w_1 and w_2 are unknown, they are estimated by

$$\hat{w}_1 = 1/E_e,$$
$$\hat{w}_2 = (bk - v)/[k(b-1)E_b - (v-k)E_e],$$

where E_e = mean intrablock error SS and E_b = mean intrablock adjusted block SS. Then, the approximate F statistic is given by

$$F = \sum_{s=1}^{v} t_s^* P_s/(v-1).$$

Further,

$$\text{var}(t_s^*) = k/wr$$

$$\text{cov}(t_s^*, t_{s'}^*) = \frac{k(w_1 - w_2)\sum_{i=1}^{m} \lambda_i h^{ij}}{wr}, \text{ if } t_s \text{ and } t_{s'} \text{ are j-th associates.}$$

Hence,

$$\text{var}(t_s^* - t_{s'}^*) = \frac{2k}{wr}\left[1 - (w_1 - w_2)\sum_{i=1}^{m} \lambda_i h^{ij}\right], s \neq s'$$

if t_s and $t_s{}'$ are j-th associates.

(vi) *Analysis of a 2-associate Class PBIBD with Recovery of Interblock Information*:

The combined intra and inter block treatment estimates are given by

$$t_s^* = k[DP_s - BS_i(P_s)]/\Delta, s = 1, 2, \ldots, v$$

where $\Delta = AD - BC$, and $A = w + \lambda_2(w_1 - w_2)$

$$B = (w_1 - w_2)(\lambda_2 - \lambda_1), C = Bp_{12}{}^2,$$
$$D = A + B\left(P_{11}{}^1 - p_{11}^2\right).$$

Then the approximate F statistic is given by

$$F = \frac{D\Sigma(kP_s)^2 - B\Sigma(kP_s)S_i(kP_s)}{k(v-1)\Delta},$$

wherein the expressions are calculated by replacing w_1 and w_2 by their estimates $\hat{w}_1$ and $\hat{w}_2$. Also,

$$\text{var}(t_s^*) = kD/\Delta$$
$$\text{cov}(t_s^*, t_{s'}^*) = -\frac{kB}{\Delta}, \text{ it } t_s \text{ and } t_{s'} \text{ are first associates}$$
$$= 0, \text{ if } t_s \text{ and } t_{s'} \text{ are second associates.}$$
$$\text{var}(t_s^* - t_{s'}^*)) = \frac{2k(B+D)}{\Delta}, \text{ if } t_s \text{ and } t_{s'} \text{ are first associates}$$
$$= \frac{2kD}{\Delta}, \text{ if } t_s \text{ and } t_{s'} \text{ are second associates.}$$

(vii) *Further Results in PBIBD*: The characteristic roots of NN′ of a 2-associate class PBIBD are given by

$$\theta_o = rk, \text{ with multiplicity } 1$$
$$\theta_i = r - \frac{1}{2}(\lambda_1 + \lambda_2) + \frac{1}{2}(\lambda_1 + \lambda_2)[r + (-1)^i\sqrt{\Delta}],$$

with multiplicity

$$\alpha_i = \frac{n_1 + n_2}{2} - (-1)^i \frac{[(n_1 - n_2) + r(n + n_2)]}{2\sqrt{\Delta}}, i = 1, 2$$

where $r = p_{12}{}^2 - p_{12}{}^1$, $\beta = p_{12}{}^1 + p_{12}{}^2$ and $\Delta = r^2 + 2\beta + 1$.

The 2-associate class PBIBD designs are classified into five types: (1) Group divisible, (2) Simple, (3) Triangular, (4) Latin square type (L_i) and, (5) Cyclic.

(1) Group divisible design: A 2-associate class PBIBD is called *group divisible* if there are $v = mn$ treatments and the treatments can be divided into m groups of n treatments each, such that any two treatments of the same group are first associates and two treatments from differenct groups are second associates.

The parameters of the GD design are $v, b, r, k, n_1 = n - 1,\ n_2 = n(m-1), \lambda_1, \lambda_2$ and

$$P_1 = \begin{bmatrix} n-2 & 0 \\ 0 & n(m-1) \end{bmatrix}, \quad P_2 = \begin{bmatrix} 0 & n-1 \\ n-1 & n(m-2) \end{bmatrix}.$$

The characteristic roots of NN' of a GD design are

$$\theta_0 = rk, \text{ with multiplicity } 1,$$
$$\theta_1 = r - \lambda_1, \text{ with multiplicity } m(n-1)$$
$$\theta_2 = rk - v\lambda_2, \text{ with multiplicity } (m-1).$$

A GD desgin is called *singular* if $r - \lambda_1 = 0$; it is called *semi-regular* if $r - \lambda_1 > 0$ and $rk - v\lambda_2 = 0$; and it is called *regular* if $r - \lambda_1 > 0$ and $rk - v\lambda_2 > 0$.

(2) Simple: A PBIB design with two associate classes is called *simple* if either (a) $\lambda_1 \neq 0, \lambda_2 = 0$, or (b) $\lambda_1 = 0, \lambda_2 \neq 0$. A simple PBIBD may belong to group divisible, triangular, Latin square type or cyclic.

(3) Triangular: A PBIB design with 2-associate classes is called *triangular* if there are $v = n(n-1)/2$ treatments and the association scheme is an array of n rows and n columns with the following properties:

(a) the positions in the principal diagonal are left blank;
(b) the $n(n-1)/2$ positions above the principal diagonal are filled by the numbers $1, 2, \ldots, n(n-1)/2$ corresponding treatments;
(c) the $n(n-1)/2$ positions below the principal diagonal are filled so that the array is symmetrical about the principal diagonal,
(d) the first associates of any treatment i are those that occur in the same row (or in the same column) as i.

The parameters of a triangular PBIBD are

$$v = n(n-1)/2, b, r, k, n_1 = 2(n-2), n_2 = (n-2)(n-3)/2, \lambda_1, \lambda_2, \text{ and}$$

$$P_1 = \begin{bmatrix} n-2 & n-3 \\ n-3 & \dfrac{(n-3)(n-4)}{2} \end{bmatrix}, \quad P_2 = \begin{bmatrix} 4 & 2n-8 \\ 2n-8 & \dfrac{(n-4)(n-5)}{2} \end{bmatrix}.$$

The characteristic roots of NN' of a triangular design are

$$\theta_0 = rk, \text{ with multiplicity } 1,$$
$$\theta_1 = r - 2\lambda_1 + \lambda_2, \text{ with multiplicity } n(n-3)/2,$$
$$\theta_2 = r + (n-4)\lambda_1 - (n-3)\lambda_2, \text{ with multiplicity } (n-1).$$

(4) Latin square type (L_i): A PBIB design with two associate classes is called a *Latin square type* with i constraints (L_i) if there are $v = s^2$ treatments and the treatments are arranged in a $s \times s$ square and $i - 2$ mutually orthogonal Latin squares are superimposed and two treatments are first associates if they occur in the same row or column or correspond to the same letter in any Latin square.

The parameters of a L_i design are

$$v = s^2, b, r, k, n_1 = i(s-1), n_2 = (s-i+1)(s-1), \lambda_1, \lambda_2, \text{ and}$$

$$P_1 = \begin{bmatrix} (i-1)(i-2)+s-2 & (s-i+1)(i-1) \\ (s-i+1)(i-1) & (s-i+1)(s-i) \end{bmatrix}$$

$$P_2 = \begin{bmatrix} i(i-1) & i(s-i) \\ i(s-i) & (s-i)(s-i-1)+s-2 \end{bmatrix}.$$

The characteristic roots of NN' of a L_i design are

$$\begin{aligned} \theta_0 &= rk, \text{ with multiplicity } 1, \\ \theta_1 &= r - i\lambda_1 + \lambda_2(i-1), \text{ with multiplicity } (s-1)(s-i+1), \\ \theta_2 &= r + \lambda_1(s-i) - \lambda_2(s-i+1), \text{ with multiplicity } i(s-1). \end{aligned}$$

(5) Cyclic design: A non-group divisible PBIB design with two associate classes is called *cyclic* if the set of first associates of the i-th treatment is $(i + d_1, i + d_2, \ldots, i + d_{n_1})$ mod v, where the d elements satisfy the following conditions:

(a) the d elements are all different and $0 < d_j < v$ for $j = 1, 2, \ldots, n_1$.

(b) among the $n_1(n_1 - 1)$ differences $d_j - d_{j'}$, each of the $d_1, d_2, \ldots, d_{n_1}$ elements occurs $p_{11}{}^1$ times and each of $e_1, e_2, \ldots, e_{n_2}$ elements occurs $p_{11}{}^2$ times, where $d_1, d_2, \ldots, d_{n_1}$ and $e_1, e_2, \ldots, e_{n_2}$ are all distinct non-zero elements of module M of v elements $0, 1, 2, \ldots, v - 1$ corresponding to the v treatments.

(c) the set $D = (d_1, d_2, \ldots, d_{n_1})$ is such that $D = (-d_1, -d_2, \ldots, -d_{n_1})$.

The cyclic association scheme has parameters $v = 4t + 1$, $n_1 = n_2 = 2t$,

$$P_1 = \begin{bmatrix} t-1 & t \\ t & t \end{bmatrix}, \quad P_2 = \begin{bmatrix} t & t \\ t & t-1 \end{bmatrix}.$$

(6) Rectangular Design: A rectangular design is a PBIBD with 3 associate classes in which $v = mn$ treatments are arranged in b blocks of k plots each such that each treatment occurs in exactly r blocks and the v treatments are arranged in a rectangle of m rows and n columns. The first associates of any treatment are the other $(n - 1)$ treatments of the same row; its second associates are the other $(m - 1)$ treatments of the same column and the remaining treatments are its third associates.

Thus, the parameters of a rectangular design are $v = mn, b, r, k, n_1 = n - 1, n_2 = m - 1, n_3 = (m - 1)(n - 1), \lambda_1, \lambda_2, \lambda_3$ and

$$P_1 = \begin{bmatrix} n-2 & 0 & 0 \\ 0 & 0 & m-1 \\ 0 & m-1 & (m-1)(n-2) \end{bmatrix},$$

$$P_2 = \begin{bmatrix} 0 & 0 & n-1 \\ 0 & m-2 & 0 \\ n-1 & 0 & (m-2)(n-1) \end{bmatrix},$$

$$P_3 = \begin{bmatrix} 0 & 1 & n-2 \\ 1 & 0 & m-2 \\ n-2 & m-2 & (n-2)(m-2) \end{bmatrix}.$$

The characteristic roots of NN' of a rectangular design are given by

$$\begin{aligned} \theta_0 &= rk, \text{ with multiplicity } 1, \\ \theta_1 &= r - \lambda_1 + (m-1)(\lambda_2 - \lambda_3), \text{ with multiplicity } (n-1) \\ \theta_2 &= r - \lambda_2 + (n-1)(\lambda_1 - \lambda_3), \text{ with multiplicity } (m-1) \\ \theta_3 &= r - \lambda_1 - \lambda_2 + \lambda_3, \text{ with multiplicity } (n-1)(m-1). \end{aligned}$$

23. **Split-plot design**

It consists of r randomized blocks and each block contains v main plots to which v main treatments are assigned at random. Further, each main plot of a block is split up into s sub-plots to which s sub-treatments are alloted at random. Let y_{ijk} denote the yield of a sub-plot to which the k-th sub-treatment is assigned and the j-th main treatment is assigned and which occurs in the i-th block, $i = 1, 2, \ldots, r$; $j = 1, 2, \ldots, v$, $k = 1, 2, \ldots, s$. The assumed model is

$$y_{ijk} = \mu + \alpha_i + t_j + \rho_k + \delta_{jk} + e_{ijk}$$

with usual meanings for the different symbols, and

$$\text{var}(e_{ijk}) = \sigma^2, \text{cov}(e_{ijk}, e_{i'j'k'}) = \rho\sigma^2,$$

for $i = i'$, $j = j'$, $k \neq k'$ and 0, otherwise.

The intra-block solutions of the normal equations are

$$\begin{aligned} \hat{\mu} &= y_{\ldots}, \hat{\alpha}_i = y_{i..} - y_{\ldots}, i = 1, 2, \ldots r, \\ \hat{t}_j &= y_{.j.} - y_{\ldots}, j = 1, 2, \ldots, v \\ \hat{\rho}_k &= y_{..k} - y_{\ldots}, k = 1, 2, \ldots, s \\ \hat{\delta}_{jk} &= y_{.jk} - y_{.j.} - y_{..k} + y_{\ldots}, \quad j = 1, 2, \ldots, v; \ k = 1, 2, \ldots, s \end{aligned}$$

For testing the significance of block difference, the F statistic is

$$F = \frac{vs \sum_{i=1}^{r} (y_{i..} - y_{\ldots})^2/(r-1)}{s \sum_{i=1}^{r} \sum_{j=1}^{u} (y_{ij.} - y_{.j.} + y_{\ldots})^2/(r-1)(v-1)}$$

with (r − 1) and (r − 1) (v − 1) d.f.

For testing the significance of differences among main treatments, the F statistic is given by

$$F = \frac{rs\sum_{j=1}^{v}(y_{.j.} - y_{...})^2/(v-1)}{s\sum_{i=1}^{r}\sum_{j=1}^{v}(y_{ij.} - y_{i..} - y_{.j.} + y_{...})^2/(r-1)(v-1)}$$

with (v − 1) and (r − 1) (v − 1) d.f.

For testing the significance of differences among the subtreatments, the F statistic is given by

$$F = \frac{rv\sum_{k=1}^{s}(y_{..k} - y_{...})^2/(s-1)}{\Sigma\Sigma\Sigma(y_{ijk} - y_{.jk} - y_{ij.} + y_{.j.})^2/v(r-1)(s-1)}$$

with (s − 1) and v(r − 1) (s − 1) d.f.

For testing the significance of the interaction between the main and sub-treatments, the F statistic is given by

$$F = \frac{r\Sigma\Sigma(y_{.jk} - y_{.j.} - y_{..k} + y_{...})^2/(v-1)(s-1)}{\Sigma\Sigma\Sigma(y_{ijk} - y_{.jk} - y_{ij.} + y_{.j.})^2/v(r-1)(s-1)}$$

with (v − 1) (s − 1) and v(r − 1) (s − 1) d.f.

The analysis of variance table of a split-plot design is as follows.

Analysis of Variance Table of a Split-Plot Design

SOURCE	SS	d.f.
Blocks	$vs\sum_{i=1}^{r}(y_{i..} - y_{...})^2$	r − 1
Main Treatments	$rs\sum_{j=1}^{v}(y_{.j.} - y_{...})^2$	v − 1
Error (a)	$s\sum_{i}(y_{ij.} - y_{i..} - y_{.j.} + y_{...})^2$	(r − 1) (v − 1)
Sub-treatments	$rv\sum_{k}(y_{..k} - y_{...})^2$	s − 1
Interaction between main and subtreatments)	$r\sum_{j}\sum_{k}(y_{.jk} - y_{.j.} - y_{..k} + y_{...})^2$	(v − 1) (s − 1)
Error (b)	$\sum_{i}\sum_{j}\sum_{k}(y_{ijk} - y_{.jk} - y_{ij.} + y_{.j.})^2$	v(r − 1) (s − 1)
Total	$\Sigma\Sigma\Sigma y_{ijk}^2 - rvsy_{...}^2$	rvs − 1

The different standard errors are given below.

(i) $S.E.(y_{..k} - y_{..k'}) = \sqrt{2E_b/rv}$

(ii) $S.E.(y_{.j.} - y_{.j'.}) = \sqrt{2E_a/rs}$

(iii) $\text{S.E.}(y_{.jk} - y_{.jk'}) = \sqrt{2E_b/r}$
(iv) $\text{S.E.}(y_{.jk} - y_{.j'k}) = \sqrt{2[E_a + (s-1)E_b]/rs}$
(v) $\text{S.E.}(y_{.jk} - y_{.j'k'}) = \sqrt{2[E_a + (s-1)E_b]/rs}$

where E_a is the mean SS for Error (a) and E_b is the mean SS for Error (b).

1.4 Two-Way Designs

24. **(i) Notations:** We consider a two-way design in which uu' plots are arranged in u rows and u' columns. Let there be v treatments, the i-th treatment being replicated r_i times, $i = 1, 2, \ldots, v$. Let the i-th treatment occur ℓ_{ij} times in the j-th row and m_{ik} times in the k-th column, $i = 1, 2, \ldots, v; j = 1, 2, \ldots, u; k = 1, 2, \ldots, u'$. Let

$L = (\ell_{ij}) =$ a $v \times u$ matrix of elements ℓ_{ij}
$M = (m_{ik}) =$ a $v \times u'$ matrix of elements m_{ik}
$E_{pq} =$ a $p \times q$ matrix with all elements unity
$I_p =$ an identity matrix of order p
$\mathbf{r}' = [r_1, r_2, \ldots r_v]$,
$\alpha_j =$ effect of the j-th row, $j = 1, 2, \ldots, u$
$\boldsymbol{\alpha}' = [\alpha_1, \alpha_2, \ldots, \alpha_u]$
$\beta_k =$ effect of the k-th column, $k = 1, 2, \ldots, u'$
$\boldsymbol{\beta}' = [\beta_1, \beta_2, \ldots, \beta_{u'}]$
$t_i =$ effect of the i-th treatment, $i = 1, 2, \ldots, v$.
$\mathbf{t}' = [t_1, t_2, \ldots, t_v]$
$F = \text{diag}(r_1, r_2, \ldots, r_v) - \frac{1}{u'}LL' - \frac{1}{u}MM' + \frac{1}{uu'}\mathbf{rr}'$
$\mathbf{R}' = [R_1, R_2, \ldots R_u] =$ row vector of row totals
$\mathbf{C}' = [C_1, C_2, \ldots C_u] =$ row vector of column totals
$\mathbf{T}' = [T_1, T_2, \ldots, T_v] =$ row vector of treatment totals
$G =$ Grand total
$\mathbf{Q} = \mathbf{T} - \frac{1}{u'}\mathbf{LR} - \frac{1}{u}\text{MC} + \frac{G}{uu'}\mathbf{r}$.

One can easily verify the following results:

$LE_{u1} = ME_{u'1} = \mathbf{r}$
$LE_{uu'} = ME_{u'u'}$
$LE_{uu} = ME_{u'u}$
$\sum_{j=1}^{u} \ell_{ij} = \sum_{k=1}^{u'} m_{ik} = r_i$
$\sum_{i=1}^{v} \ell_{ij} = u', \sum_{i=1}^{v} m_{ik} = u$
$E_{1v}L = u'E_{1u}, E_{1v}M = uE_{1u'}$,
$E_{1v}\mathbf{Q} = 0, E_{1v}F = \mathbf{0}_{1v}$

(ii) Analysis: The normal equations are

$$\begin{aligned}
G &= uu'\hat{\mu} + u'E_{1u}\hat{\boldsymbol{\alpha}} + E_{1u'}\hat{\boldsymbol{\beta}} + \mathbf{r}'\hat{\mathbf{t}}\\
\mathbf{R} &= u'\hat{\mu}E_{u1} + u'\hat{\boldsymbol{\alpha}} + E_{uu'}\hat{\boldsymbol{\beta}} + L'\hat{\mathbf{t}}\\
\mathbf{C} &= u\hat{\mu}E_{u'1} + E_{u'u}\hat{\boldsymbol{\alpha}} + u\hat{\boldsymbol{\beta}} + M'\hat{\mathbf{t}}\\
\mathbf{T} &= \hat{\mu}\mathbf{r} + L\hat{\boldsymbol{\alpha}} + M\hat{\boldsymbol{\beta}} + \text{diag}\,(r_1, r_2, \ldots, r_v)\hat{\mathbf{t}}.
\end{aligned}$$

The reduced normal equations for treatment effects are

$$\mathbf{Q} = F\hat{\mathbf{t}}.$$

A particular solution of the normal equations is taken as

$$\begin{aligned}
\hat{\mu} &= G/uu'\\
\hat{\boldsymbol{\alpha}} &= (\mathbf{R} - \tfrac{G}{u}E_{u1} - L'\hat{\mathbf{t}})/u'\\
\hat{\boldsymbol{\beta}} &= (\mathbf{C} - \tfrac{G}{u}E_{u'1} - M'\hat{\mathbf{t}}/u\\
\mathbf{Q} &= F\hat{\mathbf{t}}.
\end{aligned}$$

Adjusted treatment SS is given by $\hat{\mathbf{t}}'\mathbf{Q}$ with $(v - g)$ d.f., where

$$(v - g) = \text{rank}\,(F).$$

The error SS is given by

$$SSE = (\mathbf{y}'\mathbf{y} - \frac{G^2}{uu'}) - (\frac{1}{u'}\mathbf{R}'\mathbf{R} - \frac{G^2}{uu'}) - (\frac{1}{u}\mathbf{C}'\mathbf{C} - \frac{G^2}{uu'}) - \hat{\mathbf{t}}'\mathbf{Q}$$

with $(u - 1)(u' - 1) - (v - g)$ d.f.

For testing the significance of treatment differences, the F statistic is given by

$$F = \frac{\hat{\mathbf{t}}'\mathbf{Q}/(v - g)}{SSE/[(u-1)(u'-1) - (v-g)]}$$

with $(v - g)$ and $(u - 1)(u' - 1) - (v - g)$ d.f.

The analysis of variance table is given below.

Analysis of Variance Table For a Two-way Design

SOURCE	SS	d.f.
Rows	$\frac{1}{u'}\mathbf{R}'\mathbf{R} - \frac{G^2}{uu'}$	$u - 1$
Columns	$\frac{1}{u}\mathbf{C}'\mathbf{C} - \frac{G^2}{uu'}$	$u' - 1$
Treatments (adjusted)	$\hat{\mathbf{t}}'\mathbf{Q}$	$v - g$
Error	by subtraction	$(u - 1)(u' - 1) - (v - g)$
Total	$\mathbf{y}'\mathbf{y} - \frac{G^2}{uu'}$	$uu' - 1$

The expected values of sums of squares appearing in the above analysis of variance table are given in the following table.

Analysis of Variance Table For a Two-way Design

SOURCE	SS	E(SS)
Rows	$\frac{1}{u'}\mathbf{R}'\mathbf{R} - \frac{G^2}{uu'}$	$(u-1)\sigma^2 + u'(\boldsymbol{\alpha}'\boldsymbol{\alpha} - \frac{1}{u}\boldsymbol{\alpha}'E_{uu}\boldsymbol{\alpha}$ $+2(\boldsymbol{\alpha}'L\mathbf{t} - \frac{1}{u}\boldsymbol{\alpha}'E_{uu}L'\mathbf{t}) + \frac{1}{u'}$ $(\mathbf{t}'LL'\mathbf{t} - \frac{1}{u}\mathbf{t}'LE_{uu}L'\mathbf{t})$
Columns	$\frac{1}{u}\mathbf{C}'\mathbf{C} - \frac{G^2}{uu'}$	$(u'-1)\sigma^2 + u(\boldsymbol{\beta}'\boldsymbol{\beta} - \frac{1}{u'}\boldsymbol{\beta}'E_{u'u'}\boldsymbol{\beta})$ $+2(\boldsymbol{\beta}'M\mathbf{t} - \frac{1}{u}\boldsymbol{\beta}'E_{u'u'}M'\mathbf{t})+$ $\frac{1}{u}(\mathbf{t}'MM'\mathbf{t} - \frac{1}{u'}\mathbf{t}'ME_{u'u}M'\mathbf{t})$
Treatments (adjusted)	$\hat{\mathbf{t}}'\mathbf{Q}$	$(v-g)\sigma^2 + \mathbf{t}'F\mathbf{t}$
Error	$\mathbf{y}'\mathbf{y} - \frac{1}{u}\mathbf{R}'\mathbf{r} - \frac{1}{u}\mathbf{C}'\mathbf{C} - \mathbf{t}'\mathbf{Q} +$ G^2/uu'	$[(u-1)(u'-1) - (v-g)]\sigma^2$

25. **Latin Square Design:**

A Latin square design (LSD) is an arrangement of v^2 plots in v rows and v columns and v treatments are assigned to them such that each treatment occurs once in each row and once in each column. For a LSD,

$$L = M = E_{vv},\ F = vI_v - E_{vv}$$
$$\mathbf{Q} = \mathbf{T} - (G/v)E_{v1}$$

A particular solution for $\mathbf{t}$ is

$$\hat{\mathbf{t}} = (\mathbf{T} - v^{-1}GE_{v1})/v$$

The analysis of variance table of a LSD is given below.

Analysis of Variance Table of a LSD

SOURCE	SS	d.f.
Rows	$v^{-1}\mathbf{R}'\mathbf{R} - v^{-2}G^2$	$v-1$
Columns	$v^{-1}\mathbf{C}'\mathbf{C} - v^{-2}G^2$	$v-1$
Treatments	$v^{-1}\mathbf{T}'\mathbf{T} - v^{-2}G^2$	$v-1$
Error	by subtraction	$(v-1)(v-2)$
Total	$\mathbf{y}'\mathbf{y} - v^{-2}G^2$	v^2-1

The expected values of sums of squares in the above table are given in the following table.

Expected Values of Sums of Squares in the Analysis of Variance Table of a LSD

SOURCE	SS	$\mathcal{E}$(SS)
Rows	$\frac{1}{v}\mathbf{R}'\mathbf{R} - \frac{G^2}{v^2}$	$(v-1)\sigma^2 + v\boldsymbol{\alpha}'\boldsymbol{\alpha} - \boldsymbol{\alpha}'E_{vv}\boldsymbol{\alpha}$
Columns	$\frac{1}{v}\mathbf{C}'\mathbf{C} - \frac{G^2}{v^2}$	$(v-1)\sigma^2 + v\boldsymbol{\beta}'\boldsymbol{\beta} - \boldsymbol{\beta}'E_{vv}\boldsymbol{\beta}$
Treatments	$\frac{1}{v}\mathbf{T}'\mathbf{T} - \frac{G^2}{v^2}$	$(v-1)\sigma^2 + v\mathbf{t}'\mathbf{t} - \mathbf{t}'E_{vv}\mathbf{t}$
Error	$\mathbf{y}'\mathbf{y} - \frac{1}{v}\mathbf{R}'\mathbf{R} - \frac{1}{v}\mathbf{C}'\mathbf{C} - \frac{1}{v}\mathbf{T}'\mathbf{T} + \frac{2G^2}{v^2}$	$(v-1(v-2)\sigma^2$

The efficiency of a LD relative to a RBD with blocks as columns of a LSD is

$$[E_r + (v-1)E_e]/vE_e,$$

and the efficiency of a LSD relative to a RBD with blocks as rows of a LSD is

$$[E_c + (v-1)E_e]/vE_e,$$

where E_r, E_c, and E_e denote respectively the mean row SS, the mean column SS, and the mean error SS.

26. **Cross-over Design**

A cross-over design may be regarded as a repetition of a Latin square design. Let us consider a crosss-over design which is obtained by s repetitions of a $v \times v$ LSD. Then, for this design

$$L = s\,E_{vv},\ M = E_{v(su)}$$
$$F = svI_v - sE_{vv},\ \text{rank}\,(F) = v - 1$$
$$\mathbf{Q} = T - (G/v)E_{v1}.$$

A solution of $\mathbf{t}$ is given by

$$\hat{\mathbf{t}} = (sv)^{-1}I_v - (G/sv^2)E_{v1}.$$

The analysis of variance of table of a cross-over design is given below.

Analysis of Variance Table of a Cross-over Design

SOURCE	SS	d.f.
Rows	$(sv)^{-1}\mathbf{R}'\mathbf{R} - (sv^2)^{-1}G^2$	$v-1$
Columns	$v^{-1}\mathbf{C}'\mathbf{C} - (sv^2)^{-1}G^2$	$sv-1$
Treatments	$(sv)^{-1}\mathbf{T}'\mathbf{T} - (sv^2)^{-1}G^2$	$v-1$
Error	by subtraction	$(v-1)(sv-1)$
Total	$\mathbf{y}'\mathbf{y} - (sv^2)^{-1}G^2$	sv^2-1

27. **Graeco-Latin Square Design**

Two Latin squares of the same side s and same letters are said to be orthogonal, if when one Latin square is superimposed on the another, then each letter of the former square coincides exactly once with each letter of the second square.

A set of Latin squares of the same side s and the same letters is said to be a set of mutually orthogonal Latin squares (MOLS), if any two of them are orthogonal.

A complete set of MOLS of side s consists of atmost (s − 1) Latin squares. If s is a prime or a power prime, then the complete set of MOLS of side s consists of (s − 1) Latin squares.

If two Latin squares of side s are orthogonal and if the letters of one are written in Latin letters and those of the second are written in Greek letters and if one is superimposed on the other, then the composite square is called *Graeco-Latin square.*

A Graeco-Latin Square Design (GLSD) is an arrangement of v treatments in a $v \times v$ Latin square such that each treatment occurs exactly once in each row, exactly once in each column and exactly once with each letter of the Latin square.

The analysis of variance of a GLSD is given in the following table.

Analysis of Variance Table of a GLSD

SOURCE	SS	d.f.
Rows	$v^{-1}\Sigma R_i^2 - v^{-2}G^2$	$v-1$
Columns	$v^{-1}\Sigma C_j^2 - v^{-2}G^2$	$v-1$
Letters	$v^{-1}\Sigma A_k^2 - v^{-2}G^2$	$v-1$
Treatments	$v^{-1}\Sigma T_\ell^2 - v^{-2}G^2$	$v-1$
Error	by subtraction	$(v-1)(v-3)$
Total	$\Sigma y^2 - v^{-2}G^2$	v^2-1

28. **Youden Square Design**

A Youden square design is a rectangular arrangement of v treatments in $r = k$ rows and $b = v$ columns such that each treatment occurs exactly once in each row and the columns constitute the blocks of a symmetrically balanced incomplete block design ($v = b, r = k, \lambda$).

(i) Intrablock analysis: Here $u = r$, $u' = v$, $L = E_{vr}$, $M = N$, where N is the incidence matrix of a SBIBD ($v = b, r = k, \lambda$).

Also

$$F = (\lambda v/r)[I_v - v^{-1}E_{vv}] = C,$$

$$\mathbf{Q} = \mathbf{T} - \frac{1}{k}\mathbf{NB}$$

where C is the c-matrix of the corresponding SBIBD. Clearly, rank (F) = v − 1.

The intrablock treatment estimates are given by

$$\hat{\mathbf{t}} = (r/\lambda v)\mathbf{Q}.$$

The intrablock analysis of variance table of a Youden square design is given in the following table.

Intrablock Analysis of Variance Table of a Youden Square Design

SOURCE	SS	d.f.	SS	SOURCE
Rows	$\frac{\Sigma R_i^2}{v} - \frac{G^2}{vr}$	$r-1$	→	Rows
Blocks (unadj.)	$\frac{\Sigma B_j^2}{r} - \frac{G^2}{vr}$	$v-1$	by subtraction	Blocks (adj.)
Treatments (adj.)	$\frac{r}{\lambda v}\Sigma Q_i^2$	$v-1$	$\frac{\Sigma T_i^2}{r} - \frac{G^2}{vr}$	Treatments (unadj.)
Error (intrablock)	by subtraction	$(v-1)(r-2)$	→	Error (intrablock)
Total	$\Sigma y^2 - \frac{G^2}{vr}$	$vr-1$	→	Total

Note: "→" means carried forward.

The expected values of sums of squares in the intrablock analysis of variance table of a Youden Square are given in the following table.

Expected Values of Sums of Squares In the Intrablock Analysis of Variance Table of a Youden Square Design

SOURCE	SS	$\mathcal{E}$(SS)
Rows	$\frac{\Sigma R_i^2}{v} - \frac{G^2}{vr}$	$(r-1)\sigma^2 + v\boldsymbol{\alpha}'\boldsymbol{\alpha} - \frac{v}{r}(\Sigma\alpha)^2$
Blocks (unadj.)	$\frac{\Sigma B_j^2}{r} - \frac{G^2}{vr}$	$(v-1)\sigma^2 + r\boldsymbol{\beta}'\boldsymbol{\beta} + 2\boldsymbol{\beta}'\mathbf{N}'\mathbf{t} + \frac{1}{r}\mathbf{t}'\mathbf{NN}'\mathbf{t} - \frac{r}{v}(\Sigma\beta + \Sigma t)^2$
Treatments	$\frac{r}{\lambda v}\Sigma Q_i^2$	$(v-1)\sigma^2 + \mathbf{t}'\mathbf{C}\mathbf{t}$
Error	$(v-1)(r-2)E_e$	$(v-1)(r-2)\sigma^2$

Note: E_e = mean intrablock error SS

(ii) Analysis with recovery of interblock information. This analysis is exactly similar to that of a SBIBD except for the fact that the sum of squares due to rows is separated out from the error sum of squares. The approximate F statistic for testing the significance of treatment differences is

given by

$$F = \left[\frac{1}{r}\Sigma(T_i + \theta W_i)^2 - \frac{G^2}{vr}\right] / (v-1)E'_e$$

with (v − 1) and (v − 1)(r − 2) d.f., where

$$\begin{aligned}
\theta &= (E_b - E_e)/v(r-1)E_b \\
W_i &= (v-r)T_i - (v-1)B_{(i)} + (r-1)G \\
E_e &= \text{mean intrablock error SS} \\
E_b &= \text{mean intrablock adjusted block SS} \\
E'_e &= [1 + (v-r)\theta]E_e
\end{aligned}$$

29. **Lattice Design**

In an m-ple lattice design, there are $y = k^2$ treatments, arranged in $b = mk$ blocks of k plots each such that each treatment is replicated $r = m$ times and a pair of treatments either do not occur together in a block or occurs together once in a block.

An m-ple lattice design is constructed as follows. The $v = k^2$ treatments are arranged in a $k \times k$ square. The first set of k blocks is formed by taking k rows as blocks. The second set of k blocks is formed by taking k columns as blocks. We now take (m − 2) MOLS of side k, $(m - 2 \leq k - 1)$. We superimpose these Latin squares, in succession on the square of treatments and form blocks by taking treatments which correspond to the same letters of a Latin square. We thus get (m − 2) sets of k blocks each. Thus, we get mk blocks.

When the lattice design is obtained by taking the first two replicates, i.e., rows as blocks and columns as blocks, the lattice design is called a *simple lattice design*. A simple lattice design is a PBIBD with 2 associate classes, where the first associates of a treatment are treatments in the same row or the same column.

When $m = k + 1$, the lattice design is called the *balanced lattice design*.

The intrablock treatment estimates are given by

$$\hat{t}_s = \frac{k(m-1)Q_s + S_R(Q_s) + S_c(Q_s) + \sum_{i=1}^{m-2} S_i(Q_s)}{mk(m-1)},$$
$$s = 1, 2, \ldots, v$$

where

$$\begin{aligned}
S_R(Q_s) &= \text{sum of } Q'\text{s over treatments in the same row as } t_s \\
S_c(Q_s) &= \text{sum of } Q'\text{s over treatments in the same column as } t_s \\
S_i(Q_s) &= \text{sum of } Q'\text{s over treatments which correspond to the} \\
&\quad\ \text{same letter as } t_s \text{ in the } i-\text{th Latin square.}
\end{aligned}$$

The intrablock analysis of variance table is given below.

Analysis of Variance Table of an m-ple Lattice Design

SOURCE	SS	d.f.	SS	SOURCE
Replications	$\frac{1}{k}\Sigma R_i^2 - \frac{G^2}{mk^2}$	$m-1$	→	Replications
Blocks within replications (unadj.)	**	$m(k-1)$	**	Blocks within replications (adj.)
Blocks (unadj.)	$\frac{1}{k}\Sigma B_j^2 - \frac{G^2}{mk^2}$	$mk-1$	*	Blocks (adj.)
Treatments (adj.)	$\Sigma\hat{t}_s Q_s$	k^2-1	$\frac{1}{m}\Sigma T_i^2 - \frac{G^2}{mk^2}$	Treatments (unadj.)
Error	*	$(k-1)(mk-k-1)$	→	Error
Total	$\Sigma y^2 - \frac{G^2}{mk^2}$	mk^2-1	→	Total

Note: *obtained by subtraction, **obtained by subtraction, →carried forward

If t_s and $t_{s'}$ occur together in a block, then

$$\mathrm{var}(\hat{t}_s - \hat{t}_s) = 2\sigma^2(k+1)/mk,$$

and if t_s and $t_{s'}$ do not occur together in a block, then

$$\mathrm{var}(\hat{t}_s - \hat{t}_{s'}) = 2\sigma^2\left[\frac{1}{m} + \frac{1}{k(m-1)}\right].$$

When the lattice design is balanced, then every pair of treatment occurs once in a block and in this case

$$\mathrm{var}(\hat{t}_s - \hat{t}_{s'}) = 2\sigma^2/k.$$

For a balanced lattice design, the intrablock treatment estimate is given by $\hat{t}_s = Q_s/k$ and the adjusted treatment SS is given by $\Sigma\hat{t}_s Q_s = \Sigma Q_s^2/k$.

1.5 Factorial Designs

30. **Factorial design**

Let there be m factors denoted by $F_1, F_2, \ldots, F_m$. The factor F_i occurs at s_i levels, $i = 1, 2, \ldots, m$. The s_i levels of the factor F_i are denoted by $0, 1, 2, \ldots, s_i - 1, i = 1, 2, \ldots, m$. The treatments consist of all combinations of levels of all the m factors. The treatment combination in which F_1 occurs at x_1-th level, F_2 at x_2-th level, $\ldots, F_m$ at x_m-th level is denoted by $f_1^{x_1} f_2^{x_2} \ldots f_m^{x_m}$, $x_i = 0, 1, \ldots, s_i - 1, i = 1, 2, \ldots, m$. The total number of treatments is $N = s_1 s_2 \ldots s_m$, and this factorial design is called $s_1 \times s_2 \times \cdots \times s_m$ factorial design. If $s_1 = s_2 = \cdots = s_m = s$, then the *factorial design* is called *symmetrical*.

Treatment contrast. A linear function
$L = \sum_{x_1,\ldots,x_m} c(x_1, x_2, \ldots, x_m) f_1^{x_1} f_2^{x_2} \ldots f_m^{x_m}$, where the coefficients are not all zero, will be called a *treatment contrast*, if $\Sigma c(x_1, x_2, \ldots, x_m) = 0$, the summation being taken over all values of $x_1, \ldots, x_m$.

The following convention is followed in the the selection of coefficients $c(x_1, x_2, \ldots, x_m)$:

(i) $c(x_1, x_2, \ldots, x_m) = c(x_1).c(x_2) \ldots c(x_m)$

(ii) $c(x_i)$ will correspond to the orthogonal polynomials of various degrees based on s_i points, $i = 1, 2, \ldots, m$. There will be $(s_i - 1)$ sets of coefficients for $c(x_i)$; the t-th set of coefficients will be denoted by $c_t(x_i)$ and will correspond to the orthogonal polynomial of t-th degree, $t = 1, 2, \ldots, s_i - 1$.

The factorial experiment is assumed to be carried out in r randomized blocks. Let $T(x_1, x_2, \ldots, x_m)$ denote the total yield of the treatment $f_1^{x_1} f_2^{x_2} \ldots f_m^{x_m}$. Then, the BLUE of an estimable treatment contrast $L = \Sigma c(x_1, x_2, \ldots, x_m) f_1^{x_1} f_2^{x_2} \ldots f_m^{x_m}$ is given by

$$\hat{L} = \Sigma c(x_1, x_2, \ldots, x_m) T(x_1, x_2, \ldots, x_m)/r$$

with $\operatorname{var} \hat{L} = \sigma^2 \Sigma c^2(x_1, x_2, \ldots, x_m)/r$. Further, the SS due to L (i.e., for testing $L = 0$) is given by

$$\begin{aligned} SS\,(L) &= \sigma^2 \hat{L}^2/\operatorname{var}(\hat{L}) \\ &= \frac{[\Sigma c(x_1, x_2, \ldots, x_m) T(x_1, \ldots x_m)]^2}{r \Sigma c^2(x_1, x_2, \ldots, x)} \end{aligned}$$

and it will have 1 d.f.

Orthogonal treatment contrasts. Two treatment contrasts

$$\begin{aligned} L_1 &= \Sigma c(x_1, x_2, \ldots, x_m) f_1^{x_1} f_2^{x_2} \ldots f_m^{x_m} \\ L_2 &= \Sigma d(x_1, x_2, \ldots, x_m) f_1^{x_1} f_2^{x_2} \ldots f_m^{x_m} \end{aligned}$$

are said to be *orthogonal* if

$$\Sigma c(x_1, x_2, \ldots, x_m) d(x_1, x_2, \ldots, x_m) = 0,$$

the summation being taken over all values of $x_1, x_2, \ldots, x_m$.

A set of treatment contrasts will be said to be *mutually orthogonal* if every pair of them is orthogonal.

If $L_1, L_2, \ldots, L_{kvk}$ be a set of v mutually orthogonal treatment contrasts, then the sum of squares due to $L_1, L_2, \ldots, L_k$ is equal to sum of squares due to $L_1, L_2, \ldots, L_k^b$ and it will have k d.f.

k-factor interaction. A treatment contrast

$$L = \Sigma c(x_1, x_2, \ldots, x_m) f_1^{x_1} f_2^{x_2} \ldots f_m^{x_m},$$

with $\Sigma c(x_1, x_2, \ldots, x_m) = 0$, will be said to belong the k-factor interaction $F_{i_1} F_{i_2} \ldots F_{i_k}$ if

(i) the coefficients $c(x_1, x_2, \ldots x_m)$ depend only on the levels $x_{i_1}, x_{i_2}, \ldots, x_{i_k}$ of the factors $F_{i_1}, F_{i_2}, \ldots$, and F_{i_k}, that is

$$c(x_1, x_2, \ldots, x_m) = c(x_{i_1}) c(x_{i_2}) \ldots c(x_{i_k})$$

(ii) the sum of coefficients $c(x_{i_1}), c(x_{i_2}), \ldots, c(x_{i_k})$ over each arguments $x_{i_1}, x_{i_2}, \ldots x_{i_k}$ is zero, i.e., $\sum_{x_{i_1}} c(x_{i_1}) = 0, \sum_{x_{i_2}} c(x_{i_2}) = 0, \ldots, \sum_{x_{i_k}} c(x_{i_k}) = 0$.

An expression for a treatment contrast L belonging to the k-factor interaction $F_{i_1} F_{i_2} \ldots F_{i_k}$ can be taken as

$$L = \frac{\phi_1 \phi_2 \ldots \phi_m}{\phi_{i_1} \phi_{i_2} \ldots \phi_{i_k}} \left[\Sigma\, c(x_{i_1}) f_{i_1}^{x_{i_1}}\right] \left[\Sigma\, c(x_{i_2}) f_{i_2}^{x_{i_2}}\right] \ldots \left[\Sigma\, c(x_{i_k}) f_{i_k}^{x_{i_k}}\right]$$

where

$$\phi_i = f_i^0 + f_i + \ldots + f_i^{s_i - 1}, i = 1, 2, \ldots, m.$$

The selection of coefficients $c(x_{i_1}), c(x_{i_2}), \ldots, c(x_{i_k})$ is done by using orthogonal polynomials of different degrees based on $s_{i_1}, s_{i_2}, \ldots, s_{i_k}$ points respectively. Since, there are $(s_{i_j} - 1)$ sets of coefficients available for the selection of $c(x_{i_j})$, the total number of contrasts belonging to the k-factor interaction $F_{i_1} F_{i_2} \ldots F_{i_k}$ is equal to $(s_{i_1} - 1)(s_{i_2} - 1) \ldots (s_{i_k} - 1)$.

31. **2^m design**

There are m factors each at 2 levels called as the upper and lower levels. The factors are denoted by capital letters A, B, C, ... etc. The lower level of a factor is denoted by unity, while its upper level by its small letter. The 2^m treatments are written down as

(1)	c	d	cd	etc.
a	ac	ad	acd	
b	bc	bd	bcd	
ab	abc	abd	abcd	

The rules for writing down any factorial effect are as follows:

Rule 1: Let Z be any factorial effect in a 2^m design. Then, Z is given by

$$Z = \frac{1}{2^{m-1}} (a \pm 1)(b \pm 1) \ldots,$$

there being m brackets on the right hand side of Z, and the sign in each bracket is determined as follows. The sign in a bracket is + , if the corresponding letter is absent in Z and is −, if the corresponding letter is present in Z.

Rule 2: Let Z be any factorial effect in a 2^m design. Then, Z is given by

$$Z = \frac{1}{2^{m-1}}[\pm(1) \pm a \pm b \pm (ab) \pm \ldots.],$$

where the signs of treatments are determined as follows. If Z contains an odd (even) number of letters, then treatments having odd (even) number of letters in common with Z will have + sign and other treatments will have − sign.

Let [Z], denoted for Z, be obtained by substituting treatment yields for treatments in the expression for Z, except the divisor 2^{m-1}. Then, the BLUE of any factorial effect Z in a 2^m design is given by

$$\hat{Z} = [Z]/2^{m-1}r$$

and the SS due to Z is given by

$$SS(Z) = [Z]^2/2^m \,.\, r$$

with 1 d.f.

The sum of squares due to treatments is obtained by summing the sums of squares due to all factorial effects. The analysis of variance table for a 2^m design carried out in r randomized blocks is given below.

ANOVA Table For a 2^m Design

SOURCE	SS	d.f.
Blocks	$\frac{1}{2^m}\Sigma B_j^2 - \frac{G^2}{2^m \cdot r}$	
Treatments	$\frac{1}{r}\Sigma T_i - \frac{G^2}{2^m \cdot r}$	$2^m - 1$
A	$[A]^2/2^m \cdot r$	1
B	$[B]^2/2^m \cdot r$	1
.	.	
.	.	
.	.	
Error	by subtraction	$(r-1)(2^m-1)$
Total	$\Sigma y^2 - G^2/2^m \cdot r$	$r \cdot 2^m - 1$

Yates' method for estimating all factorial effects in a 2^m design. It consists of constructing (m + 1) columns $c_0, c_1, \ldots c_m$. The column c_0 is constructed by writing down treatment yields in the standard form. The column $c_i, i = 1, 2, \ldots, m$, is constructed from the column c_{i-1} as follows. The upper half of the column c_i is constructed by taking sums of pairs of values of the column c_{i-1}; while its lower half is constructed by taking differences (lower minus upper) of pairs of values of the column c_{i-1}. The column c_m gives the grand total G, and the totals for all factorial effects in a 2^m design in the standard form. The columns $c_0, c_1, \ldots, c_m$ constitute Yates' table. Dividing the entries of column c_m by $2^m \cdot r$, we obtain the correction term and the sums of squares due to all factorial effects.

32. **3^m design**

There are m factors $A_1, A_2, \ldots, A_m$ each at 3 levels. The three levels of the factor A_i will be denoted by $a_i^0 = 1, a_i, a_i^2, i = 1, 2, \ldots, m$. The 3^m treatments will be denoted by

$$\begin{array}{cccc} (1) & a_2 & a_2^2 & \text{etc.} \\ a_1 & a_1a_2 & a_1a_2^2 & \\ a_2^1 & a_2^1a_2 & a_2^1a_2^2 & \end{array}$$

A contrast belonging to k-factor interaction between k factors A_{i_1}, $A_{i_2}, \ldots, A_{i_k}$ is

$$Z = \left[\sum_{x_{i_1}=0}^{2} c(x_{i_1}) f_{i_1}^{x_{i_1}}\right]\left[\sum_{x_{i_2}=0}^{2} c(x_{i_2}) f_{i_2}^{x_{i_2}}\right] \cdots \left[\sum_{x_{i_k}=0}^{2} c(x_{i_k}) f_{i_k}^{x_{i_k}}\right].$$

The set of coefficients $c(x_{i_j})$ corresponding to the factor $A_i, j = 1, 2, \ldots, k$ can be selected in two different ways: one based on the orthogonal polynomial of first degree and the second based on the orthogonal polynomial of the second degree. The two sets of coefficients $c(x_{i_j})$ based on the orthogobal polynomials of first and second degree are

linear $c(x_{i_j}) : -1, 0, 1$
quadratic $c(x_{ij}) : -1, -2, 1$

The k-factor interaction is said to be linear or quadratic in the factor $A_{ij}, j = 1, 2, \ldots, k$ according as the corresponding set of coefficients $c(x_{ij})$ is based on the orthogonal polynomial of the first degree or the second degree. There are 2^k contrasts belonging to the k-factor interaction between k factors, $A_{i1}, A_{i2}, \ldots, A_{ik}$. For example, the 2-factor interaction linear in A_1 and quadratic in A_2 in a 3^3 design is given by

$$A_{1\ell}A_{2q} = (a_2^1 - 1)(a_2^2 - 2a_2 + 1)(a_3 + a^3 + 1).$$

The sum of squares due to any contrast Z is obtained by

$$SS(Z) = [Z]^2/rS(Z),$$

where [Z] = total for the contrast obtained by substituting treatment yields for treatments in the expression for Z and S(Z) = sum of squares of the coefficients which occur in the expression for Z.

1.6 CONSTRUCTION OF DESIGNS

33. **Construction of MOLS**

When s is a prime or a power prime, the complete set of (s − 1) mutually-orthogonal Latin Squares of side s is construced as follows:

(i) Write down the elements of GF(S) as $u_o = 0, u_1 = 1, u_2 = x, u_3 = x^2, \ldots, u_{s-1} = x^{s-2}$, where x is a primitive root of GF(S).

(ii) The rows and columns are numbered as $0, 1, 2, \ldots, s-1$.

(iii) The (s − 1) Latin Squares are denoted by $L_1, L_2, \ldots, L_{s-1}$.

(iv) L_1 is called the key Latin Square and is constructed as follows:

(a) Write down the 0-th row of L_1 as $0, 1, 2, \ldots, (s-1)$

(b) The first row of L_1 is constructed by filling its $(1, \beta)$-th cell by the subscript of the element $u_1 u_{1t} u_\beta = 1 + u_\beta$, $\beta = 0, 1, 2, \ldots, (s-1)$.

(c) The cells in the upper triangle of L_1 are filled by the following rule: if the cell (α, β) contains 0, then write 0 in the cell $(\alpha + 1, \beta + 1)$; if the cell (α, β) contains j, $(j = 1, 2, \ldots, s-2)$, then write $j + 1$ in the cell $(\alpha + 1, \beta + 1)$, if the cell (α, β) contains $(s-1)$, then write 1 in the cell $(\alpha + 1, \beta + 1)$.

(d) The cells in the lower triangle of L_1 are filled by noting the symmetry from the cells of the upper triangle of L_1.

(v) The Latin square L_i, $i = 2, 3, \ldots, (s-1)$, is constructed from the Latin square L_{i-1} as follows.

(a) The 0-th row of L_i is taken as $0, 1, 2, .., (s-1)$

(b) The 1st, 2nd, ..., (s − 2)th rows of Li are obtained from the 2nd, 3rd, ..., (s − 1)th rows of L_{i-1} by moving them one step up. The last row of L_i is taken as the first row of L_{i-1}.

If s is neither a prime nor a power prime, then exactly n(s) MOLS of side s can be constructd where

$$n(s) = \min(p_1^{e_1}, p_2^{e_2}, \ldots, p_m^{e_m}) - 1,$$

and $s = p_1^{e_1} p_2^{e_2}, \ldots p_m^{e_m}$, where $p_1, p_2, \ldots, p_m$ are distinct prime numbers and $e_1, e_2, \ldots, e_m$ are positive integers. We now describe the construction of n(s) MOLS in this case.

(i) Denote the elements of $GF(p_i^{e_i})$, $i = 1, 2, \ldots, m$ by

$$g_{io} = 0, g_{i1} = 1, g_{i2} = \alpha_i, g_{i3} = \alpha_i^2, \ldots, g_i p_i^{e_i - 1} = \alpha_i p_i^{e_i - 2},$$

where α_i is a primitive root of GF $(p_i^{e_i})$

(ii) Form the set {w} of elements w defined by

$$w = (g_{1\ell_1}, g_{2\ell_2}, \ldots, g_{m\ell_m}),$$

where $g_{1\ell_1} \in GF(p_1^{e_1}), g_{2\ell_2} \in GF(p_2^{e_2}), \ldots, g_{m\ell_m} \in GF(p_m^{e_m})$.

(iii) Define the operations of addition and multiplication among the elements of the set $\{w\}$ as follows:

$$w_1 + w_2 = (g_{1\ell_1} + g_{1j_1}, g_{2\ell_2} + g_{2j_2}, \ldots, g_{m\ell_m} + g_{mj_m})$$
$$w_1 w_2 = (g_{1\ell_1} \cdot g_{ij_1}, g_{2\ell_2} \cdot g_{2j_2}, \ldots, g_{m\ell_m} \cdot g_{mj_m})$$

where $w_1 = (g_{1\ell_1}, g_{2\ell_2}, \ldots, g_{m\ell_m})$ and $w_2 = (g_{1j_1}, g_{2j_2}, \ldots, g_{mj_m})$ are any two elements of the set $\{w\}$.

(iv) Number of elements of the set $\{w\}$ such that the first $n(s) + 1$ elements of the set $\{w\}$ are

$$w_j = (g_{1j}, g_{2j}, \ldots, g_{mj}), j = 0, 1, 2, \ldots, n(s)$$

while the remaining elements of the set $\{w\}$ are numbered arbitrarily.

(v) The j-th Latin Square $L_{j,j} = 1, 2, \ldots, n(s)$ is constructed by filling its (α, β)-th cell by the element (or its subscript)

$$w_j w_\alpha + w_\beta, \alpha, \beta = 0, 1, 2, \ldots, s - 1, j = 1, 2, \ldots, n(s).$$

34. **Construction of BIBDs**

(1) **Use of PG (m, s).** Identify a point of PG (m, s) with a treatment and a g-flat $(1 \leq g \leq m - 1)$ of PG (m, s) with a block. Then, a BIBD with the following parameters is constructed,

$$v = (s^{m+1} - 1)/(s - 1),$$
$$b = \phi(m, g, s)$$
$$r = \phi(m - 1, g - 1, s)$$
$$k = (s^{g+1} - 1)/(s - 1)$$
$$\lambda = \phi(m - 2, g - 2, s),$$

where

$$\phi(m, g, s) = \frac{(s^{m+1} - 1)(s^m - 1)\ldots(s^{m-g+1} - 1)}{(s^{g+1} - 1)(s^g - 1)\ldots(s - 1))}.$$

(2) **Use of EG (m, s).** Consider EG (m, s). Identify the points of EG (m, s) with the treatments and g-flats $(1 \leq g \leq m - 1)$ of EG (m, s) as blocks. Then, a BIBD with the following parameters is constructed.

$$v = s^m$$
$$b = \phi(m, g, s) - \phi(m - 1, g, s)$$
$$r = \phi(m - 1, g - 1, s)$$
$$k = s^g$$
$$\lambda = \phi(m - 2, g - 2, s).$$

(3) **Use of a complementary design of a BIBD.** Let D, a BIBD (v, b, r, k, λ), be known to exist. Then, taking its complementary design, we construct a BIBD with the following parameters:

$$v^* = v, b^* = b, r^* = b - r, k^* = v - k, \lambda^* = b - 2r + \lambda.$$

(4) **Use of a residual design of a SBIBD.** Let D, a SBIBD ($v = b$, $r = k, \lambda$), be known to exist. Then, omitting one block and all its treatments from the remaining blocks, we obtain its residual design which is a BIBD with the following parameters:

$$v_1 = v - k, b_1 = b - 1, r_1 = r, \quad k_1 = r - \lambda, \lambda_1 = \lambda.$$

(5) **Use of a derived design of a SBIBD.** Let D, a SBIBD ($v = b, r = k, \lambda$), be known to exist. Then, omitting one block and retaining its treatments in the remaining blocks, we obtain a derived design which is a BIBD with the following parameters:

$$v_2 = k, b_2 = b - 1, r_2 = r - 1, \quad k_2 = \lambda, \lambda_2 = \lambda - 1.$$

(6) **Orthogonal Series of Yates.** The two series of BIBDs due to Yates are as follows:

$$\begin{aligned} &OS1: \ v = s^2, b = s^2 + s, r = s + 1, k = s, \lambda = 1 \\ &OS2: \ v = b = s^2 + s + 1, r = k = s + 1, \lambda = 1 \end{aligned}$$

where s is a prime or a power prime.

(i) **Construction of OS 1.** Construct a complete set of MOLS of side $s : L_1, L_2, \ldots, L_{s-1}$. The $v = s^2$ treatments are arranged in a $s \times s$ square L, say. Take rows and columns of L as blocks. We thus get 2s blocks, each of size s. The blocks from rows and columns form two separate replications. Now take the Latin square $L_i, i = 1, 2, \ldots, s - 1$ and superimpose it on L and form blocks by taking treatments which correspond to the same letters of L_i. Thus, each Latin square will give s blocks forming one replication. Then, we get in all $b = 2s + s(s - 1) = s^2 + s$ blocks with $k = s, r = s + 1$ and $\lambda = 1$.

(ii) **Construction of OS 2.** Out of the $v = s^2 + s + 1$ treatments, take s^2 treatments and construct $s^2 + s$ blocks as in OS 1. To the blocks of each replication, add one treatment from the remaining treatments. Thus we get a BIBD with $v = b = s^2 + s + 1, r = k = s + 1, \lambda = 1$.

It may be noted that OS 1 can be constructed by using EG(2, s) and that OS 2 can be constructed by using PG(2, s).

(7) **The first fundamental theorem of symmetric differences.** Let M be a module of m elements and there be $v = mn$ treatments. Suppose that n treatments are associated with each element of M. The treatments associated with an element a of M are denoted by $a_1, a_2, \ldots, a_n$. Suppose that it is possible to find a set of t blocks $B_1, B_2, \ldots, B_t$ which satisfy the following conditions:

(i) Each block contains k treatments.

(ii) Of the tk treatments occuring in t blocks, exactly r belong to each class, thus $tk = nr$.

(iii) The differences arising from the t blocks are symmetrically repeated, each occurring λ times.

If a_i is any treatment and θ any element of M, then we define $a_i + \theta = (a + \theta)_i$. From each block B_i, $i = 1, 2, \ldots, t$, we form another block $B_{i\theta}$ by adding the element θ to the treatments of B_i. Form all mt blocks $B_{i\theta}$ by taking all values of i and θ. Then, these mt blocks form a BIBD with parameters

$$v = mn, b = mt, r, k, \lambda.$$

As an application of the above theorem, we get the following BIBDs.

(i) **Bose T_1 Series.**Consider a module M consisting of residues mod $(2s+1)$. Thus, $m = 2s+1$. To each element of M, associate 3 treatments, thus $n = 3$, and hence $v = mn = 6s+3$. Take the following $t = 3s+1$ blocks as the initial blocks:

$\longleftarrow$ first, s blocks $\longrightarrow$
$[1_1, (2s)_1, 0_2], [2_1, (2s-1)_1, 0_2], \ldots, [s_1, (s+1)_1, 0_2],$
$\longleftarrow$ second s blocks $\longrightarrow$
$[1_2, (2s)_2, 0_3], [1_2, (2s-1)_2, 0_3], \ldots, [s_2, (s+1)_2, 0_3],$
$\longleftarrow$ third s blocks $\longrightarrow$
$[1_3, (2s)_3, 0_1], [1_3, (2s-1)_3, 0_1], \ldots, [s_3, (s+1)_3, 0_1]$
last block $[0_1, 0_2, 0_3]$.

Then, adding the elements of M in succession, to the treatments of the above initial blocks, we get a BIBD with parameters

$$v = 6s+3, b = (3s+1)(2s+1), r = 2s+1, k = 3, \lambda = 1.$$

(ii) **Bose T_2 Series.** Let $s = 6t+1$, where t is a positive integer and s is a prime or a power prime. To each element of GF(s), associate one treatment. Thus $m = s$ and $n = 1$. Let x be a primitive root of GF(s). Take the following t blocks as the initial blocks;

$$[x^0, x^{2t}, x^{4t}], [x, x^{2t+1}, x^{4t+1}], \ldots, [x^{t-1}, x^{3t-1}, x^{5t-1}].$$

Then, adding the elements of GF(s) successively, to the treatments of the above initial blocks, we get a BIBD with parameters

$$v = 6t+1, b = t(6t+1), r = 3t, k = 3, \lambda = 1.$$

(iii) Let $s = 4t+3$ be a prime or a power prime, where t is any positive integer. To each element of GF(s), associate one treatment so that m = s and n = 1. Let x be a primitive root of GF(s). Take the following block as the initial block:

$$x^0, x^2, x^4, \ldots, x^{4t}.$$

Then, adding the elements of GF(s) successively to the treatments of the above initial block, we get a BIBD with parameters

$$v = b = 4t + 3, r = k = 2t + 1, \lambda = t.$$

(iv) Let $s = 4t + 1$ be a prime or a power prime, where t is any positive integer. Associate one treatment with each element of GF(s), so that $m = s$, $n = 1$. Let x be a primitive root of GF(s). Take the following 2 blocks as initial blocks:

$$x^0, x^2, x^4, \ldots, x^{4t-2}$$
$$x, x^3, x^5, \ldots, x^{4t-1}.$$

Then, adding the elements of GF(s) successively to the treatments of the above two initial blocks, we get a BIBD with parameters

$$v = 4t + 1, b = 8t + 2, r = 4t, k = 2t, \lambda = 2t - 1.$$

(8) **The second fundamental theorem of symmetric differences.** Let M be a module of m elements and let there be $v = mn$ treatments. Associate n treatments with each element of M. The treatments associated with an element a of M are denoted by $a_1, a_2, \ldots, a_n$. These $v = mn$ treatments are called finite treatments. We add a new treatment called an infinite treatment which will be donoted by ∞. Suppose that it is possible to find $t + s$ blocks $B_1, B_2, \ldots, B_t, B_1^*, B_2^*, \ldots, B_s^*$ satisfying the following conditions:

(i) Each block of the set $B_1, B_2, \ldots, B_t$ contains k different finite treatments, and each block of the set $B_1^*, B_2^*, \ldots, B_s^*$ contains $(k - 1)$ distinct finite treatments and the infinite treatment ∞.

(ii) Of the m finite treatments of the i-th class, $i = 1, 2, \ldots, n$, exactly $(ms - \lambda)$ treatments occur in blocks $B_1, B_2, \ldots, B_t$ and λ treatments occur in the blocks $B_1^*, B_2^*, \ldots, B_s^*$. Thus $kt = (ms - \lambda)n$ and $s(k - 1) = n\lambda$.

(iii) The differences among the finite treatments arising from the $t + s$ blocks are symmetrically repeated, each occurring λ times.

The addition of an element θ of the module M to the treatments of the blocks is defined as; $a_i + \theta = (a + \theta)_i$. Also we define $\infty + \theta = \infty$. The blocks $B_1, B_2, \ldots, B_t$ and $B_1^*, B_2^*, \ldots, B_s^*$ are called the initial blocks. Then, adding the elements of M, one by one, to the treatments of the above $(t + s)$ blocks, we get a BIBD with parameters

$$v = mn + 1, b = m(t + s), r = ms, k, \lambda.$$

As an application of the above result, we construct the following BIBD.

Let $s = 4t + 1$ be a prime or a power prime, where t is a positive integer. To each element of GF(s), associate 3 treatments. We also take one more infinite treatment denoted by ∞. Let x be a primitive root of GF(s). Take the following $(3t + 1)$ blocks as the initial blocks.

$$\begin{aligned}
&[x_1^{2i}, x_1^{2t+2i}, x_2^{2i+\alpha}, x_2^{2t+2i+\alpha}],\\
&[x_2^{2i}, x_2^{2t+2i}, x_3^{2i+\alpha}, x_3^{2t+2i+\alpha}],\\
&[x_3^{2i}, x_3^{2t+2i}, x_1^{2i+\alpha}, x_1^{2t+2i+\alpha}], \qquad i = 0, 1, 2, \ldots, (t-1)\\
&[\infty, 0_1, 0_2, 0_3],
\end{aligned}$$

where α is chosen so that $\frac{x^\alpha+1}{x^\alpha-1} = x^2, q = 1(\text{mod}2)$. Then, by adding the elements of GF(s), one by one, to the treatments of the above initial blocks, we get a BIBD with parameters

$$v = 12t + 4, b = (3t + 1)(4t + 1), r = 4t + 1, k = 4, \lambda = 1.$$

35. **Construction of a Youden Square.**

Take a SBIBD with parameters $v = b, r = k, \lambda$. Denote the v treatments by v integers 1, 2, .., v, and write the blocks of the SBIBD as columns. Let S = the set of treatments and $S_1, S_2, \ldots, S_v$, the columns (blocks) of the SBIBD be the subsets of S. The columns $S_1, S_2, \ldots, S_v$ possess a SDR, which is a permutation of integers $1, 2, \ldots, v$. Take this SDR as the first row. Delete this row from $S_1, S_2, \ldots, S_v$. Denote the new columns by $S_1^*, S_2^*, \ldots, S_v^*$. Then, these columns $S_1^*, S_2^*, \ldots, S_v^*$ possess a SDR. Take this SDR as the second row. Proceeding in this way, we get $r = k$ SDRs, which are taken as rows. These $r = k$ rows form a Youden Square.

36. **Construction of PBIBD Designs.**

(1) **Use of PG(m, s).** Consider a finite projective geometry PG(m, s). Omit one point P, say, from this geometry and take the remaining points as treatments. Take g-flats $(1 \le g \le m - 1)$, not passing through the omitted point P as blocks. Then two treatments will be called *first associates* if they occur together in the same block, if they do not occur together in the same block, they will be called *second associates*. Then, we get a PBIBD with parameters

$$\begin{aligned}
v &= s(s^m - 1)/(s - 1),\\
b &= \phi(m, g, s) - \phi(m - 1, g - 1, s)\\
r &= \phi(m - 1, g - 1, s) - \phi(m - 2, g - 2, s)\\
k &= (s^{g+1} - 1)/(s - 1)\\
n_1 &= s^2(s^{m-1} - 1)/(s - 1),\\
n_2 &= (s - 1),\\
\lambda_1 &= \phi(m - 2, g - 2, s) - \phi(m - 3, g - 3, s)\\
\lambda_2 &= 0\\
P_1 &= \begin{bmatrix} n_1 - n_2 - 1 & n_2 \\ n_2 & 0 \end{bmatrix}, P_2 = \begin{bmatrix} n_1 & 0 \\ 0 & n_2 - 1 \end{bmatrix}.
\end{aligned}$$

Another PBIBD can be constructed by using PG(m, s) as follows. Select a point P and choose t lines passing through it. Take points on these t lines other than P as treatments. Take (m − 1)-flats not passing through the point P as blocks. Two treatments are *first associates* if they lie on one of the chosen lines through P, otherwise they are *second associates*. Then, we get a PBIBD with parameters

$$v = st, b = s^m, r = s^{m-1}, k = t$$
$$n_1 = s - 1, n_2 = s(t-1), \lambda_1 = 0, \lambda_2 = s^{m-2},$$
$$P_1 = \begin{bmatrix} n_1 - 1 & 0 \\ 0 & n_2 \end{bmatrix}, P_2 = \begin{bmatrix} 0 & n_1 \\ n_1 & n_2 - n_1 - 1 \end{bmatrix}.$$

(2) **Use of EG(m, s).** Consider a finite Euclidean Geometry E(m, s). Omit one point P, say, and all g-flats ($1 \leq g \leq m - 1$) through P. take the remaining points as treatments and the g-flats, not containing the point P as blocks. Two treatments are *first associates* if they occur together in the same block, otherwise they are called *second associates*. Then, we get a PBIBD with the parameters

$$v = s^m - 1,$$
$$b = \phi(m, g, s) - \phi(m-1, g-1, s) - \phi(m-1, g, s),$$
$$r = \phi(m-1, g-1, s) - \phi(m-2, g-2, s)$$
$$k = s^g$$
$$n_1 = s^m - s, n_2 = s - 2$$
$$\lambda_1 = \phi(m-2, g-2, s) - \phi(m-3, g-3, s)$$
$$\lambda_2 = 0,$$
$$P_1 = \begin{bmatrix} n_1 - n_2 - 1 & n_2 \\ n_2 & 0 \end{bmatrix}, P_2 = \begin{bmatrix} n_1 & 0 \\ 0 & n_2 - 1 \end{bmatrix}$$

(3) **Other methods.**

(i) Consider a double triangle as shown below.

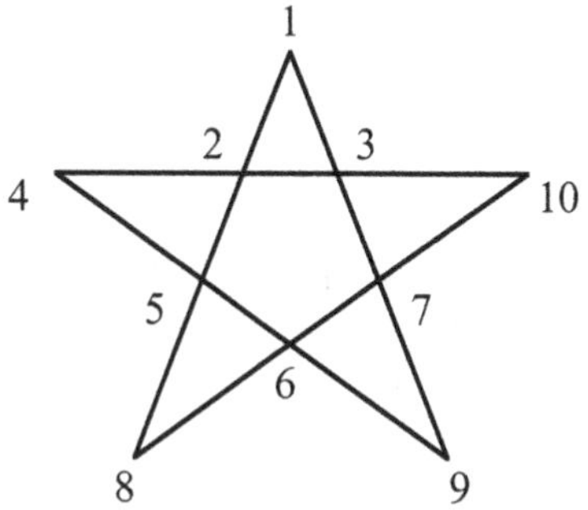

Denote the vertices by numbers 1, 2, ..., 10 and take them as treatments. Take lines as blocks. Two treatments are *first associates*

if they lie on the same line, if they do not lie on the same line, they are *second associates*. Then, we get a PBIBD with parameters $v = 10, b = 5, r = 2, k = 4, n_1 = 6, n_2 = 3, \lambda_1 = 1, \lambda_2 = 0,$

$$P_1 = \begin{bmatrix} 3 & 2 \\ 2 & 1 \end{bmatrix}, P_2 = \begin{bmatrix} 4 & 2 \\ 2 & 0 \end{bmatrix}.$$

(ii) Consider a parallelopiped as shown below

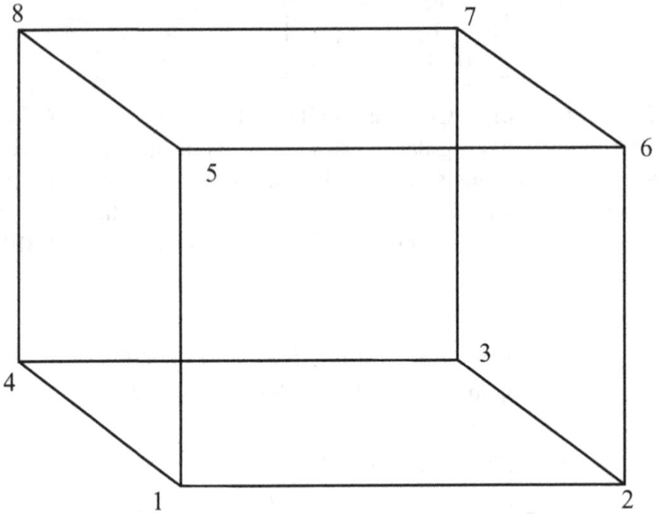

Denote the vertices of the parallelopiped by numbers 1, 2, ..., 8, and take them as treatments. Take faces as blocks. Two treatments which lie in the same face but not on the same edges are *first associates*; two treatments on the same edges are *second associates*; two treatments whichn lie on the diagonals of the parallelepiped are *third associates*. Then, we get a PBIBD with parameters

$$v = 8, b = 6, r = 3, k = 4,$$
$$n_1 = 3, n_2 = 3, n_3 = 1,$$
$$\lambda_1 = 1, \lambda_2 = 2, \lambda_3 = 0,$$

$$P_1 = \begin{bmatrix} 2 & 0 & 0 \\ 0 & 2 & 1 \\ 0 & 1 & 0 \end{bmatrix}, P_2 = \begin{bmatrix} 0 & 2 & 1 \\ 2 & 0 & 0 \\ 1 & 0 & 0 \end{bmatrix}, P_3 = \begin{bmatrix} 0 & 3 & 0 \\ 3 & 0 & 0 \\ 0 & 0 & 0 \end{bmatrix}.$$

(iii) Let $v = pq$. Arrange the $v = pq$ treatments in an array of p rows and q columns. Blocks are formed by taking each treatment and treatments in the same row and the same column as that treatment. Two treatments are *first associates* if they lie in the same row; they are *second associates* if they lie in the same column; otherwise they are *third associates*. Then, we get a PBIBD with parameters

$$v = b = pq, r = k = p + q - 1,$$
$$n_1 = q - 1, n_2 = p - 1, n_3 = (p - 1)(q - 1),$$
$$\lambda_1 = q, \lambda_2 = p, \lambda_3 = 2.$$
$$P_1 = \begin{bmatrix} q-2 & 0 & 0 \\ 0 & 0 & p-1 \\ 0 & p-1 & (p-1)(q-2) \end{bmatrix},$$
$$P_2 = \begin{bmatrix} 0 & 0 & q-1 \\ 0 & p-2 & 0 \\ q-1 & 0 & (p-2)(q-1) \end{bmatrix},$$
$$P_3 = \begin{bmatrix} 0 & 1 & q-2 \\ 1 & 0 & p-2 \\ q-2 & p-2 & (p-2(q-2)) \end{bmatrix}.$$

(iv) Let $v = pq$. Arrange the $v = pq$ treatments in an array of p rows and q columns. Blocks are formed taking treatments that occur in the same row and the same column as each treatment but excluding that treatment. Two treatments are *first associates* if they occur in the same row; they are *second associates* if they occur in the same column; otherwise they are *third associates*. Then we get a PBIBD with parameters

$$v = b = pq, r = k = p + q - 1,$$
$$n_1 = q - 1, n_2 = p - 1, n_3 = (p - 1)(q - 1),$$
$$\lambda_1 = q - 2, \lambda_2 = p - 2, \lambda_3 = 2,$$

and P_1, P_2, and P_3 are same as in (iii).

(v) Let $v = p^2$. Arrange the $v = p^2$ treatments in a $p \times p$ square. Blocks are formed by taking treatments that occur in the same row, the same column and which correspond to the same letter of a $p \times p$ Latin square as each treatment. Two treatments are *first associates* if they occur in the same row; they are *second associates* if they occur in the same column; they are *third associates* if they correspond to the same letter of the $p \times p$ Latin square; otherwise they are *fourth associates*. Then we get a PBIBD with parameters

$$v = b = p^2, r = k = (3p - 2),$$
$$n_1 = n_2 = n_3 = p - 1, n_4 = (p - 1)(p - 2),$$
$$\lambda_1 = \lambda_2 = \lambda_3 = p + 2, \lambda_4 = 6,$$

$$P_1 = \begin{bmatrix} p-2 & 0 & 0 & 0 \\ 0 & 0 & 1 & p-2 \\ 0 & 1 & 0 & p-2 \\ 0 & p-2 & p-2 & (p-2)(p-3) \end{bmatrix},$$

$$P_2 = \begin{bmatrix} 0 & 0 & 1 & p-2 \\ 0 & p-2 & 0 & 0 \\ 1 & 0 & 0 & p-2 \\ p-2 & 0 & p-2 & (p-2)(p-3) \end{bmatrix},$$

$$P_3 = \begin{bmatrix} 0 & 1 & 0 & p-2 \\ 1 & 0 & 0 & p-2 \\ 0 & 0 & p-2 & 0 \\ p-2 & p-2 & 0 & (p-2)(p-3) \end{bmatrix},$$

$$P_4 = \begin{bmatrix} 0 & 1 & 1 & p-3 \\ 1 & 0 & 1 & p-3 \\ 1 & 1 & 0 & p-3 \\ p-3 & p-3 & p-3 & p^2-6p+10 \end{bmatrix}.$$

37. **Confounded 2^m Design.**

(1) To confound one interaction X. The two blocks are constructed as follows. The first block is formed by taking treatments having even number of letters in common with X. The second block is constructed by taking treatments obtained by multiplying the treatments of the first block with a treatment not included in the first block and replacing the square of any letter in the product by unity.

(2) To confound two interactions X and Y. Construct the key block B_1 by taking treatments which have even number of letters in common with X and Y. The block B_i, $i = 2, 3, 4$ is constructed by taking the products of a treatment not included in the blocks $B_1, B_2, \ldots, B_{i-1}$ with treatments of the key block B_1, and replacing the square of any letter in the products by unity. The interactions X, Y and XY are confounded.

(3) To confound k independent interactions. We construct 2^k blocks as follows. The key block B_1 is constructed by taking treatments which have even number of letters in common with the k independent interactions $X_1, X_2, \ldots, X_k$. The block B_i, $i = 2, 3, \ldots, 2^k$ is constructed by taking the products of a treatment not included in the blocks $B_1, B_2, \ldots, B_{i-1}$ with the treatments of the key block B_1 and replacing the square of any letter in the products by unity. Each block will contain 2^{m-k} treatments. The following interactions are confounded:

(i) the k interactions X_i, $i = 1, 2, \ldots, k$

(ii) the $\binom{k}{2}$ 2-factor interactions X_iX_j, $i \neq j = 1, 2, \ldots, k$

(iii) the $\binom{k}{3}$ 3-factor interactions $X_iX_jX_k$, $i \neq j \neq k = 1, 2, \ldots, k$ and so on, the last interaction being the k-factor interaction $X_1X_2 \ldots X_k$.

38. **Confounded s^m Design.**

The treatments are identified by the points of EG(m, s). The interactions are identified by the pencils $P(a_1, a_2, \ldots, a_m)$. The pencils are written so that the first non-zero element is 1. We assume that s is a prime or a power prime.

(1) To confound the (s – 1) d.f. carried by the pencil $P(a_1, a_2, \ldots, a_m)$: Consider the equations

$$a_1x_1 + a_2x_2 + \ldots + a_mx_m = \alpha_i, i = 0, 1, 2, \ldots, s-1$$

of the pencil $P(a_1, a_2, \ldots, a_m)$. The treatments lying on each equation form a block. Thus, we get s blocks between which the $(s-1)$ d.f. carried by the pencil $P(a_1, a_2, \ldots, a_m)$ are confounded. Each block contains s^{m-1}treatments. A simple method of writing these s blocks is as follows. Write the key block B_1 by taking treatments which are the solutions of the equation

$$a_1x_1 + a_2x_2 + \ldots a_mx_m = 0.$$

The other $s-1$ blocks are obtained by adding the non-null elements to the first co-ordinates of the treatments of the key block B_1.

(2) To confound the d.f. belonging to the k independend pencils $P(a_{i1}, a_{i2}, \ldots, a_{im}), i = 1, 2, \ldots, k$: The s^k blocks are cosntructed by taking treatments which satisfy k equations, one being taken from each of the sets of the equations defining those k pencils. The d.f. carried by the $(s^k - 1)/(s-1)$ generalized pencils

$$P\left(\sum_1^k \lambda_i a_{i1}, \sum_1^k \lambda_i a_{i2}, \ldots, \sum_1^k \lambda_i a_{im}\right)$$

where $\lambda_1, \lambda_2, \ldots, \lambda_k$ are elements of GF(s), not all zero, are confounded with the s^k blocks. Each block contains s^{m-k} treatments. The above construction of s^k blocks is simplified as follows. The key block B_1 is constructed by taking treatments which satisfy the equations

$$a_{i1}x_1 + a_{i2}x_2 + \ldots + a_{im}x_m = 0, i = 1, 2, \ldots, k.$$

Write down k coordinates $(x_1x_2 \ldots x_k)$ in all possible s^k ways by assigning s values of the elements of GF(s) to each of $x_1, x_2, \ldots, x_k$. Omit the combination $(0\ 0 \ldots 0)$. The remaining $(s^k - 1)$ combinations are used to obtain the remaining $(s^k - 1)$ blocks as follows. To the first k coordinates of the treatments of the key block B_1, add the above $(s^k - 1)$ combinations, one by one, and obtain the remaining $(s^k - 1)$ blocks. These $(s^k - 1)$ blocks together with the key block form all the s^k blocks.

39. **Construction of $(s-1)^{m-1}$ replications of a (s^m, s) design achieving complete balance over the highest order interaction.**

We write down $(s-1)^{m-1}$ pencils $P(a_1, a_2, \ldots, a_m)$ belonging to the m-factor interaction, where the $a_1, a_2, \ldots, a_m$ are non-null elements of GF(s). We confound each of these pencils in one replication to construct a (s^m, s) design. Since, there are $(s-1)^{m-1}$ pencils, there will be $(s-1)^{m-1}$ replications. Each pencil belonging to the m-factor interaction is thus confounded in one replication and is unconfounded in the $(s-1)^{m-1}-1$ replications and hence the m-factor interaction is completely balanced and the relative loss of information on each d.f. belonging to the m-factor interaction is $1/(s-1)^{m-1}$.

40. **To construct $(s-1)^{m-1}$ replications of a (s^m, s^{m-1}) design, achieving complete balance over all order interaction from 1st to (m − 1)-th.** Write down distinct $(s-1)^{m-1}$ vertices of the type $(\alpha_1, \alpha_2, \ldots, \alpha_m)$ in which each $\alpha_i, i = 1, 2, \ldots, m$ is non-null element of the GF(s). Corresponding to each vertex $(\alpha_1, \alpha_2, \ldots, \alpha_m)$, take (m − 1) independent pencils $P(a_{i1}, a_{i2}, \ldots, a_{im}), i = 1, 2, \ldots, m-1$ such that $\sum_1^m \alpha_j a_{ij} = 0, i = 1, 2, \ldots, m-1$, and construct one replication of a (s^m, s^{m-1}) design by confounding these (m - 1) pencils $P(a_{i1}, a_{i2}, \ldots, a_{im}), i = 1, 2, \ldots, m-1$. Then we get $(s-1)^{m-1}$ replications of (s^m, s^{m-1}) in which no main effect is confounded and there is complete balance on 1st order, 2nd order, ..., (m − 1)-th order interactions. The relative loss of information on the (k − 1)-th order interaction is

$$[(s-1)^{k-1} - (-1)^{k-1}]/s(s-1)^{k-1}$$

41. **Construction of a Fractional Replicate of a Design.**
(1) To construct a half replicate of a 2^m design: Let the defining interaction be X. If X contains an odd (even) number of letters, then the half replicate of a 2^m design is formed by taking treatments which have an odd (even) number of letters in common with X. The alias of any interaction Y is the generalized interaction XY, in which the square of a letter is replicated by unity.
(2) To construct $(1/2^p)$-th replicate of a 2^m design: Let $X_1, X_2, \ldots, X_p$ be p independent interactions selected as defining interactions. The $(1/2^p)$-th replicate of a 2^m design is constructed by taking treatments which have an odd (even) number of letters in common with $X_i, i = 1, 2, \ldots, p$, if X_i contains odd (even) number of letters. The generalized interactions of $X_1, X_2, \ldots, X_p$ are also the defining interactions, and their number will be $(2^p - 1)$. The aliases of any interaction Y are obtained by multiplying Y with each of the $(2^p - 1)$ defining interactions and replacing the square of any letter by unity in the products.

(3) To construct the (1/s)-th replicate of a s^m design: We assume s is a prime or a power prime. Identify the points of an EG(m, s) with the treatments of a s^m design. Let the pencil $P(a_1, a_2, \ldots, a_m)$ be the defining pencil. The (1/s)-th replicate of a s^m design is constructed by taking treatments which satisfy

$$a_1x_1 + a_2x_2 + \ldots + a_mx_m = 0,$$

The constrasts belonging to the pencil $P(a_1, a_2, \ldots, a_m)$ are not estimable. The aliases of any other pencil $P(b_1, b_2, \ldots, b_m)$ are its generalized pencils with the defining pencil $P(a_1, a_2, .., a_m)$, that is, are the pencils

$$P(b_1 + \lambda a_1, b_2 + \lambda a_2, \ldots, b_m + \lambda a_m),$$

where λ is a non-null element of GF(s). Thus each alias set consists of s pencils and there will be $(s^{m-1} - 1)/(s - 1)$ alias sets.

(4) To construct $(1/s^p)$-th replicate of a s^m design: Let us select p independent pencils $P(a_{i1}, a_{i2}, \ldots, a_{im})$, $i = 1, 2, \ldots, p$ as defining pencils. Their generalized pencils

$$P\left(\sum_1^p \lambda_i a_{i1}, \sum_1^p \lambda_i a_{i2}, \ldots, \sum_1^p \lambda_i a_{im}\right),$$

where λ's are elements of GF(s), not all zero are also defining pencils and information on the d.f. carried by these pencils is lost. The $(1/s^p)$-th replicate is formed by taking treatments which satisfy the equations

$$a_{i1}x_1 + a_{i2}x_2 + \ldots + a_{im}x_m = 0\ , \ i = 1, 2, \ldots, p.$$

The aliases of any other pencil $P(b_1, b_2, \ldots, b_m)$ are the pencils

$$P\left(b_1 + \sum_1^p \lambda_i a_{i1}, b_2 + \sum_1^p \lambda_i a_{i2}, \ldots, b_m + \sum_1^p \lambda_i a_{im}\right),$$

where λ's are elements of GF(s) not all zero.

There will be $(s^p - 1)$ alias pencils of a given pencil and hence each set of alias pencils will consist of s^p pencils. The number of alias sets of pencils will be $(s^{m-p} - 1)/(s - 1)$.

42. **Confounded Fractional Replicate Designs**

(1) To confound k independent interactions in a half replicate of a 2^m design: Let the interaction X be the defining interaction. The information on X is lost. Suppose, we wish to confound k independent interactions $Y_1, Y_2, \ldots, Y_k$. The generalized interactions of $Y_1, Y_2, \ldots, Y_k$ and their aliases will also be confounded. The number of generalized interactions of $Y_1, Y_2, \ldots, Y_k$ is $2^k - 1$. Hence, there will be $(2^k - 1)$ alias sets, each consisting of 2 interactions, which will be confounded. The construction for confounding $Y_1, Y_2, \ldots, Y_k$ in a half replicate of a 2^m design obtained by selecting X as the defining interaction is as follows. We construct the half replicate by selecting treatments which have odd (even) number of letters in common with X if X contains odd (even) number of letters. Suppress one letter in these treatments so as to express them as treatments of a 2^{m-1} design in the standard form. Express the k interactions $Y_1, Y_2, \ldots, Y_k$ so that they contain letters of the factors of a 2^{m-1} design. Then, using the construction described above in 37(3), we construct 2^k blocks confounding the interactions $Y_1, Y_2, \ldots, Y_k$. Each block will contain 2^{m-k-1} treatments. Introduce the suppressed letter in the treatments in which it was suppressed.

(2) To confound k independent interactions in a $(1/2^p)$-th replicate of a 2^m design: Let $X_1, X_2, \ldots, X_p$ be p independent interactions to be used as the defining interactions for constructing the $(1/2^p)$-th replicate. Let $Y_1, Y_2, \ldots, Y_k$ be the k independent interactions which are to be confounded. Using the construction described above in 41 (2), we select the treatments of the $(1/2^p)$-th replicate. Suppress p suitably chosen letters in these treatments so that they become treatments of a 2^{m-p} design in the standard form. If any $Y_i, i = 1, 2, \ldots, k$ contains the suppressed letters, we take its alias which does not contain the suppressed letters. Thus, we express $Y_1, Y_2, \ldots, Y_k$ so that they contain letters of the factors of a 2^{m-p} design. Then, using the construction described in 37 (3), we construct 2^k blocks confounding the interactions $Y_1, Y_2, \ldots, Y_k$. Each block will contain 2^{m-p-k} treatments.

Information on $X_1, X_2, \ldots, X_p$ and their generalized interactions is lost. Thus information on $2^p - 1$ d.f. is lost.

Interactions $Y_1, Y_2, \ldots, Y_k$ and their generalized interactions and their aliases are confounded. Thus, there will be $(2^k - 1)$ alias sets of 2^p interactions each, that will be confounded.

(3) To confound k independent pencils in a $(1/s^p)$-th replicate of a s^m design: We assume here that s is a prime or a power prime. Let $P(a_{i1}, a_{i2}, \ldots, a_{im}), i = 1, 2, \ldots, p$ be p independent pencils used as the defining pencils for the construction of a $(1/s^p)$-th replicate. Suppose we wish to confound k independent pencils $P(b_{j1}, b_{j2}, \ldots, b_{jm}), j = 1, 2, \ldots, k$, none of which belongs to the set of the generalized pencils $P(a_{i1}, a_{i2}, \ldots, a_{im}), i = 1, 2, \ldots, p$. Let $S(\alpha_{i1}, \alpha_{i2}, \ldots, \alpha_{ik})$ denote the set of treatments which satisfy the equations

$$a_{i1}x_1 + a_{i2}x_2 + \ldots + a_{im}x_m = 0, i = 1, 2, \ldots, p$$
$$b_{j1}x_1 + b_{j2}x_2 + \ldots, b_{jm}x_m = \alpha_j, j = 1, 2, \ldots, k$$

where α's are elements of GF(s).

The number of equations in the above system is (p + k) and hence they determine a (m − p − k)-flat and the set $S(\alpha_{i1}, \alpha_{i2}, \ldots, \alpha_{ik})$ contains s^{m-p-k} treatments. Take the set $S(\alpha_{i1}, \alpha_{i2}, \ldots, \alpha_{ik})$ as a block. Since each $\alpha_{i1}, \alpha_{i2}, \ldots, \alpha_{ik}$ can be selected in s ways, we get s^k blocks.

The contrasts carried by the pencils

$$P\left(\sum_1^k \lambda_j b_{j1}, \sum_1^k \lambda_j b_{j2}, \ldots, \sum_1^k \lambda_j b_{jm}\right),$$

where λ's are elements of GF(s) not all zero, and their aliases are confounded. The number of these pencils is $(s^k - 1)/(s - 1)$. Hence there will be $(s^k - 1)/(s - 1)$ alias sets of s^p pencils each, which will be confounded.

1.7 Missing Plot Technique

43. **(1) Missing yields:** Let $y_1, y_2, \ldots, y_n$ be the observed yields and $x_1, x_2, \ldots, x_k$ be the missing yields. Further, let

$$E(\mathbf{y}) = A\theta, E(\mathbf{x}) = B\theta.$$

The estimates of the missing yields are obtained by minimizing the error sum of squares

$$S_e^2(\mathbf{y}, \mathbf{x}) = \mathbf{y}'\mathbf{y} + \mathbf{x}'\mathbf{x} - \hat{\theta}'(A'\mathbf{y} + B'\mathbf{x})$$

and are given by

$$\hat{\mathbf{x}} = B\hat{\theta},$$

where $\hat{\theta}$ is any solution of the normal equations

$$A'\mathbf{y} + B'\mathbf{x} = (A'A + B'B)\hat{\theta}'.$$

The actual error SS is given by

$$S_e^2(\mathbf{y}) = \min_{x} S_e^2(\mathbf{y}, \mathbf{x}).$$

For testing the hypothesis $\theta_{\ell+1} = \theta_{\ell+2} = \ldots = \theta_m = 0$, we apply the usual F test, using the estimated SS due to the hypothesis, which is calculated as follows:

Est. SS due to the hypothesis
= [Est. conditional error SS] − [min. value of error SS]

If the F value is insignificant, no correction for bias is necessary. If the F value is significant, the bias is calculated as follows:

Bias in the estimated SS due to hypothesis
= [Est. conditional error SS] − [min. value of conditional error SS].

(2) Mixed-up yields: If some yields are mixed up so that their total yield is only known then the mixed-up yields, are estimated by minimizing the error SS subject to the condition that their total yield is equal to the given total yield. The bias in the estimated SS due to hypothesis is calculated as follows:

Bias = [Est. conditional error SS]
− [min. value of conditional error SS subject to the condition that their total yield is equal to the given total yield].

1.8 Weighing Designs

44. **Chemical-Balance Weighing Design.**

Here we consider a chemical-balance weighing design. Let there be p objects to be weighed in n weighings. Define the matrix $X = (x_{ij})$ as follows.

$$\begin{aligned} x_{ij} &= 1\text{ , if the j-th object is kept in the left pan in the i-th weighing,}\\ &= -1\text{ , if the j-th object is kept in the right pan in the i-th weighing,}\\ &= 0\text{ , if the j-th object is not included in the i-th weighing}\end{aligned}$$
$$i = 1, 2, \ldots, n\ ;\ j = 1, 2, \ldots, p.$$

The $n \times p$ matrix X is called the weighing design matrix. Let $\mathbf{w}' = [w_1, w_2, \ldots, w_p]$ be the vector of the true weights of p objects and $\mathbf{y}' = [y_1, y_2, \ldots, y_p]$ be the vector of readings of the scales in n weighings.

We assume that

$$\mathbf{y} = X\mathbf{w} + \mathbf{e}$$

and $\mathbf{e} \sim N(\mathbf{0}, \sigma^2 I)$.

The least square estimates of true weights are given by $(X'X)\hat{\mathbf{w}} = X'\mathbf{y}$. If $(X'X)$ is non-singular, then

$$\hat{\mathbf{w}} = (X'X)^{-1}X'\mathbf{y}$$

with variance-covariance matrix given by

$$\text{var}(\hat{\mathbf{w}}) = \sigma^2(X'X)^{-1}.$$

The following results have been established:

(i) For any weighting design X,

$$\text{var}(\hat{w}_i) \geq \sigma^2/n,\ i = 1, 2, \ldots, p.$$

(ii) For any weighing design X, the variances of all the estimated weights are minimum if and only if $X'X = nI_p$.

(iii) If a Hadmard matrix H_n of order n exists, then by choosing any p columns of it, we can form an optimum weighing design matrix to weigh p objects.

The following three definitions of efficiency are in use.

Definition 1: Of the class of all $n \times p$ weighing designs, a weighing design X is optimal, if the average variance of all the estimated weights is minimum and the *efficiency* of any weighing design X is defined by

$$p/n\ \text{tr}\ (X'X)^{-1}.$$

Definition 2: Of the class of all $n \times p$ weighing design, a weighing design X is optimal if the generalized variance of the estimated weights, i.e. $|X'X|^{-1}$ is minimum, that is, $|X'X|$ is maximum. The *efficiency* of any weighing design X is defined by

$$\frac{\min\cdot|X'X|^{-1}}{|X'X|^{-1}} = \frac{|X'X|}{\max\cdot|X'X|}$$

Definition 3: Of the class of all $n \times p$ weighing design, a weighing design X is optimal if it has the least value of λ_{max}, where λ_{max} is the maximum characteristic root of $(X'X)^{-1}$ and the *efficiency* of any weighing design X is defined by

$$\frac{\min \lambda_{max}}{\lambda_{max}}$$

where λ_{max} is the maximum characteristic root of $(X'X)^{-1}$.

45. **Spring-Balance Weighing Design.**

In this case, the weighing design matrix $X = (x_{ij})$ is defined as

$$\begin{aligned} x_{ij} &= 1, \text{ if the j-th object is included in the i-th weighing} \\ &= 0, \text{ if the j-th object is not included in the i-th weighing.} \end{aligned}$$

If bias is present it can be assumed to be one object and its value is estimated by taking $x_{ij} = 1$,

$$i = 1, 2, \ldots, n; j = 1, 2, \ldots, p.$$

The incidence matrix N of block design in which a treatment can occur at most once in a block may be employed to construct a weighing design with a spring balance by identifying objects with treatments and blocks with weighings and N′ with X. Let there be v objects to be weighed and suppose k objects are weighed in each weighing. Suppose the total number of weighings is equal to b. Further, the weighings are such that each object is weighed exactly r times and every pair is weighed λ times. Then, we have $X = N'$, where N is the incidence matrix of a BIBD. Therefore

$$X'X = NN' = (r - \lambda)I_v + \lambda E_{vv},$$

and

$$(X'X)^{-1} = \frac{1}{r - \lambda} I_v - \frac{\lambda}{rk(r - \lambda)} E_{vv}.$$

Hence, the variance-covariance matrix of the estimated weights is given by

$$\text{var}\,(\hat{w}) = \sigma^2[I_v - \frac{\lambda}{rk} E_{vv})/(r - \lambda)].$$

Also,

$$\text{var}\,(\hat{w}_1) = \frac{(rk - \lambda)}{(r - \lambda)rk}\sigma^2, \quad i = 1, 2, \ldots, v.$$

Applying Definition 1 of Result 44, it is seen that the efficiency of the above weighing design is $k^2(r - \lambda)/(rk - \lambda)$.

When a Hadamard matrix H_{n+1} of order $n + 1$ exists, a spring balance weighing design of maximum efficiency involving n weighings of n objects can be constructed as follows. H_{n+1} is written so that the elements in its first row and first column are all $+1$. Subtract the first row from each of the remaining rows and multiply the 2nd, 3rd, …, $(n + 1)$-st rows by $-y_2$. Omit the first row and first column. Then, the resulting matrix is the matrix of spring balance weighing design having the maximum efficiency.

1.9 Some Useful Mathematical Results

46. **Galois Fields**

(1) The number of elements in a Galois field is p^n, where p is a prime number and n a positive integer. A Galois field consisting of p^n elements is denoted by $GF(p^n)$.
(2) If t is the smallest positive integer such that $\alpha^t = 1$, where $\alpha \neq 0$ is any element of $GF(p^n)$, then t is called the order of α.
(3) An element of a $GF(p^n)$ having an order $p^n - 1$ is called a primitive root of $GF(p^n)$. The primitive roots of GF(p) for various values of p are given in the following table.

Primitive Roots of GF(p)

p	3	5	7	11	13	17	19	23
primitive root	2	2	3	2	2	3	2	5

(4) If p is a prime number, then residues mod p constitute a GF(p).
(5) If F(x) is an irreducible polynomial of degree n in GF(p), then the residues mod F(x) constitute the $GF(p^n)$.
(6) The residues mod a minimum function of a $GF(p^n)$ constitute the $GF(p^n)$. Minimum functions of $GF(p^n)$ for various values of p and n are given in the following table.

Minimum Functions of $GF(p^n)$

p	n	minimum functions
2	2	$x^2 + x + 1$
	3	$x^3 + x^2 + 1,\ x^3 + x + 1$
	4	$x^4 + x^3 + 1,\ x^4 + x + 1$
	5	$x^5 + x^2 + 1$
3	2	$x^2 + x + 2,\ x^2 + 2x + 2$
	3	$x^3 + 2x + 1$
	4	$x^4 + x^3 + x^2 + 2x + 2$
5	2	$x^2 + 2$
7	2	$x^2 + 1$

(7) The roots of a minimum function of a $GF(p^n)$ are primitive roots of $GF(p^n)$.
(8) If α is a primitive root of $GF(p^n)$, then the elements of $GF(p^n)$ are taken as $0, 1, \alpha, \alpha^2, \ldots, \alpha^{p^n-2}$.

47. **Finite projective Geometry PG(m, s).**

Let $s = p^n$, where p is a prime number and n a positive integer.
(1) A point in PG(m, s) is taken as $(x_0, x_1, \ldots, x_m)$, where $x_0, x_1, \ldots, x_m$ are elements of GF(s), not all zero. Two points $(x_0, x_1, \ldots, x_m)$ and

$(\rho x_0, \rho x_1, \ldots, \rho x_m)$ represent the same point, where ρ is a non-null element of GF(s).

(2) The total number of points in PG(m, s) is equal to $(s^{m+1} - 1)/(s - 1)$.

(3) **g-flat:** All points of PG(m, s) which satisfy a set of (m − g) linear independent homogeneous equations

$$a_{i0}x_0 + a_{i1}x_1 + \cdots + a_{im}x_m = 0, i = 1, 2, \ldots, m - g$$

will be said to form a g-flat. Alternatively, a g-flat is determined by a set of (g + 1) linearly independent points of PG(m, s). A 0-flat is a point in PG(m, s). An 1-flat is a line in PG(m, s).

(4) The number of points on a g-flat of a PG(m, s) is equal to $(s^{(g+1)} - 1)/(s - 1)$.

(5) The number of g-flats contained in a PG(m, s) is equal to

$$\phi(m, g, s) = \frac{(s^{m+1} - 1)(s^m - 1)\ldots(s^{m-g+1} - 1)}{(s^{g+1} - 1)(s^g - 1)\ldots(s - 1)}.$$

(6)(i) $\phi(m, g, s) = \phi(m, m - g - 1, s)$

(ii) $\phi(m, g, s) = 1, \text{ if } g = -1$
$= 0, \text{ if } g \leq -2.$

(7) The number of g-flats in a PG(m, s) which contain a given t-flat $(0 \leq t < g \leq m)$ is equal to $\phi(m - t - 1, g - t - 1, s)$.

(8) The number of g-flats in a PG(m, s) which contain a given point is equal to $\phi(m - 1, g - 1, s)$.

(9) The number of g-flats in a PG(m, s) which contain a given pair of points is equal to $\phi(m - 2, g - 2, s)$.

48. **Finite Euclidean Geometry EG(m, s)**

Let $s = p^n$, where p is a prime number and n a positive integer.

(1) A point in EG(m, s) is taken as $(x_1, x_2, \ldots, x_m)$, where $x_1, x_2, \ldots, x_m$ are elements of GF(s). Two points $(x_1, x_2, \ldots, x_m)$ and $(x'_1, x'_2, \ldots, x'_m)$ are same if and only if $x_i = x'_i$ for $i = 1, 2, \ldots, m$.

(2) The total number of points in an EG(m, s) is s^m.

(3) **g-flat:** All points of EG(m, s) which satisfy the m − g consistent and independent equations

$$a_{i0} + a_{i1}x_1 + a_{i2}x_2 + \ldots + a_{im}x_m = 0, i - 1, 2, \ldots, m - g$$

will be said to form a g-flat.

(4) The number of points on a g-flat of EG(m, s) is s^g.

(5) The number of g-flats in an EG(m, s) is

$$\phi(m, g, s) - \phi(m - 1, g, s).$$

(6) The number of g-flats of an EG(m, s) which contain a given point is

$$\phi(m - 1, g - 1, s)$$

(7) The number of g-flats of an EG(m, s) which contain a given pair of points is

$$\phi(m-2, g-2, s).$$

49. **Hadamard Matrix**

(1) **Definition:** A square matrix of order n, H_n will be called a hadamard matrix if it has the elements +1 and −1 and is such that $H_n'H_n = nI_n$.

(2) A Hadamard matrix of order 2 is

$$H_2 = \begin{bmatrix} 1 & 1 \\ 1 & -1 \end{bmatrix}.$$

(3) If H_m and H_n are Hadamard matrices of orders m and n respectively, then their Kronecker product $H_m \otimes H_n$ is a Hadamard matrix of order mn.

(4) A Hadamard matrix or order 2^n can always be constructed.

(5) A Hadamard matrix can be written so that the elements in its first row and first column are all +1.

(6) Let H_n be a Hadamard matrix of order n, written so that all elements in its first row and first column are +1. Let D be the matrix of order (n − 1) obtained from H_n by omitting its first row and first column. If the elements −1 in D are replaced by 0, then we get the incidence matrix of a SBIBD with parameters

$$v = b = n - 1, \ \gamma = k = \frac{n}{2} - 1, \ \lambda = \frac{n}{4} - 1.$$

(7) If we subtract the first row from the other rows of H_n, and multiply the 2nd, 3rd, ..., nth rows by $-y_2$, and omit the first row and first column, then we get the incidence matrix of a SBIBD. Equivalently, if we subtract the first row from the other rows of H_n and omit the first row and first column and replace the non-zero element by +1, then the resulting matrix is the incidence matrix of a SBIBD.

50. **Systems of Distinct Representatives (SDR)**

(1) **Definition :** Let S be a set of v elements 1, 2, ..., v. Let $T_1, T_2, \ldots, T_n$ be a non-null subsets of S, not necessarily disjoint. Then, a set $(a_1, a_2, \ldots, a_n)$ where $a_i \neq T_i$ for every i and $a_i \neq a_j$ for every $i \neq j$, is defined to be a system of distinct representatives (SDR) for the sets $T_1, T_2, \ldots, T_n$.

(2) A necessary and sufficient condition for the existence of a SDR for the sets $T_1, T_2, \ldots, T_n$ is that for every integral $k, i_1, i_2, \ldots, i_k$ satisfying $1 \leq k \leq n$ and $1 \leq i_1 < i_2 < \ldots < i_k \leq n$,

$$|T_{i_1} \cup T_{i_2} \cup \ldots \cup T_{i_k}| \geq k,$$

where |T| denotes the cardinality of the set T.

(3) Let |T| denote the cardinality of the set T and $R_n(T_1, T_2, \ldots, T_n)$ denote the number of SDRs for the sets $T_1, T_2, \ldots, T_n$. If $|T_i| \geq s$ for every i, then

$$\begin{aligned} R_n(T_1, T_2, \ldots, T_n) &\geq s!, && \text{if } s \leq n \\ &\geq s!/(s-n)!, && \text{if } s \geq n \end{aligned}$$

51. **Symmetrically Repeated Differences**

Let M be a module containing exactly m elements. We associate exactly n symbols with each element of M. The symbols corresponding to an element a of M are denoted by $a_1, a_2, \ldots, a_n$. There will be mn symbols in all. Symbols with the same subscript i will be said to belong to the class i, $i = 1, 2, \ldots, n$. There will be n classes.

Consider a set S containing k symbols. Let S contain p_i symbols from the i-th class, $i = 1, 2, \ldots, n$. Thus, $k = \sum_{i=1}^{n} p_i$. Let p_i symbols contained in the set S from the i-th class be donoted by $a_i^1, a_i^2, \ldots, a_i^{p_i}$, $i = 1, 2, \ldots, n$. The $p_i(p_i - 1)$ differences $a_i{}^u - a_i{}^w$, $u \neq w = 1, 2, \ldots, p_i$ are called pure differences of the type [i, i]. Clearly, there are n types of different pure differences. Let the p_i symbols contained in the set S from the j-th class be donoted by $b_j{}^1, b_j{}^2, \ldots, b_j^{p_j}$. The $p_i p_j$ differences $a_i{}^u - b_j{}^w$, $i \neq j = 1, 2, \ldots, n$, $u = 1, 2, \ldots, p_i$, $w = 1, 2, \ldots, n_j$ are called the mixed differences of the type [i, j]. Clearly, there are $n(n-1)$types of mixed differences. Let $S_1, S_2, \ldots, S_t$ be t sets satisfying the following conditions:

(i) Each set contains k symbols.

(ii) Let $p_{i\ell}$ denote the number of symbols from the i-th class contained in the ℓ-th set.

(iii) Among the $\sum_{\ell=1}^{t} pi\ell(pi\ell - 1)$ pure differences of the type [i, i], arising from the t sets, every non-zero element of M is repeated exactly λ times independently of i.

(iv) Among the $\sum_{\ell=1}^{t} pi\ell pj\ell$ mixed differences of the type [i, j], arising from the t sets, every element of M is repeated exactly λ times, independently of i and j.

Then, we say that in the t sets $S_1, S_2, \ldots, S_t$, the differences are symmetrically repeated, each occuring λ times.

CHAPTER 2

EXERCISES

1. For the model $\mathcal{E}(\mathbf{y}) = A\boldsymbol{\theta}$, $\text{var}(\mathbf{y}) = \sigma^2 I$, show that the parametric functions (i) $\boldsymbol{\ell}'A\boldsymbol{\theta}$, (ii) $\boldsymbol{\ell}'A'A\boldsymbol{\theta}$ and (iii) $\boldsymbol{\ell}'GA'A\boldsymbol{\theta}$, where G is a g-inverse of $A'A$, are estimable for any vector $\boldsymbol{\ell}$ of appropriate order. Find their BLUEs and their variances.

2. Show that a necessary and sufficient condition for the estimability of $\boldsymbol{\ell}'\boldsymbol{\theta}$ is that

$$\text{rank}\,(A'A) = \text{rank}\,(A'A,\ \boldsymbol{\ell}),$$

where it is assumed that $\mathcal{E}(\mathbf{y}) = A\boldsymbol{\theta}$.

3. Let y_1, y_2, and y_3 be indepenent random variables with a common variance σ^2 and expectations given by $\mathcal{E}(y_1) = \theta_1 + \theta_3$, $\mathcal{E}(y_2) = \theta_2 + \theta_3$ and $\mathcal{E}(y_3) = \theta_1 + \theta_3$. Prove that $b_1 + b_2\theta_2 + b_3\theta_3$ is estimable if and only if $b_3 = b_1 + b_2$. If $b_1\theta_1 + b_2\theta_2 + b_3\theta_3$ is estimable, find its BLUE and its variance. Find an unbiased estimator of σ^2.

4. Let $y_i, i = 1, 2, \ldots, 6$, be independent random variables with a common variance σ^2 and $\mathcal{E}(y_1) = \theta_1 + \theta_4$, $\mathcal{E}(y_2) = \theta_2 + \theta_4$, $\mathcal{E}(y_3) = \theta_3 + \theta_5$, $\mathcal{E}(y_4) = \theta_4 + \theta_5$, $\mathcal{E}(y_5) = \theta_1 + \theta_6$, and $\mathcal{E}(y_6) = \theta_2 + \theta_6$. Show that $\theta_1 - \theta_2$ is estimable and find its BLUE and its variance. Find an unbiased estimator of σ^2. Show that $\theta_1 + \theta_2$ is not estimable.

5. Let $y_i, i = 1, 2, \ldots, 6$ be independent observations with a common variance σ^2 and expectations given by $\mathcal{E}(y_i) = \theta_1\cos(2\pi i/6) + \theta_2\sin(2\pi i/6)$, $i = 1, 2, \ldots, 6$. Find the BLUES of θ_1 and θ_2 and their variances and the covariance between them.

6. Let x_i, y_i and $z_i, i = 1, 2, \ldots, n$ be 3n independent observations with a common variance σ^2 and expectations given by

$$\mathcal{E}(x_i) = \theta_1,\ \mathcal{E}(y_i) = \theta_2,\ \mathcal{E}(z_i) = \theta_1 + \theta_2, i = 1, 2, \ldots, n.$$

Find the BLUES of θ_1 and θ_2 and their variance-covariance matrix. Find the unbiased estimator of σ^2 based on the total available degrees of freedom.

7. Let x_i, y_i, and z_i be 3n independent observation with a common variance σ^2 and expectations given by $\mathcal{E}(x_i) = \theta_1$, $\mathcal{E}(y_i) = \theta_2$, $\mathcal{E}(z_i) = \theta_1 - \theta_2$, $i = 1, 2, \ldots, n$. Find the BLUES of θ_1 and θ_2 and their variance-covariance

matrix. Find the unbiased extimator σ^2 based on the total variable degrees of freedom.

8. Let $y_{ij}, j = 1, 2, \ldots, n, i = 1, 2, 3$ be independent observations with a common variance σ^2 and expectations given by $\mathcal{E}(y_{1j}) = \theta_1 + \theta_2 + \theta_3$; $\mathcal{E}(y_{2j}) = \theta_1 + \theta_3$; and $\mathcal{E}(y_{3j}) = -2\theta_2 + \theta_3, j = 1, 2, \ldots, n$. Find the BLUEs of θ_1, θ_2, and θ_3 and their variance-covariance matrix. Find the unbiased estimator of σ^2 based on the total available degrees of freedom.

9. At a post office, three parcels are weighed singly, in pairs and all together. All the weighings are independent and have equal accuracy. The weights are denoted by $y_{100}, y_{010}, y_{001}, y_{110}, y_{101}, y_{011}$ and y_{111}, where the suffix 1 indicates the presence of a particular parcel and the suffix 0 indicates its absence. Obtain the BLUEs of the weights of the parcels and their variance-covariance matrix. Find the unbiased estimator of σ^2, where σ^2 is the variance of each observation.

10. Let $y_{ij}, j = 1, 2, \ldots, n; i = 1, 2, 3$ be independent observations with a common variance σ^2 and expectations given by $\mathcal{E}(y_{1j}) = \theta_1 + 2\theta_2 + 3\theta_3$, $\mathcal{E}(y_{2j}) = 2\theta_1 + 3\theta_2 + \theta_3$ and $\mathcal{E}(y_{3j}) = 3\theta_1 + \theta_2 + 2\theta_3; j = 1, 2, \ldots, n$. Find the BLUEs of θ_1, θ_2 and θ_3 and their variance-covariance matrix. Also, find the unbiased estimator of σ^2 based on the total available degrees of freedom.

11. Assuming normality of observations, derive suitable test statistics for testing the following hypotheses

(i) H_0: $\theta_1 = \theta_2$ in Exercise 6.
(ii) H_0: $\theta_1 = a\theta_2$ in Exercise 7, where a is a known constant.
(iii) H_0: $\theta_3 = 0$ in Exercise 8.
(iv) H_0: $\theta_1 + \theta_2 + \theta_3 = 0$ in Exercise 10.
(v) H_0: $\theta_2 = \theta_3 = 0$ in Exercise 10.

12. Let $y_i, i = 1, 2, \ldots, n$ be independent normal variates with a common variance σ^2 and expectations given by $\mathcal{E}(y_i) = \alpha + \beta(x_i - \bar{x}), i = 1, 2, \ldots, n$, where α and β are unknown constants and $\bar{x} = \Sigma x_i/n$. Obtain the BLUEs of α and β and their variance-covariance matrix. Derive appropriate statistics for testing (i) $\alpha = 0$ and (ii) $\beta = 0$.

13. Let $y_i, i = 1, 2, \ldots, n$ be independent normal variates with a common variance σ^2 and expectations given by $\mathcal{E}(y_i) = \alpha + \beta_1(x_{1i} - \overline{x}_1) + \beta_2(x_{2i} - \overline{x}_2) + \ldots + \beta_k(x_{ki} - \overline{x}_k)$ $i = 1, 2, \ldots, n$, where $\alpha, \beta_1, \beta_2, \ldots \beta_k$ are unknown parameters and x_{ij}'s are known constants and $\overline{x}_p = \Sigma x_{pi}/n, p = 1, 2, \ldots, k$. Obtain the BLUEs of $\alpha, \beta_1, \beta_2, \ldots, \beta_k$ and their variance-covariance matrix. Derive tests for testing (i) $\beta_i = 0$, (ii) $\beta_i = \beta_j, i \neq j = 1, 2, \ldots k$.

14. Let $y_{ij}, j = 1, 2, \ldots, n_i;\ i = 1, 2, \ldots, k$ be independent normal variates with a common variance σ^2 and expectations given by $\mathcal{E}(y_{ij}) = \mu + t_i, j = 1, 2, \ldots, n_i,\ i = 1, 2, \ldots,\ k$, where μ and t's are unknown parameters. Derive tests for testing (i) $n\mu + \Sigma n_i t_i = 0$, (ii) $t_1 = t_2 = \ldots = t_k = 0$ and

(iii) $t_1 = t_2 = \ldots = t_k$. Show that the test statistics in (ii) and (iii) are same.

15. Let y_{ij}, $i = 1, 2, \ldots, r, j = 1, 2, \ldots,$ s be independent normal observations with a common variance σ^2 and expectations given by $\mathcal{E}(y_{ij}) = \mu + \alpha_i + \beta_j, i = 1, 2, \ldots, r$ and $j = 1, 2, \ldots, s$. Derive appropriate statistics for testing (i) $\alpha_i = 0, i = 1, 2, \ldots, r$ and (ii) $\beta_j = 0, j = 1, 2, \ldots,$ s.

16. Let $y_{ij}, j = 1, 2, \ldots, n_i$ and $i = 1, 2, \ldots, k$ be independent normal observations with a common variance σ^2 and expectations given by $\mathcal{E}(y_{ij}) = \mu + t_i + \beta x_{ij}, j = 1, 2, \ldots, n_i, j = 1, 2, \ldots, k$, where $\mu, t_1, t_2, \ldots, t_k$ and β are unknown numbers and x_{ij}'s are known constants. Derive suitable statistics for testing (i) $\beta = 0$ and (ii) $t_1 = t_2 = \ldots = t_k = 0$.

17. Let y_{ij}, $i = 1, 2, \ldots, r; j = 1, 2, \ldots,$ s be independent normal observations with a common variance σ^2 and expectations given by $\mathcal{E}(y_{ij}) = \mu + \alpha_j + t_i + \beta x_{ij}, i = 1, 2, \ldots, s$, where μ, α's, t's and β are unknown parameters and x_{ij}'s are known numbers. Derive statistics for testing (i) $\beta = 0$ and (ii) $t_1 = t_2 = \ldots = t_r = 0$.

18. Given that $\mathcal{E}(\mathbf{y}) = A\theta$, $\text{var}(\mathbf{y}) = \sigma^2 I$, and $A'A$ is a 3×3 non-singular matrix and the following

$$\hat{\sigma}^2 = 200, \hat{\theta}_1 = 3, \hat{\theta}_2 = 5, \hat{\theta}_3 = 2,$$

$$\text{var}(\hat{\theta}) = \frac{3}{10}\begin{bmatrix} 3 & 2 & 1 \\ 2 & 4 & 2 \\ 1 & 2 & 3 \end{bmatrix}.$$

Show that the value of the F-statitsic for testing $\theta_1 = \theta_2 = \theta_3 = 0$ is $F = 65/9$.

19. Let $y_i, i = 1, 2, \ldots, n$ be independent random variables with $\mathcal{E}(y_i) = \alpha + \beta x_i, i = 1, 2, \ldots, n$ and $\text{var}(y) = \text{diag}\,(d_1, d_2, \ldots, d_n)$, where α and β are unknown parameters and x's, $d_1, d_2, \ldots, d_n$ are known numbers. Find the BLUEs of α and β and their variance-covariance matrix. Discuss the cases when (i) $x_i = i$ and $d_i = \sigma^2 i$, (ii) $x_i = i$ and $d_i = \sigma^2 i^2, i = 1, 2, \ldots, n$.

20. Let y_1, y_2 and y_3 be 3 observations with $\mathcal{E}(y_1) = \theta_1 + a_1\theta_2$, $\mathcal{E}(y_2) = \theta_2$ and $\mathcal{E}(y_3) = \theta_1 + a_3\theta_2$ and variance-covariance matrix

$$V = \frac{3}{20}\begin{bmatrix} 3 & 2 & 1 \\ 2 & 4 & 2 \\ 1 & 2 & 3 \end{bmatrix}, \text{ and } a_1 = -\sqrt{3/2},\ a_2 = \sqrt{3/2}.$$

Find the BLUEs of θ_1 and θ_2 and their variance-covariance matrix.

21. Let $y_i, i = 1, 2, \ldots, n$ be n independent random observations with $\mathcal{E}(y_i) = \mu$ and variance-covariance matrix V given by $V = [\sigma^2/(N-1)].[NI_n - E_{nn}]$, where E_{nn} is an $n \times n$ matrix with all elements unity. Obtain the BLUE of μ and its variance. Also, obtain an estimator of the variance of the BLUE of μ.

22. Let $y_i, i = 1, 2, \ldots, n$ be n independent random variables with $\mathcal{E}(y_i) = \mu + \theta x_i, i = 1, 2, \ldots, n$ where μ and θ are unknown parameters and $x_i = \sqrt{3}(2i - n - 1)/(n + 1), i = 1, 2, \ldots, n$, and variance-covariance matrix $V = \sigma^2(a_{ij})$, where $a_{ij} = 12i(n - j + 1)/(n + 1)^2(n + 2)$ for $i \leq j$. Show that the BLUEs of μ and θ are given by $\hat{\mu} = (y_1 + y_n)/2$ and $\hat{\theta} = (n + 1)(y_n - y_1)/2\sqrt{3}(n - 1)$. Further, show that

$$\text{var}(\hat{\mu}) = 6\sigma^2/(n + 1)(n + 2)$$
$$\text{var}(\hat{\theta}) = 2\sigma^2/(n - 1)(n + 2)$$
$$\text{cov}(\hat{\mu}, \hat{\theta}) = 0.$$

$$\text{Hint: } V^{-1} = \frac{(n + 1)(n + 2)}{12\sigma^2}\begin{bmatrix} 2 & -1 & 0 & 0 & \ldots & 0 & 0 \\ -1 & 2 & -1 & 0 & \ldots & 0 & 0 \\ 0 & -1 & 2 & -1 & \ldots & 0 & 0 \\ \cdot & \cdot & \cdot & \cdot & \ldots & \cdot & \cdot \\ \cdot & \cdot & \cdot & \cdot & \ldots & \cdot & \cdot \\ \cdot & \cdot & \cdot & \cdot & \ldots & \cdot & \cdot \\ \cdot & \cdot & \cdot & \cdot & \ldots & \cdot & \cdot \\ 0 & 0 & 0 & 0 & \ldots & -1 & 2 \end{bmatrix}$$

23. Given that $\mathcal{E}(\mathbf{y}) = A\boldsymbol{\theta}$, $\text{var}(\mathbf{y}) = V$, V being a non-singular matrix, A is an $n \times 3$ matrix with rank 3 and the following,

$$[SSE/(n - 3)] = 100, \hat{\boldsymbol{\theta}}' = (6, 10, 4)$$
$$\text{var}(\hat{\boldsymbol{\theta}}) = \frac{1}{20}\begin{bmatrix} 3 & 2 & 1 \\ 2 & 4 & 2 \\ 1 & 2 & 3 \end{bmatrix}.$$

Show that the value of F for testing $\theta_1 = \theta_2 = \theta_3 = 0$ is given by $F = 1.73$

24. In Exercise 20, find the F statistic for testing the hypothesis $\theta_2 = 0$. Also give the F statistic for testing the hypothesis $\theta_1 = \theta_2 = 0$.

25. In Exercise 22, obtain the values of F statistic for testing the hypotheses (i) $\mu = \theta = 0$ and (ii) $\theta = 0$.

26. Given below is the incidence matrix of a block design. Find its C matrix, the degrees of freedom associated with the adjusted treatment sum of squares and the degrees of freedom for the error sum of squares.

$$N = \begin{bmatrix} 1 & 0 & 0 & 0 \\ 1 & 0 & 1 & 0 \\ 1 & 0 & 1 & 0 \\ 0 & 1 & 0 & 1 \\ 0 & 1 & 0 & 1 \\ 0 & 0 & 0 & 1 \end{bmatrix}$$

27. Determine which of the following designs are connected.

Design	Blocks							
	1	2	3	4	5	6	7	8
I	C, A, B	C, B, D	A, C, D	A, B, D	D, F, G	E, F, G	D, E, F	D, E, G
II	A, B, G	A, F, G	A, E, F	A, D, E	A, C, D	A, B, C		
III	B, E, G	A, F, G	C, E, F	A, D, E	C, D, G	B, D, F	A, B, C	

28. Below is the incidence matrix of a design. Show that the design is not connected but is balanced.

$$N = \begin{bmatrix} 1 & 0 \\ 0 & 1 \\ 0 & 1 \\ 1 & 0 \end{bmatrix}$$

Hint: Here

$$K = \begin{bmatrix} 2 & 0 \\ 0 & 2 \end{bmatrix}, \quad R = \begin{bmatrix} 1 & 0 & 0 & 0 \\ 0 & 1 & 0 & 0 \\ 0 & 0 & 1 & 0 \\ 0 & 0 & 0 & 1 \end{bmatrix}.$$

Hence

$$\begin{aligned} C &= R - NK^{-1}N' \\ &= \begin{bmatrix} 1 & 0 & 0 & 0 \\ 0 & 1 & 0 & 0 \\ 0 & 0 & 1 & 0 \\ 0 & 0 & 0 & 1 \end{bmatrix} - \begin{bmatrix} 1/2 & 0 & 0 & 1/2 \\ 0 & 1/2 & 1/2 & 0 \\ 0 & 1/2 & 1/2 & 0 \\ 1/2 & 0 & 0 & 1/2 \end{bmatrix} \\ &= \begin{bmatrix} 1/2 & 0 & 0 & -1/2 \\ 0 & 1/2 & -1/2 & 0 \\ 0 & -1/2 & 1/2 & 0 \\ -1/2 & 0 & 0 & 1/2 \end{bmatrix} \end{aligned}$$

The fourth column of C is -1 times its first column and its third column is -1 times its second column. Hence

$$\text{rank}\ (C) = 2.$$

Thus, rank (C) $\neq v - 1 = 3$. Hence the design is not connected.

We now find the characteristic roots of the C matrix. We have

$$|C - \lambda I| = \lambda^2(\lambda - 1)^2.$$

Hence the two non-zero characteristic roots of C are each equal to 1. Therefore the design is balanced.

29. Show that the design with incidence matrix $N = aE_{vb}$, where a is some positive integer, is connected, balanced and orthogonal.

30. If N is the incidence matrix of an equi-replicate binary regular design and $\overset{*}{N} = E_{vb} - N$, then show that the design whose incidence matrix is $\overset{*}{N}$ is an

equi-replicate binary regular design. Determine the C matrix of the design $\overset{*}{N}$ in terms of the C matrix of design N. Further if the design N is connected balanced, show that the design $\overset{*}{N}$ is also connected balanced.

31. Let N be the incidence matrix of an orthogonal design (v, b, r, k) with $v = 2k$. Consider $\overset{*}{N} = E_{vb} - N$. Then show that the design $\overset{*}{N}$ is orthogonal.

32. For a connected balanced design, show that its C-matrix is given by

$$C = \theta(I_V - \frac{1}{V}E_{vv}),$$

where θ is the non-zero characteristic root of the matrix C.

33. Let N be the incidence matrix of an equi-replicate, regular connected balanced design. Show that

$$NN' = k[(r - \theta)I_v + (\theta/v)E_{vv}],$$

where θ is the non-zero characteristic root of its C-matrix, k is the block size and r is the number of replications of a treatment, that is, show that the characteristic roots of NN' are rk and $k(r - \theta)$ with respective multiplicities 1 and $v - 1$.

34. Show that the average variance of BLUEs of elementary treatment constrasts in a connected design is

$$\frac{2\sigma^2}{(v-1)}\sum_{i=1}^{v-1}(1/\theta_i),$$

where $\theta_1, \theta_2, \ldots, \theta_{v-1}$ are the $(v - 1)$ non-zero characteristic roots of the matrix C.

35. Prove that the average variance of the BLUEs of the elementary treatment contrasts in a connected design lies between $2\sigma^2/\theta_{min}$ and $2\sigma^2/\theta_{max}$ where θ_{min} and θ_{max} are respectively the minimum and the maximum characteristic roots of the C-matrix of the design.

36. Consider a connected design and let $\boldsymbol{\ell}'\mathbf{t}$ be any treatment contrast. Show that the variance of the BLUE of $\boldsymbol{\ell}'\mathbf{t}$ lies between $\boldsymbol{\ell}'\boldsymbol{\ell}\,\sigma^2/\theta_{max}$ and $\boldsymbol{\ell}'\boldsymbol{\ell}\,\sigma^2/\theta_{min}$, where θ_{max} and θ_{min} are respectively the maximum and the minimum non-zero characteristic roots of the matrix C of the design.

37. Prove that the efficiency of a binary connected design is given by

$$E \leq \frac{v(n-b)}{n(v-1)},$$

where n = total number of plots in the design.

38. Show that the non-zero characteristic roots of the C-matrix of a connected balanced design is $(n - b)/(v - 1)$, where n = total number of plots in the design.

39. For a binary connected equi-replicated incomplete block design, show that $b \geq v$.

40. Show that the incidence matrix N of an equi-replicated regular connected balanced design satisfies the relation

$$NN' = (r - \lambda)I_v + \lambda E_{vv},$$

where $\lambda(v - 1) = r(k - 1)$.

41. Prove that the most efficient connected design is balanced.

42. Prove that the average variance of the BLUEs of elementary treatment contrasts in a connected balanced design is given by

$$\overline{V} = 2\sigma^2(v - 1)/(n - b)$$

and that its efficiency is $E = v(n - b)/n(v - 1)$.

43. Let the incidence matrix of a design be

$$N = \begin{bmatrix} 1 & 1 & 1 & 0 \\ 1 & 1 & 0 & 1 \\ 1 & 0 & 1 & 1 \\ 0 & 1 & 1 & 1 \end{bmatrix}.$$

Show that (i) the design is connected balanced, and (ii) its efficiency factor is $E = (8/9)$.

44. Suppose mk treatments are divided into m sets of k each and the treatments of each set are assigned to blocks of size k. Further, let there be r repetitions of these m blocks. Show that the design so obtained is orthogonal.

45. Show that

$$\begin{vmatrix} C & -N \\ N' & K \end{vmatrix} = k_1 k_2 \ldots k_b . r_1 r_2 \ldots r_v$$

46. Show that the following designs are connected balanced. Find the non-zero characteristic root of the C matrix in each case. Find also the variance of the BLUE of an elementary contrast in each case.

$$\text{(i) } N = E_{vb} \quad \text{(ii) } C = aI - (a/v)E_{vv}.$$

47. Show that in a one-way design, the BLUE of an estimable treatment contrast $\boldsymbol{\ell}'\mathbf{t}$ is given by $\mathbf{p}'\mathbf{Q}$, where $\mathbf{C}\mathbf{p} = \boldsymbol{\ell}$ and that its variance is given by $\sigma^2\mathbf{p}'\boldsymbol{\ell}$. Hence show that if $C = aI - (a/v)\ E_{vv}$, the BLUE of a treatment contrast $\boldsymbol{\ell}'\mathbf{t}$ is $\boldsymbol{\ell}'\mathbf{Q}/a$ with variance $\sigma^2\boldsymbol{\ell}'\boldsymbol{\ell}/a$. As a particular case, when $C = aI_v - (a/v)E_{vv}$, deduce that the BLUE of an elementary treatment contrast $t_i - t_j$ is $(Q_i - Q_j)/a$ with variance $2\sigma^2/a$.

48. Show that in a connected design $C + \mathbf{rr}'/n$ is non-singular and that $(C + \mathbf{rr}'/n)^{-1}\mathbf{r} = E_{v1}$, where $\mathbf{r}' = (r_1, r_2, \ldots, r_v)$ and $n = \mathbf{r}'E_{v1}$. Also prove that $(C + \mathbf{rr}'/n)^{-1}$ is a g-inverse of C.

49. Prove that in a connected design, a set of solutions of normal equations in the intrablock analysis is given by

$$\hat{\mu} = G/n$$
$$\hat{\alpha} = K^{-1}\mathbf{B} - (G/n)E_{b1} - K^{-1}N'[C + \mathbf{rr}'/n]^{-1}\mathbf{Q}$$
$$\hat{\mathbf{t}} = [C + \mathbf{rr}'/n]^{-1}\mathbf{Q}$$

Also show that

(i) $\mathrm{var}(\hat{\mathbf{t}}) = \sigma^2[(C + \mathbf{rr}'/n)^{-1} - E_{vv}/n]$

(ii) if $\boldsymbol{\ell}'\mathbf{t}$ is any treatment contrast, then its intrablock estimate is $\boldsymbol{\ell}'(C + \mathbf{rr}'/n)^{-1}\mathbf{Q}$, with variance $\sigma^2\boldsymbol{\ell}'(C + \mathbf{rr}'/n)^{-1}\boldsymbol{\ell}$.

50. Show that in the intrablock analysis of variance of a one-way design, the adjusted block sum of squares is obtained as

$$\text{Adj. Block SS} = (\text{Total SS}) - (\text{Error SS}) - (\text{Unadj. Treatment SS}).$$

51. Derive the expected values of different sums of squares which occur in the intrablock analysis of variance of a one-way design.

52. In the intrablock analysis of variance of a one-way design, show that the variance-covariance matrices of **Q** and **P** are given by

$$\mathrm{var}(\mathbf{Q}) = \sigma^2 C$$
$$\mathrm{var}(\mathbf{P}) = \sigma^2 D$$
$$\mathrm{cov}(\mathbf{Q}, \mathbf{P}) = -\sigma^2 CR^{-1}N = -\sigma^2 NK^{-1}D.$$

53. Show that the equations for obtaining t_s, $s = 1, 2, \ldots, v$ in the combined inter and intrablock analysis of an equiblock-size one-way design can be obtained from the corresponding equations in the intrablock analysis by replacing Q_s, r_s and $\lambda_{ss'}$, $(s \neq s' = 1, 2, \ldots, v)$ by P_s, R_s and $\Lambda_{ss'}$, where $P_s = w_1 Q_s + w_2 Q'_s$, $\quad Q'_s = T_s - Q_s - (G/bk)\, r_s$, $R_s = r_s\, w_1 + \dfrac{w_2}{R-1}$, $\Lambda_{ss'} = (w_1 - w_2)\lambda_{ss'}$ and setting $\mathbf{r}'\mathbf{t}_s = 0$.

54. Show that, in the analysis with recovery of interblock information,

$$\mathrm{var}(\mathbf{Q}) = C/w_1,\ \mathrm{var}(\mathbf{Q}_1) = C_1/w_2$$
$$\mathrm{cov}(\mathbf{Q}, \mathbf{Q}_1) = \mathbf{0}$$
$$\mathrm{var}(w_1\mathbf{Q} + w_2\mathbf{Q}_1) = w_1 C + w_2 C_1$$

where $w_1 = 1/\sigma_e^2$ and $w_2 = 1/(\sigma_e^2 + k\sigma_b^2)$.

55. Show that in a connected design in which each block is of the same size k and NN′ is non-singular, the intrablock, interblock and the combined intra and inter block estimates of a treatment contrast $\boldsymbol{\ell}'\mathbf{t}$ are respectively given by

(i) $\boldsymbol{\ell}'(C + \mathbf{rr}'/n)^{-1}\mathbf{Q}$ with variance $w\, \boldsymbol{\ell}'(C + \mathbf{rr}'/n)^{-1}\boldsymbol{\ell}$

(ii) $\boldsymbol{\ell}'(NN')^{-1}N\,\mathbf{B}$ with variance $(k/w_2)\, \boldsymbol{\ell}'(NN')^{-1}\boldsymbol{\ell}$

(iii) $\boldsymbol{\ell}'(w_1 C + \frac{w_2}{R}NN')^{-1}(w_1\mathbf{Q} + \frac{w_2}{R}N\mathbf{B})$ with variance $\boldsymbol{\ell}'(w_1 C + \frac{w_1}{R}NN')^{-1}\boldsymbol{\ell}$

Obtain the linear combination of the estimates (i) and (ii) which has the least variance and show that it is equal to the estimate (iii) if

$$(C + \frac{\mathbf{r}\mathbf{r}'}{n})^{-1}\mathbf{Q} = (NN')^{-1}N\mathbf{B}.$$

56. Show that in a connected design with equal block size, the average variance of the combined intra and inter block estimates of elementary treatment contrasts is

$$\frac{2\,v\,\mathrm{tr}(w_1C + \frac{w_2}{k}NN')^{-1} - E_{1v}(w_1C + \frac{w_2}{k}NN')^{-1}E_{v1}}{v(v-1)}.$$

STANDARD DESIGNS

57. Derive the analysis of variance of a design whose incidence matrix is $N = aE_{vb}$; where a is some positive integer and compare its efficiency with that of a design with incidence matrix $N = E_{vb}$.

58. Suppose $v = pq$ treatments $t_{ij}, i = 1, 2, \ldots, p, j = 1, 2, \ldots, q$ are arranged at random in b blocks of k plots each such that (i) every treatment occurs at most once in each block, (ii) every treatment occurs in exactly r blocks and (iii) a pair of treatments t_{ij} and $t_{i'j'}$ occurs in exactly $\lambda_{(ij)(i'j')}$ blocks, where

$$\begin{aligned}\lambda_{(ij)(i'j')} &= r_1 \text{ if } i = i',\ j = j'\\ &= \lambda_{01} \text{ if } i = i',\ j \neq j'\\ &= \lambda_{10} \text{ if } i \neq i',\ j = j'\\ &= \lambda_{11} \text{ if } i \neq i',\ j \neq j'\end{aligned}$$

Prove that

(i) $r(k-1) = (q-1)\lambda_{01} + (p-1)\lambda_{10} + (p-1)(q-1)\lambda_{11}$.

(ii) an intrablock solution of t_{ij} is given by

$$\begin{aligned}\hat{t}_{ij} &= \frac{k}{a}[Q_{ij} + BQ_{i.} + CQ_{.j}]\,, \text{ where}\\ Q_{ij} &= \text{ the adjusted total for } t_{ij}\,,\\ Q_{i.} &= \sum_{j=1}^{q} Q_{ij}\,,\ Q_{.j} = \sum_{i=1}^{p} Q_{ij}\,,\\ B &= (\lambda_{01} - \lambda_{11})/[a - q(\lambda_{01} - \lambda_{11})]\,,\\ C &= (\lambda_{10} - \lambda_{11})/[a - p(\lambda_{10} - \lambda_{11})]\,,\\ a &= r(k-1) + \lambda_{01} + \lambda_{10} - \lambda_{11}.\end{aligned}$$

(iii) the BLUEs of $\sum_{j=1}^{q} t_{ij} - \sum_{j=1}^{q} t_{i'j}$, $\sum_{i=1}^{p}, t_{ij} - \sum_{i=1}^{p}, t_{ij'}$ and $t_{ij} - t_{ij'} - t_{i'j} + t_{i'j'}$ are respectively given by

$k(Q_{i.} - Q_{i'.})/[a - q(\lambda_{01} - \lambda_{11})]$,
$k(Q_{.j} - Q_{.j'})/[a - p(\lambda_{10} - \lambda_{11})]$, and
$k(Q_{ij} - Q_{ij'} - Q_{i'j} + Q_{i'j'})/a$
with respective variances
$2kq\sigma^2(1 + qB)/a$, $2kp\sigma^2(1 + pC)/a$, and $4k\sigma^2/a$,
where a, B and C are defined in (ii).

59. Let there be v treatments and $n = n_1 + n_2 + \ldots + n_v$ plots, there being n_i plots receiving the i-th treatment. The yields are assumed to be independent normal variates with a common variance σ^2 and expected values given by the effect of the treatment applied to the plot. Derive the analysis of variance for testing the hypothesis of equality of treatment effects. Prove that the test is most sensitive when $n_i = n/v$, $i = 1, 2, \ldots, v$.

60. In a randomized block experiment originally planned with v treatments and r replications, it was later on found that there was not enough material of treatment 1 and there was excess of material of treatment 2. Hence treatment 1 was applied once only in blocks $1, 2, \ldots, r_1$ and treatment 2 was applied once in blocks $1, 2, \ldots, r_1$ and twice in blocks $r_1 + 1, r_1 + 2, \ldots, r$. Derive the analysis of variance for this modified design. Obtain variances of the BLUEs of different elementary treatment comparisons and the loss of efficiency due to the above modification of the design.

61. In a randomized block design, with v treatments and r blocks, treatment 2 in Block 1 was interchanged with treatment 1 in Block 2 through mistake. Further, if the design is equi-replicate, then show that the above expression of the average variance becomes

$$2[\mathrm{tr}(w_1 C + \frac{w_2}{k} NN')^{-1} - \frac{1}{w_2 r}]/(v - 1).$$

62. In a randomized experiment, treatment 1 was used twice in the first two blocks and consequently treatment 2 was not used in these blocks. Show that

(i) the sum of squares for testing $\mathbf{t} = \mathbf{0}$ is

$$\frac{k}{\Delta}[(r - 2)(rk - 2)Q_1^2 + (r + 2)(rk + 2)Q_2^2 - 4(r - 2)Q_1Q_2] + \sum_{3}^{v} Q_{i'}{}^2/r,$$

(ii) $\mathrm{var}(\hat{t}_1 - \hat{t}_2) = 2k\sigma^2 r(rk + 2)/\Delta$

(iii) $\mathrm{var}(\hat{t}_1 - \hat{t}_i) = \sigma^2 \left[\frac{k(r - 2)(rk - 2)}{\Delta} + \frac{1}{r}\right]$, $i = 3, 4, \ldots, v$

(iv) $\mathrm{var}(\hat{t}_2 - \hat{t}_i) = \sigma^2 \left[\frac{k(r+2)(rk+2)}{\Delta} + \frac{1}{r} \right], \ i = 3, 4, \ldots, v$

(v) $\mathrm{var}\,(\hat{t}_i - \hat{t}_j) = 2\sigma^2/r, \ i \neq j = 3, 4, \ldots, v$
where $\Delta = (r-2)(r^2k^2 + 2rk^2 - 8)$.

63. For testing v new varieties, an experimenter divides his field into v blocks and each block into (gv + 1) plots. Each new variety is replicated g times in each block and a standard variety is assigned to the (gv + 1)st plot in each block. Derive an appropriate analysis of variance for this design.

64. For a SBIBD, show that $NN' = N'N$.

65. If N is the incidence matrix of a BIBD, then show that

(i) the characteristic roots of NN' are rk and $r - \lambda$ with multiplicities 1 and $v - 1$ respectively and the characteristic roots of $N'N$ are 0, rk and $r - \lambda$ with multiplicities $b - v$, 1, and $v - 1$ respectively.

(ii) $|N'N| = 0$, when the BIBD is non-symmetrical

(iii) $\mathrm{tr}\,(N'N) = vr$.

66. If in a BIBD, $b = 4(r - \lambda)$, then prove that $2k = v \pm \sqrt{v}$.

67. Show that the parameters of an affine resolvable BIBD with $v = nk$, $b = nr$, r, k, λ can be expressed as

$$v = n^2(nt - t + 1), \ b = n(n^2t + n + 1),$$
$$r = n^2t + n + 1, \ k = n(nt - t + 1),$$
$$\lambda = nt + 1,$$

where $n \geq 2$ and $t \geq 0$.

68. In a BIBD with parameters v, b, r, k, λ, a control treatment is added to each block, so that the block size is now (k + 1). Derive the analysis of variance, the BLUEs of various treatment comparisons and their variances.

69. Show that in a BIBD, the necessary and sufficient condition that there be the same number of treatments common between any two blocks is that $b = v$.

70. Show that in a BIBD, $(b > v + 2)$, x, the number of common treatments between any two blocks satisfies the inequality

$$-(r - \lambda - k) \leq x \leq [2\lambda k + r(r - \lambda - k)]/r.$$

71. Show that in a BIBD, x, the number of common treatments between any two blocks satisfies the inequality

$$\frac{k}{(b-1)}[(r-1) - T(b-r)] \leq x \leq \frac{k}{(b-1)}[(r-1) + T(b-r)],$$

where $T = [(b-2)(b-v)/b(v-1)]^{1/2}$.

72. Let M be the matrix obtained from the incidence matrix N of a BIBD (v, b, r, k, λ) by replacing 0 by 1. Prove that $MM' = 4(r - \lambda)I_v + [b - 4(r - \lambda)]E_{vv}$.

73. A BIBD (v, b, r, k, λ) is said to belong to a family A of designs if $b = 4(r - \lambda)$. Let N_1 and N_2 be the incidence matrices of BIBDs $(v_1, b_1, r_1, k_1, \lambda_1)$ and $(v_2, b_2, r_2, k_2, \lambda_2)$ belonging to the family A. Prove that the design whose incidence matrix N is given by

$$N = N_1 \otimes N_2 + \overset{*}{N}_1 \otimes \overset{*}{N}_2$$

is a BIBD belonging to the family A and find its parameters, where $\otimes$ denotes the Kronecker product, $\overset{*}{N}_1$ and $\overset{*}{N}_2$ denote the complementary designs of N_1 and N_2.

74. An incomplete block design is said to be a Linked Block design if (i) each block has the same number k of plots, (ii) each treatment occurs in exactly the same number r of blocks and (iii) any two blocks have the same number of treatments in common. Show that the design obtained by interchanging the roles of treatments and blocks in a BIBD $(\overset{*}{v}, \overset{*}{b}, \overset{*}{r}, \overset{*}{k}, \overset{*}{\lambda})$ is a Linked block design. Obtain the parameters of this new design. Obtain the C and D matrices of this design, and find the rank and characteristic roots of the D matrix.

75. Derive the expected values of different sums of squares in the intrablock analysis of variance of a BIBD.

76. A BIBD with parameters v, $b = v(v - 1)/2$, $k = 2$, $r = v - 1$ and $\lambda = 1$ is formed by taking $v(v - 1)/2$ combinations of v treatments taken 2 at a time as blocks. Interchange the roles of blocks and treatments in this design. Show that this new design is a PBIBD and obtain the parameters of this design.

77. Let $v = mk$ treatments be arranged in m sets of k each and each set be taken as a block. There are r such repetitions, so that the total number of blocks is equal to $b = mr$. Show that this design is a PBIBD with two associate classes and obtain the parameters of this design. Show that this design is not connected and that by selecting r suitably, the number of blocks can be made equal to, greater than or less than the number of treatments.

78. The association matrices of the asssociation scheme of an m-associate class PBIBD are defined by $v \times v$ matrices $B_i = (b_{\alpha\beta}{}^i)$, $i = 1, 2, \ldots, m$, where $b_{\alpha\beta}{}^i$, the element in the cell (α, β) of B_i is defined by

$$b_{\alpha\beta}{}^i = 1, \text{ if } \alpha \text{ and } \beta \text{ are i-th associates}$$
$$= 0, \text{ otherwise.}$$

Further, every treatment is defined to be the o-th associate of itself, so that

$$B_o = I_v, n_o = 1, \ \lambda_o = r, p_{ij}{}^o = n_i\delta_{ij}, p_{ok}{}^i = \delta_{ik}$$

where δ_{ij} is that Kronecker delta, which is defined to be 1, if $i = j$ and 0, if $i \neq j$. Show that

(i) $E_{1v}B_i = n_iE_{1v}, \ B_iE_{v1} = n_iE_{v1}, i = 0, 1, 2, \ldots, m$

(ii) $\sum_{i=0}^{m} B_i = E_{vv}$

(iii) $B_o, B_1, \ldots, B_m$ are linearly independent

(iv) $B_jB_k = \sum_{i=0}^{m} p_{jk}{}^{i}B_i$, $B_jB_k = B_kB_j$, $j, k = 0, 1, 2, \ldots, m$.

(v) $\sum_{u} p_{jk}{}^{u}p_{iu}{}^{t} = \sum_{u} p_{ij}{}^{u}p_{uk}{}^{t}$

79. Define P_i , $i = 0, 1, 2, \ldots$, m matrices of a PBIBD as follows:

$$P_i = \begin{bmatrix} p_{oi}{}^{O} & p_{oi}{}^{1} & \cdots & p_{oi}{}^{m} \\ p_{1i}{}^{O} & p_{1i}{}^{1} & \cdots & p_{1i}{}^{m} \\ \cdot & \cdot & \cdots & \cdot \\ \cdot & \cdot & \cdots & \cdot \\ \cdot & \cdot & \cdots & \cdot \\ p_{mi}{}^{O} & p_{mi}{}^{1} & \cdots & p_{mi}{}^{m} \end{bmatrix}.$$

Show that

(i) $$\sum_{i=0}^{m} P_i = \begin{bmatrix} n_o & n_o & \cdots & n_o \\ n_1 & n_1 & \cdots & n_1 \\ n_2 & n_2 & \cdots & n_2 \\ \cdot & \cdot & \cdots & \cdot \\ \cdot & \cdot & \cdots & \cdot \\ n_m & n_m & \cdots & n_m \end{bmatrix}$$

(ii) $P_0, P_1, \ldots, P_m$ matrices are linearly independent

(iii) $P_jP_k = \sum_{i=0}^{m} p_{jk}{}^{i}P_i$, $j, k = 0, 1, 2, \ldots$, m

$P_jP_k = P_kP_j$,

(iv) $P_0, P_1, \ldots, P_m$ matrices provide a regular representation in $(m + 1) \times (m + 1)$ matrices of the algebra given by B-matrices of the association scheme of a PBIBD, which are $v \times v$ matices.

80. Show that NN' of an m-associate class of a PBIBD is given by

$$NN' = \sum_{i=0}^{m} \lambda_i B_i.$$

81. With the help of B_i-matrices of the association scheme of an m-associate class of a PBIBD, show that

(i) $\sum_{i=0}^{m} n_i = v$, (ii) $\sum_{i=0}^{m} n_i\lambda_i = rk$,

(iii) $\sum_{k=0}^{m} p_{jk}{}^{i} = n_j$, (iv) $n_ip_{jk}{}^{i} = n_jp_{ik}{}^{j}$

82. The distinct characteristic roots of $B = \sum_{i=0}^{m} c_iB_i$ are the same. Hence deduce that the distinct characteristic roots of NN' of an m-associate class PBIBD are the same as those of $P = \sum_{i=0}^{m} \lambda_iP_i$. Hence, prove that the distinct characteristic

roots of NN' are given by rk and the distinct characteristic roots of the matrix $L = (\ell_{ij})$, where

$$\ell_{ii} = r - n_i\lambda_i + \sum_{t=i}^{m} \lambda_t p_{it}{}^{i},\ i = 1, 2, \ldots, m$$

$$\ell_{ij} = \sum_{t=i}^{m} \lambda_t p_{it}{}^{j} - n_i\lambda_i\ ,\ i \neq j = 1, 2, \ldots,\ m$$

83. If NN' of an incomplete block design (v, b, r, k) has only one zero characteristic root with multiplicity u, show that $b \geq v - u$. Further if this design is resolvable with $v = nk$, $b = nr$, then show that $b \geq v + (r - 1) - u$.

84. Show that the distinct characteristic roots of NN' of a 2-associate class PBIBD are given by

$$\theta_o = rk,\ \text{with multiplicity } 1$$
$$\theta_i = r - \frac{1}{2}(\lambda_1 + \lambda_2) + \frac{1}{2}(\lambda_1 - \lambda_2)[p + (-1)^i\sqrt{\Delta}],$$

with multiplicity

$$\alpha_i = \frac{n_1 + n_2}{2} - (-1)^i \left[\frac{(n_1 - n_2) + r(n_1 + n_2)}{2\sqrt{\Delta}}\right],\ i = 1, 2$$

where $r = p_{12}{}^2 - p_{12}{}^1$, $\beta = p_{12}{}^1 + p_{12}{}^2$ and $\Delta = r^2 + 2\beta + 1$.

85. Show that the values and the multiplicities of the distinct characteristic roots of NN' of a group divisible design are

$$\theta_o = rk,\ \text{with multiplicity } 1$$
$$\theta_1 = r - \lambda_1,\ \text{with multiplicity } m(n - 1)$$
$$\theta_2 = rk - v\lambda_2,\ \text{with multiplicity } (m - 1)$$

86. Show that the values of the multiplicities of the distinct characteristic roots of NN' of an L_i (Latin square type design with i constraints) design are given by

$$\theta_o = rk \text{ with multiplicity } 1$$
$$\theta_1 = r - i\lambda_1 + \lambda_2(i - 1) \text{ with multiplicity } (s - 1)(s - i + 1)$$
$$\theta_2 = r + \lambda_1(s - i) - \lambda_2(s - i + 1) \text{ with multiplicity } i(s - 1).$$

87. Show that the values and the multiplicities of the distinct characteristic roots of NN' of a triangular design are given by

$$\theta_o = rk,\ \text{with multiplicity } 1$$
$$\theta_1 = r - 2\lambda_1 + \lambda_2,\ \text{with multiplicity } n(n - 3)/2$$
$$\theta_2 = r + (n - 4)\lambda_1 - (n - 3)\lambda_2,\ \text{with multiplicity } (n - 1).$$

88. Prove that in an m-associate class PBIBD with $k > r$, $|(\ell_{ij})| = 0$, where

$$\ell_{ij} = \sum_{t=i}^{m} \lambda_t p_{it}^{\ j} - n_i\lambda_i, \ i \neq j = 1, 2, \dots, m$$

$$\ell_{ii} = r + \sum_{t=i}^{m} \lambda_t p_{it}^{\ j} - n_i\lambda_i, \ i = 1, 2, \dots, m.$$

Hence, deduce that for a 2-associate class PBIBD with $k > r$,

$$(r - \lambda_1)(r - \lambda_2) + (\lambda_1 - \lambda_2)[p_{12}^{\ 2}(r - \lambda_1) - p_{12}^{\ 1}(r - \lambda_2)] = 0$$

89. Prove that a given block in a BIBD (v, b, r, k, λ) can never have more than $b - 1 - [(r - 1)^2 k/(r - \lambda - k + k\lambda)]$ blocks disjoint with it and if some block has that many, then $(r - \lambda - k + k\lambda)/(r - 1)$ is a positive integer and each of the non-disjoint blocks has $(r - \lambda - k + k\lambda)/(r - 1)$ treatments in common with it.

90. Show that by replacing each treatment of a BIBD $(v^*, b^*, r^*, k^*, \lambda^*)$ by a group of n treatments, we get a singular group divisible design with parameters $v = nv^*$, $b = b^*$, $r = r^*$, $k = nk^*$, $n_1 = n - 1$, $n_2 = n(v^* - 1)$, $\lambda_1 = r^*$, $\lambda_2 = \lambda^*$,

$$P_1 = \begin{bmatrix} n-2 & 0 \\ 0 & n(v^* - 1) \end{bmatrix}, \ P_2 = \begin{bmatrix} 0 & n-1 \\ n-1 & n(v^* - 2) \end{bmatrix}.$$

91. Prove that in a singular group divisible design, a given block cannot have more than

$$b - 1 - \left[\frac{k(\lambda_1 - 1)^2}{n(\lambda_1 - 1) + (k - n)(\lambda_2 - 1)}\right]$$

blocks disjoint with it and if some block has that many disjoint blocks, then $n + [(k - n)(\lambda_2 - 1)/(\lambda_1 - 1)]$ is an integer and each non-disjoint block has $n + [(k - n)(\lambda_2 - 1)/(\lambda_1 - 1)]$ treatments common with that block.

92. Prove that in a singular group divisible design, the necessary and sufficient condition that a block will have the same number of treatments in common with each of the remaining blocks is that (i) $b = m$, and (ii) $k(r - 1)/(m - 1)$ is an integer.

93. Prove that for a resolvable singular group divisible design, $b \geq m + r - 1$. Further, prove that a necessary and sufficient condition for a resolvable singular group divisible design to be affine resolvable is that (i) $b = m + r - 1$, and (ii) k^2/v is an integer.

94. Prove that for a singular group divisible design, $b \geq m$, and further if it is resolvable, then $b \geq m + r - 1$.

95. Prove that for a semi-regular GD(group divisible) design, $b \geq v - m + 1$ and further if it is resolvable, then $b \geq v - m + r$.

96. Prove that in a semi-regular GD design, k is divisible by m and every block contains (k/m) treatments from each group.

97. Prove that a given block of a semi-regular group divisible design cannot have more than

$$b - 1 - \frac{v(v-m)(r-1)^2}{(v-k)(b-r) - (v-rk)(v-m)}$$

blocks disjoint with it and if some block has that many disjoint blocks, then

$$\frac{k[(v-k)(b-r) - (v-rk)(v-m)]}{v(v-m)(r-1)}$$

is an integer and each non-disjoint block has

$$\frac{k[(v-k)(b-r) - (v-rk)(v-m)]}{v(v-m)(r-1)}$$

treatments common with that block.

98. Prove that in a semi-regular group divisible design, the necessary and sufficient condition that a block will have the same number of treatments in common with each of the remaining blocks is that (i) $b = v - m + 1$, and (ii) $k(r-1)/(v-m)$ is an integer.

99. Prove that for a resolvable semi-regular group divisible design, $b \geq v - m + r$ and that the necessary and sufficient condition for it to be affine resolvable then is that (i) $b = v - m + r$, and (ii) k^2/v is an integer.

100. Prove that in a triangular design,

(i) if $r - 2\lambda_1 + \lambda_2 = 0$, then $b \geq n$ and further if the design is resolvable $b \geq n + r - 1$.;

(ii) if $r + (n-4)\lambda_1 - (n-3)\lambda_2 = 0$, then $b \geq v - n + 1$, and further if the design is resolvable, then $b \geq v - n + r$.

101. Prove that if in a triangular design $r + (n-4)\lambda_1 - (n-3)\lambda_2 = 0$, then 2k is divisible by n, and every block contains 2k/n treatments from each of the n rows of the association scheme.

102. Prove that in a triangular design with $r + (n-4)\lambda_1 - (n-3)\lambda_2 = 0$, a given block cannot have more than

$$b - 1 - \frac{v(v-n)(r-1)^2}{[(b-r)(v-k) - (v-rk)(v-n)]}$$

blocks disjoint with it and if some block has that many disjoint blocks, then $k[(b-r)(v-k) - (v-rk)(v-n)]/v(v-n)(r-1)$ is an integer and each non-disjoint block has $k[(b-r)(v-k) - (v-rk)(v-n)]/n(v-n)(r-1)$ treatments in common with that block.

103. Prove that in a triangular design with $r + (n-4)\lambda_1 - (n-3)\lambda_2 = 0$, the necessary and sufficient condition that a block will have the same number of treatments in common with each of the remaining blocks is that (i) $b = v - n + 1$ and (ii) $k(r-1)/(v-n)$ is an integer.

104. Prove that for a resolvable triangular design with $r + (n-4)\lambda_1 - (n-3)\lambda_2 = 0$, $b \geq v - n + r$ and that the necessary

and sufficient condition for it to be affine resolvable is that (i) $b = v - n + r$ and (ii) k^2/v is an integer.

105. Prove that in a Latin-square type design with i constraints (L_i)

(i) if $r - i\lambda_1 + \lambda_2(i-1) = 0$, then $b \geq 1 + i(s-1)$ and further if the design is resolvable, then $b \geq r + i(s-1)$

(ii) if $r + \lambda_1(s-i) - \lambda_2(s-i+1) = 0$, then $b \geq 1 + (s-1)(s-i+1)$ and further if the design is resolvable, then $b \geq r + (s-1)(s-i+1)$.

106. Prove that in a L_2 design, if $r + (s-2)\lambda_1 - (s-1)\lambda_2 = 0$, then k is divisible by s and every block contains k/s treatments from each of the s rows (or columns) of the association scheme.

107. Prove that in a L_2 design with $r + (s-2)\lambda_1 - (s-1)\lambda_2 = 0$, a given block cannot have more than

$$b - 1 - \frac{v(r-1)^2(s-1)^2}{(b-r)(v-k) - (v-rk)(s-1)^2}$$

blocks disjoint with it and if some block has that many, then $k[(b-r)(v-k) - (v-rk)(s-1)^2]/v(r-1)(s-1)^2$ is an integer and each non-disjoint block has $k[(b-r)(v-k) - (v-rk)(s-1)^2]/v(r-1)(s-1)^2$ treatments in common with that given block.

108. Prove that in a L_2 design with $r + (s-2)\lambda_1 - (s-1)\lambda_2 = 0$, the necessary and sufficient condition that a block will have the same number of treatments in common with each of the remaining blocks is that (i) $b = 1 + (s-1)^2$ and (ii) $k(r-1)/(s-1)^2$ is an integer.

109. Prove that for a resolvable L_2 design with $r + (s-2)\lambda_1 - (s-1)\lambda_2 = 0$, $b \geq r + (s-1)^2$ and that the necessary and sufficient condition for it to be affine resolvable is that (i) $b = r + (s-1)^2$ and (ii) k^2/v is an integer.

110. If $rk > \mu_o > \mu_1 > \cdots > \mu_p$ be the distinct non-zero characteristic roots of NN' of an incomplete block design (v, b, r, k), then x, the number of common treatments between any two blocks of this design satisfies the inequality

$$\max[0, 2k - v, k - \mu_o] \leq x \leq \min[k, \mu_o - k + 2(rk - \mu_o)b^{-1}]$$

111. Using Exercise 108, prove that x, the number of common treatments between any two blocks of a BIBD satisfies the inequality

$$\max[0, 2k - v, k - r + \lambda] \leq x \leq \min[k, r - \lambda - k + 2(\lambda k/r)].$$

112. Prove that in a singular group divisible design, x, the number of common treatments between any two blocks satisfies the inequality

$$\max[0, 2k - v, -k(b - m - r + 1)/(m - 1)] \leq x$$
$$\leq \min[k, k\{n(b - m - r - 1) + 2k\}/(v - n)].$$

113. Prove that in a semi-regular group divisible design, x, the number of common treatments between any two blocks satisfies the inequality

$$\max[0, 2k - v, -k(b - v + m - r)/(v - m)] \leq x$$
$$\leq \min[k, k\{(b - r)(b - 2) - (v - m)(b - 2r)\}/b(v - m)].$$

114. Prove that in a triangular design,

(i) if $r + (n - 4)\lambda_1 - (n - 3)\lambda_2 = 0$, then x, the number of common treatments between any two blocks satisfies the inequality

$$\max[0, 2k - v, -k(b - v - r + n)/(v - n)] \leq x$$
$$\leq \min[k, k\{(b - r)(b - 2) - (v - n)(b - 2r)\}/b(v - n)]$$

and

(ii) if $r + (n - 4)\lambda_1 - (n - 3)\lambda_2 = 0$, then x, the number of common treatments between any two blocks satisfies the inequality

$$\max[0, 2k - v, -k(b - v - r + n)/(v - n)] \leq x$$
$$\leq \min[k, k\{(b - r)(b - 2) - (v - n)(b - 2r)\}/b(v - n)]$$

115. In a triangular design, prove that x, the nubmer of common treatments between any two blocks satisfies the inequality

$$\max[0, 2k - v, k - \theta_i] \leq x \leq \min[k, \theta_i - k + 2(rk - \theta_i)b^{-1}],$$

where $i = 1$ if $\lambda_1 > \lambda_2$ and $i = 2$, if $\lambda_1 < \lambda_2$, and $\theta_1 = r + (n - 4)\lambda_1 - (n - 3)\lambda_2$, $\theta_2 = r - 2\lambda_1 + \lambda_2$.

116. In a L_i design, prove that x, the number of common treatments between any two blocks satisfies the inequality

$$\max[0, 2k - v, k - \theta_p] \leq x \leq \min[k, \theta_p - k + 2(rk - \theta_p)b^{-1}],$$

where $p = 1$ if $\lambda_1 > \lambda_2$ and $p = 2$ if $\lambda_1 < \lambda_2$ and

$$\theta_1 = r + (s - i)\lambda_1 - (s - i + 1)\lambda_2, \qquad \theta_2 = r - i\lambda_1 + (i - 1)\lambda_2.$$

117. In a L_2 design with $r + (s - 2)\lambda_1 - (s - 1)\lambda_2 = 0$, prove that x, the number of common treatments between any two blocks satisfies the inequality

$$\max[0, 2k - v, -k\{(b - r) - (s - 1)^2\}/(s - 1)^2] \leq x$$
$$\leq \min\left[k, \frac{k\{(b - r)(b - 2) - (s - 1)^2(b - 2r)\}}{b(s - 1)^2}\right]$$

118. Prove that in a semi-regular group divisible design, x, the number of common treatments between any two blocks satisfies the inequality

$$\frac{[k(r - 1) - A]}{(b - 1)} \leq x \leq \frac{[k(r - 1) + A]}{(b - 1)},$$

where $A^2 = k^2(b - 2)(b - r)(v - k)(b - v + m - 1)/v(v - m)$.

119. Prove that in a triangular design with $r + (n-4)\lambda_1 - (n-3)\lambda_2 = 0$, x, the number of common treatments between any two blocks satisfies the inequality

$$\frac{k(r-1) - A}{b-1} \leq x \leq \frac{k(r-1) + A}{(b-1)},$$

where $A^2 = k^2(b-2)(b-r)(v-k)(b-v+n-1)/v(v-n)$.

120. Prove that in a L_2 design with $r + (s-2)\lambda_1 - (s-1)\lambda_2 = 0$, the number of common treatments between any two blocks satisfies the inequality

$$\frac{k(r-1) - A}{b-1} \leq x \leq \frac{k(r-1) + A}{(b-1)},$$

where $A^2 = \dfrac{k^2(b-2)(b-r)(v-k)\{b-1-(s-1)^2\}}{v(s-1)^2}$.

121. Prove that a necessary condition for the existence of a group divisible design is that $rk - v\lambda_2 \geq 0$.

122. Prove that a necessary condition for the existence of a triangular design is that $rk - v\lambda_1 \leq n(r - \lambda_1)/2$.

123. Prove that a necessary condition for the existence of a L_2 design is that $rk - v\lambda_1 \leq s(r - \lambda_1)$.

124. In a $v \times v$ Latin square, the v treatments represented by the numbers $1, 2, \ldots, v$ occur in the first row in their natural order. Suppose, now, due to accident, treatments 1 and 2 get interchanged. Obtain the BLUEs of the various elementary treatment comparisons and their variances. Also obtain the loss in efficiency due to this interchange.

125. Show that the efficiency of an m-ple Lattice design, using intrablock analysis is given by

$$E = 1 - \frac{m}{(m-1)(k+1) + m}.$$

Deduce that, if the Lattice design is balanced, then $E = k/(k+1)$.

126. Show that the efficiency of an m-ple Lattice design utilizing the analysis with recovery of interblock information is given by

$$E = 1 - \frac{m(w_1 - w_2)}{(k+1)\{(m-1)w_1 + w_2\} + m(w_1 - w_2)}.$$

Deduce that if the Lattice design is balanced, then $E = kw_1/[(k+1)w_1 - w_2]$.

127. Let $X_o(x) = 1, X_1(x) = x - \dfrac{s_1 - 1}{2}$,

$$X_t(x) = X_1(x)X_{t-1}(x) - \frac{(t-1)^2\{s_1^2 - (t-1)^2\}X_{t-2}(x)}{4(2t-1)(2t-3)}$$

be orthogonal polynomials for the set of values $x = 0, 1, \ldots, s_1 - 1$. For $t = 1, 2, \ldots, s_1 - 1$, define

$$F_{1t} = \left[\sum_{x=0}^{s_1-1} f_1^x X_t(x)\right] \phi_2\phi_3 \ldots \phi_m$$

as the linear, quadratic, cubic, ... effect of F_1, where

$$\phi_i = f_i^1 + f_i^2 + \cdots + f_i^{s_1-1}, i = 1, 2, \ldots, m.$$

Prove that the sum of squares due to the above effect in a randomized block experiment with r replications is

$$[\hat{F}_{1t}]^2 \Big/ \left[\frac{s_1 s_2 \ldots s_m (t!)^4 (s_1^2 - 1^2) \ldots (s_1^2 - t^2)}{r(2t)!(2t+1)!}\right],$$

where $\hat{F}_{1t}$ denotes the estimate of F_{1t} obtained by replacing treatments in F_{1t} by their mean yields.

128. In a 2^m factorial design, denote the factors by $A_1, A_2, \ldots, A_m$. The upper level of A_i is denoted by a_i and its lower level by 1, $i = 1, 2, \ldots, m$. Let

$$S_i = (a_i + 1)(a_{i-1} + 1) \ldots (a_1 + 1),$$
$$X_o = 1, X_i' = [X_{i-1}' a_i X_{i-1}'], i = 1, 2, \ldots, m$$
$$H_o = 1, H_i = \begin{bmatrix} H_{i-1} & H_{i-1} \\ -H_{i-1} & H_{i-1} \end{bmatrix}, i = 1, 2, \ldots, m$$
$$Y_o = 1, Y_i' = \begin{bmatrix} (a_i + 1) & Y_{i-1} \\ (a_i - 1) & Y_{i-1} \end{bmatrix}, i = 1, 2, \ldots, m.$$

Prove that

(i) $$X_i = [{}^{1}_{a_i}] \otimes [{}^{1}_{a_{i-1}}] \otimes \ldots \otimes [{}^{1}_{a_1}]$$

(ii) $$Y_i = [{}^{a_i+1}_{a_i+1}] \otimes [{}^{a_{i-1}+1}_{a_{i-1}-1}] \otimes \ldots \otimes [{}^{a_1+1}_{a_1-1}],$$

(iii) $$H_i = H_1 \otimes H_1 \otimes \ldots \otimes H_1,$$

where $\otimes$ denotes Kronecker product and hence establish Yates' method, $i = 1, 2, \ldots, m$. Further, prove that

$$H_i^{-1} = H_i'/2^i, i = 1, 2, \ldots m,$$

and hence establish inverse Yates' method.

129. Consider a 3^m factorial design. The factors are denoted by $A_1, A_2, \ldots, A_m$ and three levels of A_i by $1, a_i, a_i^2$, $i = 1, 2, \ldots, m$. Let

$$S_i = (a_i^2 + a_i + 1)(a_{i-1}^2 + a_{i-1} + 1) \ldots (a_1^2 + a_1 + 1)$$
$$X_o = 1, X_1' = [1\ a_1\ a_1^2]$$
$$X_i' = [X_{i-1}'\ a_i X_{i-1}'\ a_i^2 X_{i-1}'], i = 1, 2, \ldots, m$$
$$H_o = 1$$

$$H_1 = \begin{bmatrix} 1 & 1 & 1 \\ -1 & 0 & 1 \\ 1 & -2 & 1 \end{bmatrix}$$

$$H_i = \begin{bmatrix} H_{i-1} & H_{i-1} & H_{i-1} \\ -H_{i-1} & 0 & H_{i-1} \\ H_{i-1} & -2H_{i-1} & H_{i-1} \end{bmatrix}, i = 1, 2, \ldots, m$$

$$Y_o = 1, \ Y_1 = \begin{bmatrix} a_1^2 + a_1 + 1 \\ a_1^2 - 1 \\ a_1^2 - 2a_1 + 1 \end{bmatrix}$$

$$Y_i = \begin{bmatrix} (a_i^2 + a_i + 1)Y_{i-1} \\ (a_i^2 - 1)Y_{i-1} \\ (a_i^2 - 2a_i + 1)Y_{i-1} \end{bmatrix}, i = 1, 2, \ldots, m.$$

Prove that for $i = 1, 2, \ldots, m$

(i) $$X_i = \begin{bmatrix} 1 \\ a_i \\ a_i^2 \end{bmatrix} \otimes \begin{bmatrix} 1 \\ a_{i-1} \\ a_i^2 - 1 \end{bmatrix} \otimes \ldots \otimes \begin{bmatrix} 1 \\ a_1 \\ a_1^2 \end{bmatrix}$$

(ii) $$Y_i = \begin{bmatrix} a_i^2 + a_i + 1 \\ a_i^2 - 1 \\ a_i^2 - 2a_i + 1 \end{bmatrix} \otimes \begin{bmatrix} a_{i-1}^2 + a_{i-1} + 1 \\ a_{i-1}^2 - 1 \\ a_{i-1-2a_{i-1}}^2 + 1 \end{bmatrix} \otimes \ldots$$
$$\otimes \begin{bmatrix} a_1^2 + a_1 + 1 \\ a_1^2 - 1 \\ a_1^2 - 2a_1 + 1 \end{bmatrix}$$

(iii) $H_i = H_1 \otimes H_1 \otimes \ldots \otimes H_1$.

Hence, establish (extended Yates' rule) $H_i X_i = Y_i$.

130. In a randomized block experiment with v treatments and b replications, the yield corresponding to ith treatment in the jth blocks is missing. Show that

(a) the estimate of the missing yield is given by

$$\hat{x} = \frac{(b\, B_j + vT_i - G)^2}{(b - 1)(v - 1)}$$

(b) the bias in estimated treatment sum of square is given by

$$\text{bias} = (B_j + v\, T_i - G)^2 / v(b - 1)^2(v - 1)$$

(c) the loss in efficiency due to missing yield is $[1 + (v - 1)(b - 1)]^{-1}$.

131. In a BIBD (v, b, r, k, λ), the yield corresponding to the ith treatment in the jth block is missing. Show that

(a) the estimate of the missing yield is given by

$$\hat{x} = \frac{\lambda v B_j - kQ'_j + k^2 Q_j}{(k-1)(\lambda v - k)}$$

where Q'_j is the sum of Q's over treatments of the jth block,

(b) the bias in the estimated adjusted treatment sum of squares is given by

$$\text{bias} = \frac{k(B_j - Q'_j + kQ_j)^2}{(k-1)(\lambda v - k)^2}.$$

132. In a binary block design with parameters $v, k_1, k_2, \ldots k_v, r_1, r_2, \ldots, r_b$, the yield corresponding to the ith treatment in the jth block is missing. Show that the estimate of the missing yield is given by

$$\hat{x} = \frac{B_j - \sum_{p=1}^{v}\sum_{s=1}^{v} n_{pj} h_{ps} Q_s + k_j \sum_{s=1}^{v} h_{is} Q_s}{k_j - 1 - k_j h_{ii} + \sum_{s=1}^{v} h_{is} n_{sj} + \sum_{p=1}^{v} n_{pj}\left(h_{pi} - \frac{1}{k_j}\sum_{s=1}^{v} h_{ps} n_{sj}\right)}$$

and that the bias in the estimated adjusted treatment sum of squares is given by

$$\text{bias} = \frac{(k_j - 1)}{k_j}\left[\hat{x} - \frac{B_j}{k_j - 1}\right]^2$$

where $N = [n_{ij}]$ is the incidence matrix of the design and (h_{ij}) is any g-inverse of the C-matrix of the design. Hence deduce the corresponding results for a RBD and BIBD.

133. Suppose the yield corresponding to the ith treatment in the jth row and k-th column in a $v \times v$ Latin square is missing. Show that

(a) the estimate of the missing yield is given by

$$\hat{x} = \frac{vR_j + vC_k + vT_i - 2G}{(v-1)(v-2)}$$

(b) bias in the estimated treatment SS is given by

$$\text{bias} = \frac{[R_j + C_k + (v-1)T_i - G]^2}{(v-2)(v-1)^2}$$

(c) the loss of efficiency due to the missing yield is

$$[1 + (v-1)(v-2)]^{-1}.$$

134. Consider a Youden Square design whose columns form blocks of a SBIBD $(v = b, r = k, \lambda)$. Suppose the yield $x_{ijk} = x$ corresponding to the ith treatment in the jth row and kth column is missing. Show that

(a) the estimate of the missing yield is given by

$$\hat{x} = \frac{\lambda(rR_j + vC_k - G) + r^2Q_i - rQ'_k}{r(r-1)(r-2)},$$

(b) the bias in the estimated adjusted treatment SS is given by

$$\text{bias} = \frac{(v-1)[\lambda(rR_j + vC_k - G) + r(r-1)(rQ_i - Q'_k)]^2}{vr^3(r-1)^3(r-2)^2}$$

when R_j, C_k, G, Q_i have usual meaning and Q'_k = sum of Q's over treatments in the kth column.

135. In a v × v Latin square design, the first row contains treatments in such way that the ith treatment occurs in the ith column, $i = 1, 2, \ldots, v$, and the first row is missing. Showing that the yields in the first row are estimated by

$$\hat{x}_i = \frac{v(c_i + T_i) - 2G}{v(v-2)}, i = 1, 2, \ldots, v$$

and that the bias in the estimated treatment sum of squares is given by

$$\text{bias} = [c_i + (v-1)T_i - G]^2/v(v-1)(v-2)^2.$$

136. In a randomized block design with r blocks and v treatments, the yields of treatments 1 and 2 in the first block are mixed up and their total yield u is only known. Estimate the mixed up yields and show that the bias in the estimated treatment SS is given by

$$\text{bias} = (T_1 - T_2)^2/2(r-1)^2$$

and that the loss in efficiency due to mixing of the yields is

$$[1 + (v-1)(r-1)]^{-1}.$$

137. Obtain the efficiencies of the following designs, the efficiency being defined by $p/n\ \mathrm{tr}(X'X)^{-1}$, where p = no. of objects to be weighted, n = no. of weighings and X is the weighing design matrix:

$$\text{(a)}\quad X = \begin{bmatrix} 1 & 1 \\ 1 & 1 \\ 1 & 1 \\ 1 & -1 \\ 1 & -1 \end{bmatrix}, \quad \text{(b)}\quad X = \begin{bmatrix} 1 & 1 & 1 & 0 \\ 1 & 1 & 0 & 1 \\ 1 & 0 & 1 & 1 \\ 0 & 1 & 1 & 1 \end{bmatrix}$$

138. Let H_{n-1} be the matrix of (n − 1) rows and p columns with elements +1 or −1 such that $H'_{n-1}H_{n-1} = (n-1)I_p$. Prove that the efficiency of design $X = [\begin{smallmatrix} E_{1p} \\ H_{n-1} \end{smallmatrix}]$ is $(n-1)(n-1+p)/n(n-2+p)$.

139. Let H_{n-1} be the matrix of (n − 1) rows and p columns with elements +1 or −1 such that $H'_{n-1}H_{n-1} = (n-1)I_p$. Prove that the efficiencies of the

weighing design

$$\text{(i)}\quad X = \begin{bmatrix} E_{rp} \\ H_{n-1} \end{bmatrix} \quad \text{and} \quad \text{(ii)}\quad X = \begin{bmatrix} E_{r1}0 \\ H_{n-1} \end{bmatrix}$$

are respectively given by

$$(n-1)(n-1+p)/(n-1+r)(n-1-r+pr)$$

and

$$p(n-1)/[p(n-1)+r(p-1)].$$

140. For any weighing design X involving n weighings of p objects, prove that $\text{var}(\hat{w}_i) \geq \sigma^2/n$.

141. For a weighing design X involving n weighings of p objects, prove that the variances of all the estimated weights are minimum if and only if $X'X = nI_p$.

142. In a spring-balance weighing design X, there are v objects and the number of weighings are b. Let k objects weighed in each weighing so that each object is exactly included r times and each pair of objects is included λ times. Show that the variance-covariance matrix of the estimated weights is given by

$$\text{var}\,(\hat{w}) = \left[\frac{1}{(r-\lambda)} I_v - \frac{\lambda}{(r-\lambda)rk} E_{vv}\right]\sigma^2.$$

Defining the efficiency of a weighing design X as $p/n\ \text{tr}(X'X)^{-1}$, where p = no. of objects weighed, n = no. of weighings, show that the efficiency of the above design is $k^2(r-\lambda)/(rk-\lambda)$.

143. A weighing design X is said to be optimal, i.e., one having the maximum efficiency if the value of $|X'X|$ is maximum. Show how to construct a spring-balance weighing design involving n weighings of n objects of maximum efficiency in the above sense with the help of a Hadamard matrix H_{n+1} of order $n+1$. Hence, construct a spring-balance design involving 3 weighings of 3 objects of maximum efficiency.

144. Consider a spring-balance weighing design $X = N'$, where N is the incidence matrix of a BIBD (v, b, r, k, λ). Show that the variance of the best linear unbiased estimate of the total weight of all objects is $\sigma^2 v/rk$.

145. Let $s = p^n$, where p is a prime and n a positive integer and denote the elements of GF (s) by $u_o = 0, u_1 = 1, u_3 = x^2, \ldots, u_{s-1} = x^{s-2}$, where x is a primitive root of GF (s). Construct the square $L_i, i = 1, 2, \ldots, s-1$ by filling its (α, β)th cell by the subscript of the element $u_i u_\alpha + u_\beta, \alpha, \beta = 0, 1, 2, \ldots, s-1$. Prove that

(i) L_i is a Latin square, $i = 1, 2, \ldots, s-1$, and

(ii) the Latin squares L_i and L_j, $i \neq j = 1, 2, \ldots, s-1$ are orthogonal.

146. (Continuation). Prove that

(i) the α-th row of L_{i+1} is the same as the $(\alpha+1)$-st row of L_i and the last row of L_{i+1} is the same as the first row of $L_i, i = 1, 2, .., s-2, \alpha = 1, 2, \ldots, s-2$.

(ii) If in the Latin square $L_i, i = 1, 2, \ldots, s-1$ the subscript in the cell (α, β), $1 \le \alpha, \beta \le s-1$ is j, then the subscript in the cell $(\alpha + 1, \beta + 1)$ is

$$0, \text{ if } j = 0$$
$$j + 1, \text{ if } j = 1, 2, \ldots, s-2$$
$$1, \text{ if } j = s-1.$$

147. Construct the key Latin squares in the set of mutually orthogonal Latin Squares of sides (i) 5, (ii) 7, (iii) 8, (iv) 9, (v) 16, (vi) 25.

148. Let $s = \prod_{i=1}^{m} {p_i}^{e_i}$, where $p_1, p_2, \ldots, p_m$ are distinct prime numbers and $e_1, e_2, \ldots, e_m$ are positive integers. Let $n(s) = \min({p_1}^{e_1}, {p_2}^{e_2}, \ldots, {p_m}^{(e_m)})$. Show that exactly n(s) mutually orthogonal Latin squares of side s can be constructed.

149. Construct 2 mutually orthogonal Latin Squares of side 12.

150. Prove that a BIBD with the following values of parameters can always be constructed:

$$v = (s^{m+1} - 1)/(s-1)$$
$$b = \frac{s^{m+1} - 1)(s^m - 1)\ldots(s^{m-g+1} - 1)}{(s^{g+1} - 1)(s^g - 1)\ldots(s-1)}$$
$$r = \frac{(s^m - 1)(s^{m-1} - 1)\ldots(s^{m-g+1} - 1)}{(s^g - 1)(s^{g-1} - 1)\ldots(s-1)}$$
$$k = (s^{g+1} - 1)/(s-1)$$
$$\lambda = \frac{(s^{m-1} - 1)(s^{m-2} - 1)\ldots(s^{m-g+1} - 1)}{(s^{g-1} - 1)(s^{g-2} - 1)\ldots(s-1)},$$

where s is a prime or a power prime and m and g are positive integers and $1 \le g \le m-1$.

151. Construct a BIBD with parameters $v = b = 13, r = k = 4, \lambda = 1$.

152. Prove that a BIBD with the following values of the parameters can always be constructed:

$$v = s^m,$$
$$b = \frac{s^{m-g}(s^m - 1)(s^{m-1} - 1)\ldots(s^{m-g+1} - 1)}{(s^g - 1)(s^{g-1} - 1)\ldots(s-1)}$$
$$k = s^g$$
$$r = \frac{(s^m - 1)(s^{m-1} - 1)\ldots(s^{m-g+1} - 1)}{(s^g - 1)(s^{g-1} - 1)\ldots(s-1)}$$
$$\lambda = \frac{(s^{m-1} - 1)(s^{m-2} - 1)\ldots(s^{m-g+1} - 1)}{(s^{g-1} - 1)(s^{g-2} - 1)\ldots(s-1)}$$

where s is a prime or a power prime and m and g are positive integers and $1 \leq g \leq m-1$.

153. Let B_i, $i = 1, 2, \ldots, b$ denote the blocks of a BIBD (v, b, r, k, λ). Form the blocks B_i^*, $i = 1, 2, \ldots, b$ such that B_i^* contains treatments which do not occur in B_i. Show that the design formed by the blocks B_i^*, $i = 1, 2, \ldots, b$ is a BIBD with parameters

$$v^* = v, b^* = b, r^* = b - r, k^* = v - k, \lambda = b - 2r + \lambda.$$

154. Let D be the given SBIBD with parameters $v = b, r = k, \lambda$. Consider the residual design D_1 obtained from D by omitting one block from D and all its treatments from the remaining blocks of D. Show that D_1 is BIBD with parameters

$$v_1 = v - k, b_1 = b - 1, r_1 = r, k_1 = k - \lambda, \lambda_1 = \lambda.$$

155. Let D be the given SBIBD with parameters $v = b, r = k, \lambda$. Consider the derived design D_2 obtained from D by omitting one block from D and retaining all its treatments in the remaining blocks of D. Show that D_2 is a BIBD with parameters $v_2 = k, b_2 = b - 1, r_2 = r - 1, k_2 = \lambda, \lambda_2 = \lambda - 1$.

156. Give the constructions of the following BIBDs:

(i) $v = s^2, b = s^2 + s, r = s + 1, k = s, \lambda = 1,$
(ii) $v = b = s^2 + s + 1, r = k = s + 1, \lambda = 1,$

where s is a prime or a power prime.

157. Prove that a BIBD with parameters $v = 6s + 3, b = (3s + 1)(2s + 1)$, $r = 3s + 1, k = 3, \lambda = 1$ can always be constructed where s is any positive integer.

158. Prove that a BIBD with parameters $v = 6t + 1, b = t(6t + 1)$, $r = 3t k = 3, \lambda = 1$, where $6t + 1$ is a prime or a power prime and t a positive integer, can always be constructed.

159. Prove that a BIBD with parameters $v = b = 4t + 3, r = k = 2t + 1, \lambda = t$, where $4t + 3$ is a prime or a power prime and t a positive integer, can always be constructed.

160. Prove that a BIBD with parameters $v = 4t + 1, b = 8t + 2, r = 4t$, $k = 2t, \lambda = 2t - 1$, where $4t + 1$ is a prime or a power prime and t a positive integer, can always be constructed.

161. Prove that a BIBD with parameters $v = 12t + 4, b = (3t + 1)(4t + 1)$, $r = 4t + 1, k = 4, \lambda = 1$, where $4t + 1$ is a prime or a power prime and t a positive integer, can always be constructed.

162. Construct the following BIBDs:

(i) $v = 9, b = 12, r = 4, k = 3, \lambda = 1$
(ii) $v = b = 13, r = k = 4, \lambda = 1$
(iii) $v = b = 61, r = k = 5, \lambda = 2$
(iv) $v = 15, b = 35, r = 7, k = 3, \lambda = 1$

(v) $v = 13, b = 26, r = 12, k = 6, \lambda = 5$
(vi) $v = 8, b = 14, r = 7, k = 4, \lambda = 3$
(vii) $v = 28, b = 63, r = 9, k = 4, \lambda = 1$

163. Prove that a Youden square can always be constructed from a SBIBD.

164. Prove that the number of Youden squares that can be constructed from a SBIBD with parameters $v = b, r = k, \lambda$, is greater than or equal to $k!(k-1)!\ldots!2!1!$

165. Prove that a PBIBD with the following values of the parameters can always be constructed:

$$v = [(s^{m+1} - 1)/(s-1)] - 1 = s(s^m - 1)/(s-1)$$
$$b = \phi(m, g, s) - \phi(m-1, g-1, s)$$
$$r = \phi(m-1, g-1, s) - \phi(m-2, g-2, s)$$
$$k = (s^{g+1} - 1)/(s-1)$$
$$n_1 = s^2(s^{m-1} - 1)/(s-1), n_2 = s-1,$$
$$\lambda_1 = \phi(m-2, g-2, s) - \phi(m-3, g-3, s),$$
$$\lambda_2 = 0$$
$$P_1 = \begin{bmatrix} n_1 - n_2 - 1 & n_2 \\ n_2 & 0 \end{bmatrix}, \qquad P_2 = \begin{bmatrix} n_1 & 0 \\ 0 & n_2 - 1 \end{bmatrix}$$

where s is a prime or a power prime and m and g are positive integers with $1 \leq g \leq m-1$.

166. Prove that a PBIBD with the following values of the parameters can always be constructed.

$$v = st, b = s^m, r = s^{m-1}, k = t,$$
$$\lambda_1 = 0, \lambda_2 = s^{m-2}, n_1 = s-1, n_2 = s(t-1)$$
$$P_1 = \begin{bmatrix} n_1 - 1 & 0 \\ 0 & n_2 \end{bmatrix}, \quad P_2 = \begin{bmatrix} 0 & n_1 \\ n_1 & n_2 - n_1 - 1 \end{bmatrix}$$

where s is a prime or a power prime and m and t are positive integers and $1 \leq t \leq (s^m - 1)/(s-1)$.

167. Prove that PBIBD with the following values of the parameters can always be constructed:

$$v = s^m - 1,$$
$$b = \phi(m, g, s) - \phi(m-1, g-1, s) - \phi(m-1, g, s)$$
$$k = s^g$$
$$r = \phi(m-1, g-1, s) - \phi(m-2, g-2, s)$$
$$n_1 = s^m - s, n_2 = s-2$$
$$\lambda_1 = \phi(m-2, g-2, s) - \phi(m-3, g-3, s)$$
$$\lambda_2 = 0$$

$$P_1 = \begin{bmatrix} n_1 - n_2 - 1 & n_2 \\ n_2 & 0 \end{bmatrix}, \quad P_2 = \begin{bmatrix} n_1 & 0 \\ 0 & n_2 - 1 \end{bmatrix}$$

where s is a prime or a power prime and m and g are positive integers with $1 \leq g \leq m - 1$.

168. Prove that a PBIBD with the following values of the parameters can always be constructed.

$$v = b = pq, r = k = p + q - 1.$$
$$n_1 = q - 1, n_2 = p - 1, n_3 = (p - 1)(q - 1)$$
$$\lambda_1 = q, \lambda_2 = p, \lambda_3 = 2,$$
$$P_1 = \begin{bmatrix} q-2 & 0 & 0 \\ 0 & 0 & (p-1) \\ 0 & p-1 & (p-1)(q-2) \end{bmatrix}$$
$$P_2 = \begin{bmatrix} 0 & 0 & q-1 \\ 0 & p-2 & 0 \\ q-1 & 0 & (p-2)(q-1) \end{bmatrix}$$
$$P_3 = \begin{bmatrix} 0 & 1 & q-2 \\ 1 & 0 & p-2 \\ q-2 & p-2 & (p-2)(q-2) \end{bmatrix}$$

where p and q are integers greater than 2.

169. Prove that a PBIBD with the following values of the parameters can always be constructed.

$$v = b = pq, r = k = p + q - 2$$
$$n_1 = q - 1, n_2 = p - 1, n_3 = (p - 1)(q - 1)$$
$$\lambda_1 = q - 2, \lambda_2 = p - 2, \lambda_3 = 2$$

P_1, P_2, P_3 having the same values as in Exercise 165, where p and q are integers greater than 2.

170. Prove that a PBIBD with the following values of the parameters can always be constructed.

$$v = p = p^2, r = k = 3p - 2$$
$$n_1 = n_2 = n_3 = p - 1, n_4 = (p - 1)(p - 2)$$
$$\lambda_1 = \lambda_2 = \lambda_3 = p + 2, \lambda_4 = 6.$$
$$P_1 = \begin{bmatrix} p-2 & 0 & 0 & 0 \\ 0 & 0 & 1 & p-2 \\ 0 & 1 & 0 & p-2 \\ 0 & p-2 & p-2 & (p-2)(p-3) \end{bmatrix},$$

$$P_2 = \begin{bmatrix} 0 & 0 & 1 & p-2 \\ 0 & p-2 & 0 & 0 \\ 1 & 0 & 0 & p-2 \\ p-2 & 0 & p-2 & (p-2)(p-3) \end{bmatrix}$$

$$P_3 = \begin{bmatrix} 0 & 1 & 0 & p-2 \\ 1 & 0 & 0 & p-2 \\ 0 & 0 & p-2 & 0 \\ p-2 & p-2 & 0 & (p-2)(p-3) \end{bmatrix},$$

$$P_4 = \begin{bmatrix} 0 & 1 & 1 & p-3 \\ 1 & 0 & 1 & p-3 \\ 1 & 1 & 0 & p-3 \\ p-3 & p-3 & p-3 & p^2-6p+10 \end{bmatrix}.$$

171. Construct the following PBIBDs:

(i) $v = 10, b = 5, r = 2, k = 4,$

$n_1 = 6, n_2 = 3, \lambda_1 = 1, \lambda_2 = 0$

$P_1 = [\begin{smallmatrix} 3 & 2 \\ 2 & 1 \end{smallmatrix}], \quad P_2 = [\begin{smallmatrix} 4 & 2 \\ 2 & 0 \end{smallmatrix}]$

(ii) $v = 8, b = 6, r = 3, k = 4$

$n_1 = 3, n_2 = 3, n_3 = 1$

$\lambda_1 = 1, \lambda_2 = 2, \lambda_3 = 0$

$$P_1 = \begin{bmatrix} 2 & 0 & 0 \\ 0 & 2 & 1 \\ 0 & 1 & 0 \end{bmatrix}, \quad P_2 = \begin{bmatrix} 0 & 2 & 1 \\ 2 & 0 & 0 \\ 1 & 0 & 0 \end{bmatrix},$$

$$P_3 = \begin{bmatrix} 0 & 3 & 0 \\ 3 & 0 & 0 \\ 0 & 0 & 0 \end{bmatrix}.$$

(iii) $v = 12, b = 9, r = 3, k = 4.$

$n_1 = 9, n_2 = 2, \lambda_1 = 1, \lambda_2 = 0$

$P_1 = [\begin{smallmatrix} 6 & 2 \\ 2 & 0 \end{smallmatrix}], \quad P_2 = [\begin{smallmatrix} 9 & 0 \\ 0 & 1 \end{smallmatrix}]$

(iv) $v = b = 8, r = k = 3, n_1 = 6,$

$n_2 = 1, \lambda_1 = 1, \lambda_2 = 0,$

$P_1 = [\begin{smallmatrix} 4 & 1 \\ 1 & 0 \end{smallmatrix}], \quad P_2 = [\begin{smallmatrix} 6 & 0 \\ 0 & 0 \end{smallmatrix}]$

(v) $v = b = 9, r = k = 5, n_1 = 2, n_2 = 2,$

$n_3 = 4, \lambda_1 = 3, \lambda_2 = 3, \lambda_3 = 2.$

$$P_1 = \begin{bmatrix} 1 & 0 & 0 \\ 0 & 0 & 2 \\ 0 & 2 & 2 \end{bmatrix}, \quad P_2 = \begin{bmatrix} 0 & 0 & 2 \\ 0 & 1 & 0 \\ 2 & 0 & 2 \end{bmatrix}, \quad P_3 = \begin{bmatrix} 0 & 1 & 1 \\ 1 & 0 & 1 \\ 1 & 1 & 1 \end{bmatrix}.$$

(vi) $v = b = 9, r = k = 4, n_1 = n_2 = 2, n_3 = 4,$
$\lambda_1 = \lambda_2 = 1, \lambda_3 = 2$ and P_1, P_2, P_3 same as in (v).

(vii) $v = b = 9, r = k = 7, n_1 = n_2 = n_3 = n_4 = 2,$
$\lambda_1 = \lambda_2 = \lambda_3 = 5, \lambda_4 = 6$ and

$$P_1 = \begin{bmatrix} 1 & 0 & 0 & 0 \\ 0 & 0 & 1 & 1 \\ 0 & 1 & 0 & 1 \\ 0 & 1 & 1 & 0 \end{bmatrix}, \quad P_2 = \begin{bmatrix} 0 & 0 & 1 & 1 \\ 0 & 1 & 0 & 0 \\ 1 & 0 & 0 & 1 \\ 1 & 0 & 1 & 0 \end{bmatrix}$$

$$P_3 = \begin{bmatrix} 0 & 1 & 0 & 1 \\ 1 & 0 & 0 & 1 \\ 0 & 0 & 1 & 0 \\ 1 & 1 & 0 & 0 \end{bmatrix}, \quad P_4 = \begin{bmatrix} 0 & 1 & 1 & 0 \\ 1 & 0 & 1 & 0 \\ 1 & 1 & 0 & 0 \\ 0 & 0 & 0 & 1 \end{bmatrix}.$$

172. Construct 8 blocks of 4 plots each confounding the interactions ACB, BCE and ABDE in a 2^5 design. Which other interactions are also confounded?

173. Construct 3 blocks of 9 plots each confonding the interaction AB^2C^2 in a 3^3 design.

174. Construct a $(3^3, 3^2)$ design confounding the interactions AB and BC.

175. Construct a $(3^2, 3)$ design achieving a complete balance over the 2-factor interaction.

176. Construct 4 replications of $(3^3, 3^2)$ design achieving complete balance over first order and second order interactions.

177. Construct a (1/4)-th replicate of a 2^5 design with factors A, B, C, D and E. Write down the different alias sets of factorial effects. Take ABC and ACDE as defining interactions.

178. Construct a half replicate of a 2^4 design in blocks of 4 plots each, confounding the interaction AD and using ABCD as the defining interaction.

179. Construct a half replicate of a 2^5 design in blocks of 4 plots each, confounding the interactions BE and CDE and using ABCDE as the defining interaction.

180. Construct (1/4)-th replicate of a 2^6 design, confounding the interactions ACE and ACDF and taking the interactions ABC and ADE as the defining interactions.

181. Construct a (1/3)-rd replicate of a 3^3 design, taking P(111) as the defining pencil. Write down all the alias sets of pencils.

182. Construct $(1/3^2)$-th replicate of a 3^5 design, using P(11111) and P(10011) as the defining pencils. Write down the pencils on which information is lost. Write down the aliases of the pencil P(11101).

183. Construct a (1/3)-rd replicate of a 3^4 design in blocks of 9 plots each, using P(1112) as the defining pencil and confounding the pencil P(1011). Which other pencils are confounded?

184. Construct a (1/3)-rd replicate of a 3^4 design in blocks of 3 plots each, using P(1110) as the defining pencil and confounding the pencils P(1011) and P(1101). Write down the alias sets of pencils which are also confounded.

CHAPTER 3

SOLUTIONS

1. Consider $\boldsymbol{\ell}'\mathbf{y}$, $\boldsymbol{\ell}'\text{A}'\mathbf{y}$, $\boldsymbol{\ell}'\text{GA}'\mathbf{y}$. Since

$$\mathcal{E}(\boldsymbol{\ell}'\mathbf{y}) = \boldsymbol{\ell}'\text{A}\boldsymbol{\theta},$$
$$\mathcal{E}(\boldsymbol{\ell}'\text{A}'\mathbf{y}) = \boldsymbol{\ell}'\text{A}'\text{A}\boldsymbol{\theta},$$
$$\mathcal{E}(\boldsymbol{\ell}'\text{GA}'\mathbf{y}) = \boldsymbol{\ell}'\text{GA}'\text{A}\boldsymbol{\theta},$$

it follows that $\boldsymbol{\ell}'\text{A}\boldsymbol{\theta}$, $\boldsymbol{\ell}'\text{A}'\text{A}\boldsymbol{\theta}$ and $\boldsymbol{\ell}'\text{GA}'\text{A}\boldsymbol{\theta}$ are estimable.

Let $\hat{\boldsymbol{\theta}}$ be any solution of

$$\text{A}'\mathbf{y} = \text{A}'\text{A}\hat{\boldsymbol{\theta}},$$

then the BLUEs of $\boldsymbol{\ell}'\text{A}\boldsymbol{\theta}$, $\boldsymbol{\ell}'\text{A}'\text{A}\boldsymbol{\theta}$ and $\boldsymbol{\ell}'\text{GA}'\text{A}\boldsymbol{\theta}$ are given by $\boldsymbol{\ell}'\text{A}\hat{\boldsymbol{\theta}}$, $\boldsymbol{\ell}'\text{A}'\text{A}\hat{\boldsymbol{\theta}}$ and $\boldsymbol{\ell}'\text{GA}'\text{A}\hat{\boldsymbol{\theta}}$ respectively. Let $\hat{\boldsymbol{\theta}} = \text{GA}'\mathbf{y}$, where G is a g-inverse of A′A. Then,

(i) The BLUE of $\boldsymbol{\ell}'\text{A}\boldsymbol{\theta}$ is $\boldsymbol{\ell}'\text{AGA}'\mathbf{y}$.

$$\begin{aligned}\text{var}(\boldsymbol{\ell}'\text{AGA}'\mathbf{y}) &= \sigma^2\boldsymbol{\ell}'\text{AGA}'.\ \text{AG}'\text{A}'\boldsymbol{\ell}\\ &= \sigma^2\boldsymbol{\ell}'(\text{AGA}')^2\boldsymbol{\ell},\ \text{since AGA}' \text{ is symmetric}\\ &= \sigma^2\boldsymbol{\ell}'\text{AGA}'\boldsymbol{\ell},\ \text{since AGA}' \text{ is idempotent.}\end{aligned}$$

(ii) The BLUE of $\boldsymbol{\ell}'\text{A}'\text{A}\boldsymbol{\theta}$ is $\boldsymbol{\ell}'\text{A}'\text{AGA}'\mathbf{y} = \boldsymbol{\ell}'\text{A}'\mathbf{y}$, since AG is a g-inverse of A′, and var $(\boldsymbol{\ell}'\text{A}'\mathbf{y}) = \sigma^2\boldsymbol{\ell}'\text{A}'\text{A}\boldsymbol{\ell}$. Note here $\boldsymbol{\ell}$ is an $\text{m} \times 1$ vector.

(iii) The BLUE of $\boldsymbol{\ell}'\text{GA}'\text{A}\boldsymbol{\theta}$ is $\boldsymbol{\ell}'\text{GA}'\text{AGA}'\mathbf{y} = \boldsymbol{\ell}'\text{GA}'\mathbf{y}$, since AG is a g-inverse of A′, and var $(\boldsymbol{\ell}'\text{GA}'\mathbf{y}) = \sigma^2\boldsymbol{\ell}'\text{GA}'\text{AG}'\boldsymbol{\ell}$.

2. Clearly,

$$\text{rank}\ (\text{A}'\text{A}) \leq \text{rank}\ (\text{A}'\text{A}, \boldsymbol{\ell}). \tag{1}$$

Now,

$$\text{rank}\ (\text{A}'\text{A}, \boldsymbol{\ell}) = \text{rank}\ (\text{A}', \boldsymbol{\ell})\begin{bmatrix}\text{A} & 0\\ 0 & \text{I}\end{bmatrix} \leq \text{rank}\ (\text{A}', \boldsymbol{\ell}) \tag{2}$$

But, a necessary and sufficient condition for the estimability of $\boldsymbol{\ell}'\boldsymbol{\theta}$ is that rank (A′) = rank $(\text{A}', \boldsymbol{\ell})$. Thus from (2), we get

$$\text{rank}\ (\text{A}'\text{A}, \boldsymbol{\ell}) \leq \text{rank}\ (\text{A}') = \text{rank}\ (\text{A}'\text{A}). \tag{3}$$

Then, the result follows from (1) and (3).

3. Here $\mathcal{E}(\mathbf{y}) = A\boldsymbol{\theta}$, where

$$A = \begin{bmatrix} 1 & 0 & 1 \\ 0 & 1 & 1 \\ 1 & 0 & 1 \end{bmatrix}, \quad \text{and} \quad \boldsymbol{\theta} = \begin{bmatrix} 1 \\ 2 \\ 3 \end{bmatrix}.$$

Now $b_1\theta_1 + b_2\theta_2 + b_3\theta_3$ is estimable iff

$$\text{rank}\,(A') = \text{rank}\,(A', \mathbf{b}),$$

i.e., there exists $\mathbf{k}' = (k_1, k_2, k_3)$ such that $\mathbf{b} = A'\mathbf{k}$, i.e. iff

$$\mathbf{b} = \begin{bmatrix} 1 & 0 & 1 \\ 0 & 1 & 0 \\ 1 & 1 & 1 \end{bmatrix} \begin{bmatrix} k_1 \\ k_2 \\ k_3 \end{bmatrix}$$

i.e. iff $b_1 = k_1 + k_3$, $b_2 = k_2$, $b_3 = k_1 + k_2 + k_3$, i.e. iff $b_3 = b_1 + b_2$. Suppose $b_1\theta_1 + b_2\theta_2 + b_3\theta_3$ is estimable. Then its BLUE is given by

$$b_1\hat{\theta}_1 + b_2\hat{\theta}_2 + b_3\hat{\theta}_3,$$

where $\hat{\theta}_1$, $\hat{\theta}_2$, and $\hat{\theta}_3$ are any solutions of $A'\mathbf{y} = A'A\hat{\boldsymbol{\theta}}$. Now,

$$\begin{bmatrix} 1 & 0 & 1 \\ 0 & 1 & 0 \\ 1 & 1 & 1 \end{bmatrix} \begin{bmatrix} y_1 \\ y_2 \\ y_3 \end{bmatrix} = \begin{bmatrix} 2 & 0 & 2 \\ 0 & 1 & 1 \\ 2 & 1 & 3 \end{bmatrix} \begin{bmatrix} 1 \\ 2 \\ 3 \end{bmatrix}$$

$$y_1 + y_3 = 2\hat{\theta}_1 + 2\hat{\theta}_3$$
$$y_2 = \hat{\theta}_2 + \hat{\theta}_3$$
$$y_1 + y_2 + y_3 = 2\hat{\theta}_1 + \hat{\theta}_2 + 3\hat{\theta}_3$$

A solution of the above equation is

$$\hat{\theta}_3 = 0,\ \hat{\theta}_2 = y_2,\ \hat{\theta}_1 = (y_1 + y_3)/2.$$

Hence, the BLUE of the estimable $b_1\theta_1 + b_2\theta_2 + b_3\theta_3$ is $(b_1y_1 + 2b_2y_2 + b_1y_3)/2$ and its variance is $\sigma^2(b_1^2 + 2b_2^2)/2$.

An unbiased estimator of σ^2 is given by $(\mathbf{y}'\mathbf{y} - \hat{\boldsymbol{\theta}}'A'\mathbf{y})/(n - r)$, where $r = \text{rank}\,(A)$. Here $n = 3$, $r = 2$; hence an unbiased estimator of σ^2 is given by

$$\begin{aligned}
&\mathbf{y}'\mathbf{y} - \hat{\boldsymbol{\theta}}'A'\mathbf{y} \\
&= \sum_1^3 y_i^2 - (\hat{\theta}_1, \hat{\theta}_2, \hat{\theta}_3) \begin{bmatrix} y_1 + y_3 \\ y_2 \\ y_1 + y_2 + y_3 \end{bmatrix} \\
&= \sum y_i^2 - \hat{\theta}_1(y_1 + y_3) - \hat{\theta}_2 y_2\,, \text{ since } \hat{\theta}_3 = 0 \\
&= (y_1^2 + y_2^2 + y_3^2) - \frac{(y_1 + y_3)^2}{2} - y_2^2 \\
&= (2y_1^2 + 2y_2^2 + 2y_3^2 - y_1^2 - 2y_1y_3 - y_3^2 - 2y_2^2)/2 \\
&= (y_1^2 - 2y_1y_3 + y_3^2)/2 = (y_1 - y_3)^2/2.
\end{aligned}$$

4. If we can find a vector $\mathbf{c}$ such that $\mathbf{b} = A'\mathbf{c}$, then $\mathbf{b}'\boldsymbol{\theta}$ is estimable and $\mathbf{c}'\mathbf{y}$ is an unbiased estimator of $\mathbf{b}'\boldsymbol{\theta}$. Consider $\mathbf{b}'\boldsymbol{\theta} = \theta_1 - \theta_2$, i.e. $\mathbf{b}' = (1, -1, 0, 0, 0, 0)$. Hence from $\mathbf{b} = A'\mathbf{c}$, we get

$$\begin{bmatrix} 1 \\ -1 \\ 0 \\ 0 \\ 0 \\ 0 \end{bmatrix} = \begin{bmatrix} c_1 + c_5 \\ c_2 + c_6 \\ c_3 \\ c_1 + c_2 + c_4 \\ c_3 + c_4 \\ c_5 + c_6 \end{bmatrix}.$$

This gives $c_1 + c_5 = 1, c_3 = 0, c_4 = 0, c_1 = -c_2, c_5 = -c_6$, and $c_2 + c_6 = -1$. Thus, we get a solution $c_1 = k, c_2 = -k, c_3 = 0, c_4 = 0, c_5 = 1 - k, c_6 = k - 1$, where k is any constant. Incidently we see that there are many unbiased estimators of $\theta_1 - \theta_2$. Thus an unbiased estimator of $\theta_1 - \theta_2$ is given by

$$ky_1 - ky_2 + (1-k)\,y_5 + (k-1)\,y_6,$$

for any constant k. The variance of this estimator is given by

$$\begin{aligned} V &= \sigma^2[2k^2 + 2(k-1)^2] \\ &= 2\sigma^2(2k^2 - 2k + 1). \end{aligned}$$

Then,

$$\frac{dV}{dk} = 2\sigma^2(4k - 2) = 0, \text{ which gives } k = 2.$$

Since $\frac{d^2V}{dk^2} = 8\sigma^2 > 0$, V is minimum at $k = 2$. Thus, the BLUE of $\theta_1 - \theta_2$ is given by

$$2y_1 - 2y_2 - y_5 + y_6$$

and its variance is $10\sigma^2$.

Now let us consider $\theta_1 + \theta_2 = \mathbf{b}'\theta$, where $\mathbf{b}' = (1, 1, 0, 0, 0, 0)$. Here $\theta_1 + \theta_2$ will be estimable if we can find $\mathbf{c}$ such that $\mathbf{b} = A'\mathbf{c}$, i.e.

$$\begin{bmatrix} 1 \\ 1 \\ 0 \\ 0 \\ 0 \\ 0 \end{bmatrix} = \begin{bmatrix} c_1 + c_5 \\ c_2 + c_6 \\ c_3 \\ c_1 + c_2 + c_4 \\ c_3 + c_4 \\ c_5 + c_6 \end{bmatrix},$$

i.e., $c_1 + c_5 = 1, c_2 + c_6 = 1, c_3 = 0, c_4 = 0, c_5 + c_6 = 0$, and $c_1 + c_2 = 0$. We then get

$$\begin{aligned} c_1 + c_5 &= 1 \\ -c_1 - c_5 &= 1 \end{aligned}$$

i.e., $c_1 + c_5 = -1$, which is not possible. Hence, we cannot find $\mathbf{c}$ such that $\mathbf{b} = A'\mathbf{c}$. Thus $\theta_1 + \theta_2$ is not estimable.

Here

$$A = \begin{bmatrix} 1 & 0 & 0 & 1 & 0 & 0 \\ 0 & 1 & 0 & 1 & 0 & 0 \\ 0 & 0 & 1 & 0 & 1 & 0 \\ 0 & 0 & 0 & 1 & 1 & 0 \\ 1 & 0 & 0 & 0 & 0 & 1 \\ 0 & 1 & 0 & 0 & 0 & 1 \end{bmatrix}$$

Denote the column vectors of A by $\boldsymbol{\alpha}_1, \boldsymbol{\alpha}_2, \boldsymbol{\alpha}_3, \boldsymbol{\alpha}_4, \boldsymbol{\alpha}_5$ and $\boldsymbol{\alpha}_6$. Then, we can easily see that $\boldsymbol{\alpha}_1 + \boldsymbol{\alpha}_2 + \boldsymbol{\alpha}_5 = \boldsymbol{\alpha}_3 + \boldsymbol{\alpha}_4 + \boldsymbol{\alpha}_6$. Hence rank of A is 5. Further $A'\mathbf{y} = A'A\hat{\boldsymbol{\theta}}$ gives

$$y_1 + y_5 = 2\hat{\theta}_1 + \hat{\theta}_4 + \hat{\theta}_6 \tag{1}$$
$$y_2 + y_6 = 2\hat{\theta}_2 + \hat{\theta}_4 + \hat{\theta}_6 \tag{2}$$
$$y_3 = \hat{\theta}_3 + \hat{\theta}_5 \tag{3}$$
$$y_1 + y_2 + y_4 = \hat{\theta}_1 + \hat{\theta}_2 + 3\hat{\theta}_4 + \hat{\theta}_5 \tag{4}$$
$$y_3 + y_4 = \hat{\theta}_3 + \hat{\theta}_4 + 2\hat{\theta}_5 \tag{5}$$
$$y_5 + y_6 = \hat{\theta}_1 + \hat{\theta}_2 + 2\hat{\theta}_6. \tag{6}$$

Put $\hat{\theta}_6 = 0$, then from (1), (2) and (6), we get

$$\hat{\theta}_1 = (y_1 - y_2 + 3y_5 + y_6)/4$$
$$\hat{\theta}_2 = (-y_1 + y_2 + y_5 + 3y_6)/4.$$

Substituting the values of $\hat{\theta}_1$ in (1), we get

$$\hat{\theta}_4 = (y_1 + y_2 - y_5 - y_6)/2.$$

From (3) and (5), we get $y_4 = \hat{\theta}_4 + \hat{\theta}_5$, which gives

$$\hat{\theta}_5 = (-y_1 - y_2 + 2y_4 + y_5 + y_6)/2.$$

Further, from (3), we get

$$\hat{\theta}_3 = (y_1 + y_2 + 2y_3 - 2y_4 - y_5 - y_6)/2.$$

Then, SS due to regression when θ's are fitted is equal to

$$\begin{aligned} SSR &= \hat{\theta}_1(y_1 + y_5) + \hat{\theta}_2(y_2 + y_6) + \hat{\theta}_3 y_3 \\ &\quad + \hat{\theta}_4(y_1 + y_2 + y_4) + \hat{\theta}_5(y_3 + y_4) + \hat{\theta}_6(y_5 + y_6) \\ &= (y_1 - y_2 + 3y_5 + y_6)(y_1 + y_5)/4 \\ &\quad + (-y_1 + y_2 + y_5 + 3y_6)(y_2 + y_6)/4 \\ &\quad + (y_1 + y_2 + 2y_3 - 2y_4 - y_5 - y_6)y_3/2 \\ &\quad + (y_1 + y_2 - y_5 - y_6)(y_1 + y_2 + y_4)/2 \\ &\quad + (-y_1 - y_2 + 2y_4 + y_5 + y_6)(y_3 + y_4)/2 \end{aligned}$$

$$= [3y_1^2 + 3y_2^2 + 4y_3^2 + 4y_4^2 + 3y_5^2 + 3y_6^2 + 2y_1y_2 + 2y_1y_5 - 2y_1y_6 - 2y_2y_5 + 2y_2y_6 + 2y_5y_6]/4.$$

Hence an unbiased estimator of σ^2 is given by

$$(\Sigma\, y_i^2 - \text{SSR})/(6-5) = \Sigma\, y^2 - \text{SSR} = (-y_1 + y_2 + y_5 - y_6)^2/4.$$

5. Here

$$A = \begin{bmatrix} \cos\frac{2\pi}{6} & \sin\frac{2\pi}{6} \\ \cos\frac{4\pi}{6} & \sin\frac{4\pi}{6} \\ \vdots & \\ \cos 2\pi & \sin 2\pi \end{bmatrix}$$

Hence

$$A'A = \begin{bmatrix} \cos\frac{2\pi}{6} \cdots \cos 2\pi \\ \\ \sin\frac{2\pi}{6} \cdots \sin 2\pi \end{bmatrix} \begin{bmatrix} \cos\frac{2\pi}{6} & \sin\frac{2\pi}{6} \\ \vdots & \vdots \\ \cos 2\pi & \sin 2\pi \end{bmatrix}$$
$$= \begin{bmatrix} \sum_i \cos^2\frac{2\pi i}{6} & \sum_i \cos^2\frac{2\pi i}{6}\sin\frac{2\pi i}{6} \\ \sum_i \cos\frac{2\pi i}{6}\sin\frac{2\pi i}{6} & \sum_i \sin^2\frac{2\pi i}{6} \end{bmatrix}$$
$$= \begin{bmatrix} 3 & \sqrt{3} \\ 3 & 3 \end{bmatrix},$$

and

$$(A'A)^{-1} = \frac{1}{2\sqrt{3}}\begin{bmatrix} \sqrt{3} & -1 \\ -1 & \sqrt{3} \end{bmatrix}.$$

Also,

$$A'\mathbf{y} = \begin{bmatrix} \Sigma\, y_i \cos\frac{2\pi i}{6} \\ \Sigma\, y_i \sin\frac{2\pi i}{6} \end{bmatrix}.$$

Thus, the BLUE of $\boldsymbol{\theta}$ is given by

$$\hat{\boldsymbol{\theta}} = (A'A)^{-1}A'\mathbf{y} = \frac{1}{2\sqrt{3}}\begin{bmatrix} \sqrt{3} & -1 \\ -1 & \sqrt{3} \end{bmatrix}\begin{bmatrix} \Sigma\, y_i \cos\dfrac{2\pi i}{6} \\ \Sigma\, y_i \sin\dfrac{2\pi i}{6} \end{bmatrix}$$

$$= \frac{1}{2\sqrt{3}}\begin{bmatrix} \sqrt{3}\Sigma\, y_i \cos\dfrac{2\pi i}{6} - \Sigma\, y_i \sin\dfrac{2\pi i}{6} \\ \sqrt{3}\Sigma\, y_i \sin\dfrac{2\pi i}{6} - \Sigma\, y_i \cos\dfrac{2\pi i}{6} \end{bmatrix}.$$

Further,

$$\text{var}\,(\hat{\boldsymbol{\theta}}) = \sigma^2(A'A)^{-1} = \sigma^2\frac{1}{2\sqrt{3}}\begin{bmatrix} \sqrt{3} & -1 \\ -1 & \sqrt{3} \end{bmatrix}.$$

Hence, var $(\hat{\theta}_1) =$ var $(\hat{\theta}_2) = \sigma^2/2$ and cov $(\hat{\theta}_1, \hat{\theta}_2) = -\sigma^2/2\sqrt{3}$.

6. Let $\mathbf{x}' = \{x_1, x_2, \ldots, x_n\}$, $\mathbf{y}' = \{y_1, y_2, \ldots, y_n\}$ and $\mathbf{z}' = \{z_1, z_2, \ldots, z_n\}$. Then

$$\mathcal{E}\begin{bmatrix} \mathbf{x} \\ \mathbf{y} \\ \mathbf{z} \end{bmatrix} = \begin{bmatrix} E_{n1} & \mathbf{0} \\ \mathbf{0} & E_{n1} \\ E_{n1} & E_{n1} \end{bmatrix}\begin{bmatrix} \theta_1 \\ \theta_2 \end{bmatrix},$$

so that $A = \begin{bmatrix} E_{n1} & \mathbf{0} \\ \mathbf{0} & E_{n1} \\ E_{n1} & E_{n1} \end{bmatrix}$. Hence,

$$A'A = n\begin{bmatrix} 2 & 1 \\ 1 & 2 \end{bmatrix}, \quad \text{rank}\,(A'A) = 2$$

and

$$(A'A)^{-1} = \frac{1}{3n}\begin{bmatrix} 2 & -1 \\ -1 & 2 \end{bmatrix}.$$

Therefore the BLUE of $\boldsymbol{\theta}$ is given by

$$\hat{\boldsymbol{\theta}} = (A'A)^{-1}A'\begin{bmatrix} \mathbf{x} \\ \mathbf{y} \\ \mathbf{z} \end{bmatrix}$$

$$= \frac{1}{3n}\begin{bmatrix} 2 & -1 \\ -1 & 2 \end{bmatrix}\begin{bmatrix} E_{1n} & \mathbf{0} & E_{1n} \\ \mathbf{0} & E_{1n} & E_{1n} \end{bmatrix}\begin{bmatrix} \mathbf{x} \\ \mathbf{y} \\ \mathbf{z} \end{bmatrix}$$

$$= \frac{1}{3n}\begin{bmatrix} 2 & -1 \\ -1 & 2 \end{bmatrix}\begin{bmatrix} X+Z \\ Y+Z \end{bmatrix}$$

$$= \frac{1}{3n}\begin{bmatrix} 2X-Y+Z \\ -X+2Y+Z \end{bmatrix},$$

where $X = \Sigma x_i$, $Y = \Sigma y_i$, $Z = \Sigma z_i$. The variance-covariance matrix of $\hat{\boldsymbol\theta}$ is given by

$$\mathrm{var}(\hat{\boldsymbol\theta}) = \sigma^2(A'A)^{-1} = \frac{\sigma^2}{3n}\begin{bmatrix} 2 & -1 \\ -1 & 2 \end{bmatrix}.$$

Now,

$$\begin{aligned}
SSR &= \hat{\boldsymbol\theta}'A'\begin{bmatrix} \mathbf{x} \\ \mathbf{y} \\ \mathbf{z} \end{bmatrix} = \frac{1}{3n}[(2X - Y + Z) - X + 2Y + Z)]\begin{bmatrix} X+Z \\ Y+Z \end{bmatrix} \\
&= \frac{1}{3n}[(2X - Y + Z)(X + Z) + (-X + 2Y + Z)(Y + Z)] \\
&= \frac{1}{3n}[2X^2 - YX + ZX + 2XZ - YZ + Z^2 - XY + 2Y^2 \\
&\quad + ZY - XZ + 2YZ + Z^2] \\
&= \frac{1}{3n}[2X^2 + 2Y^2 + 2Z^2 - 2XY + 2XZ + 2YZ] \\
&= \frac{2}{3n}[X^2 + Y^2 + Z^2 - XY + YZ + XZ], \\
SSE &= \Sigma\, x_i^2 + \Sigma\, y_i^2 + \Sigma\, z_i^2 - \frac{2}{3n}(X^2 + Y^2 + Z^2 - XY + YZ + XZ),
\end{aligned}$$

and an unbiased estimator of σ^2 is given by

$$\begin{aligned}
\hat{\sigma}^2 &= \frac{SSE}{3n-2} \\
&= \frac{[\Sigma\, x^2 + \Sigma\, y^2 + \Sigma\, z^2 - \frac{2}{3n}(X^2 + Y^2 + Z^2 - XY + YZ + XZ)]}{3n-2}.
\end{aligned}$$

7. Let $\mathbf{x}' = \{x_1, x_2, \ldots, x_n\}$, $\mathbf{y}' = \{y_1, y_2, \ldots, y_n\}$ and $\mathbf{z}' = \{z_1, z_2, \ldots, z_n\}$. Then

$$\mathcal{E}\begin{bmatrix} \mathbf{x} \\ \mathbf{y} \\ \mathbf{z} \end{bmatrix} = \begin{bmatrix} E_{n1} & \mathbf{0} \\ \mathbf{0} & E_{n1} \\ E_{n1} & -E_{n1} \end{bmatrix}\begin{bmatrix} \theta_1 \\ \theta_2 \end{bmatrix}.$$

Hence

$$A = \begin{bmatrix} E_{n1} & \mathbf{0} \\ \mathbf{0} & E_{n1} \\ E_{n1} & -E_{n1} \end{bmatrix},$$

$$A'A = n\begin{bmatrix} 2 & -1 \\ 1 & 2 \end{bmatrix},\quad (A'A)^{-1} = \frac{1}{3n}\begin{bmatrix} 2 & 1 \\ 1 & 2 \end{bmatrix}$$

and rank $(A'A) = 2$. Therefore

$$\hat{\boldsymbol{\theta}} = (A'A)^{-1}A'\begin{bmatrix}\mathbf{x}\\ \mathbf{y}\\ \mathbf{z}\end{bmatrix}$$

$$= \frac{1}{3n}\begin{bmatrix}2 & 1\\ 1 & 2\end{bmatrix}\begin{bmatrix}E_{1n} & \mathbf{0} & E_{1n}\\ \mathbf{0} & E_{1n} & E_{1n}\end{bmatrix}\begin{bmatrix}\mathbf{x}\\ \mathbf{y}\\ \mathbf{z}\end{bmatrix}$$

$$= \frac{1}{3n}\begin{bmatrix}2X + Y + Z\\ X + 2Y - Z\end{bmatrix},$$

where $X = \Sigma x_i$, $Y = \Sigma y_i$, and $Z = \Sigma z_i$. Further,

$$\text{var}\,(\hat{\boldsymbol{\theta}}') = \sigma^2(A'A)^{-1}$$

$$= \frac{\sigma^2}{3n}\begin{bmatrix}2 & 1\\ 1 & 2\end{bmatrix},$$

$$\text{SSR} = \hat{\boldsymbol{\theta}}'A'\begin{bmatrix}\mathbf{x}\\ \mathbf{y}\\ \mathbf{z}\end{bmatrix} = \frac{1}{3n}[2X - Y + Z, X + 2Y + Z]\begin{bmatrix}X + Z\\ Y - Z\end{bmatrix}$$

$$= \frac{2}{3n}[X^2 + Y^2 + Z^2 + XY - YZ + XZ]$$

and

$$\text{SSE} = \Sigma\, x_i^2 + \Sigma\, y_i^2 + \Sigma\, z_i^2 - \text{SSR}.$$

Hence, an unbiased estimator of σ^2 is given by

$$\sigma^2 = \frac{[\Sigma\, x_i^2 + \Sigma\, y_i^2 + \Sigma\, z_i^2 - \frac{2}{3n}(X^2 + Y^2 + Z^2 + XY - YZ + XZ)]}{(3n - 2)}.$$

8. Let $\mathbf{y}_1' = \{y_{11}, y_{12}, \ldots, y_{1n}\}; \mathbf{y}_2' = \{y_{21}, y_{22}, \ldots y_{2n}\}; \mathbf{y}_3' = \{y_{31}, y_{32}, \ldots, y_{3n}\}$, and $\mathbf{z}' = \{\mathbf{y}_1, \mathbf{y}_2, \mathbf{y}_3\}$. Then, we have

$$\mathcal{E}(\mathbf{z}) = A\boldsymbol{\theta},$$

where

$$A = \begin{bmatrix}E_{n1} & E_{n1} & E_{n1}\\ E_{n1} & \mathbf{0} & E_{n1}\\ \mathbf{0} & -2E_{n1} & E_{n1}\end{bmatrix}, \boldsymbol{\theta} = \begin{bmatrix}\theta_1\\ \theta_2\\ \theta_3\end{bmatrix}.$$

We find

$$A'A = n\begin{bmatrix}2 & 1 & 2\\ 1 & 5 & -1\\ 2 & -1 & 3\end{bmatrix}, (A'A)^{-1} = \frac{1}{n}\begin{bmatrix}14 & -5 & -11\\ -5 & 2 & 4\\ -11 & 4 & 9\end{bmatrix}$$

and

$$A'\mathbf{z} = \begin{bmatrix} Y_1 + Y_2 \\ Y_1 - 2Y_3 \\ Y_1 + Y_2 + Y_3 \end{bmatrix},$$

where $Y_1 = \Sigma\, y_{1j}$, $Y_2 = \Sigma y_{2j}$, and $Y_3 = \Sigma y_{3j}$. Therefore, the BLUE of $\boldsymbol{\theta}$ is given by

$$\hat{\boldsymbol{\theta}} = (A'A)^{-1}A'\mathbf{z} = \frac{1}{n}\begin{bmatrix} 2Y_1 + 3Y_2 - Y_3 \\ Y_1 - Y_2 \\ 2Y_1 - 2Y_2 + Y_3 \end{bmatrix}$$

and the variance-covariance matrix of $\hat{\boldsymbol{\theta}}$ is

$$\text{var}\,(\hat{\boldsymbol{\theta}}) = \frac{\sigma^2}{n}\begin{bmatrix} 14 & -5 & -11 \\ -5 & 2 & 4 \\ -11 & 4 & 9 \end{bmatrix}.$$

Further,

$$SSR = \hat{\boldsymbol{\theta}}'A'\mathbf{z} = \frac{1}{n}(5Y_1^2 + Y_2^2 + Y_3^2 + 4Y_1Y_2)$$

and

$$SSE = \Sigma\Sigma\, y_{ij}^2 - \frac{1}{n}(5Y_1^2 + Y_2^2 + Y_3^2 + 4Y_1Y_2).$$

Hence an unbiased estimator of σ^2 is

$$\begin{aligned}\hat{\sigma}^2 &= SSE/(3n-3) \\ &= [\Sigma\Sigma\, y_{ij}^2 - \frac{1}{n}(5Y_1^2 + Y_2^2 + Y_3^2 + 4Y_1Y_2)]/3(n-1),\end{aligned}$$

since rank $(A'A) = 3$.

9. Let $\mathbf{y}$ denote the vector of observations. Then clearly

$$A = \begin{bmatrix} 1 & 0 & 0 \\ 0 & 1 & 0 \\ 0 & 0 & 1 \\ 1 & 1 & 0 \\ 1 & 0 & 1 \\ 0 & 1 & 1 \\ 1 & 1 & 1 \end{bmatrix}, \text{ and } \boldsymbol{\theta} = \begin{bmatrix} \theta_1 \\ \theta_2 \\ \theta_3 \end{bmatrix}.$$

Then, we find

$$A'A = 2\begin{bmatrix} 2 & 1 & 1 \\ 1 & 2 & 1 \\ 1 & 1 & 2 \end{bmatrix}, \quad (A'A)^{-1} = \frac{1}{8}\begin{bmatrix} 3 & -1 & -1 \\ -1 & 3 & -1 \\ -1 & -1 & 3 \end{bmatrix}$$

and rank (A′A) = 3. Hence, the BLUE of $\boldsymbol{\theta}$ is given by

$$\hat{\boldsymbol{\theta}} = \frac{1}{8}\begin{bmatrix} 3 & -1 & -1 \\ -1 & 3 & -1 \\ -1 & -1 & 3 \end{bmatrix} A'\mathbf{y}$$

$$= \frac{1}{8}\begin{bmatrix} 3Y_1 - Y_2 - Y_3 \\ -Y_1 + 3Y_2 - Y_3 \\ -Y_1 - Y_2 + 3Y_3 \end{bmatrix},$$

where $Y_1 = \sum_{i,j=0}^{1} y_{1ij}$, $Y_2 = \sum_{i,j=0}^{1} Y_{i1j}$, and $Y_3 = \sum_{i,j=0}^{1} y_{ij1}$.
Also, the variance-covariance matrix of $\hat{\boldsymbol{\theta}}$ is given by

$$\text{var}\,(\hat{\boldsymbol{\theta}}) = \frac{\sigma^2}{8}\begin{bmatrix} 3 & -1 & -1 \\ -1 & 3 & -1 \\ -1 & -1 & 3 \end{bmatrix}.$$

Further,

$$\text{SSR} = \hat{\boldsymbol{\theta}}'\, A'\mathbf{y} = \frac{1}{8}[3Y_1 - Y_2 - Y_3, -Y_1 + 3Y_2 - Y_3,$$

$$-Y_1 - Y_2 + 3Y_3]\begin{bmatrix} Y_1 \\ Y_2 \\ Y_3 \end{bmatrix}$$

$$= \frac{1}{8}[3Y_1^2 + 3Y_2^2 + 3Y_3^2 - 2Y_1Y_2 - 2Y_1Y_3 - 2Y_2Y_3]$$

and

$$\text{SSE} = (Y_{100}^2 + y_{010}^2 + y_{001}^2 + y_{110}^2 + y_{101}^2 + y_{111}^2)$$
$$- \frac{1}{8}(3Y_1^2 + 3Y_2^2 + 3Y_3^2 - 2Y_1Y_2 - 2Y_1Y_3 - 2Y_2Y_3).$$

Therefore, an unbiased estimator of σ^2 is

$$\hat{\sigma}^2 = [y_{100}^2 + y_{010}^2 + y_{001}^2 + y_{110}^2 + y_{101}^2 + y_{011}^2 + y_{111}^2$$
$$- \frac{1}{8}(3Y_1^2 + 3Y_2^2 + 3Y_3^2 - 2Y_1Y_2 - 2Y_1Y_2 - 2Y_2Y_3)]/4.$$

10. Let $\mathbf{y}_1' = \{y_{11}, y_{12}, \ldots, y_{1n}\}$, $\mathbf{y}_2' = \{y_{21}, y_{22}, \ldots, y_{2n}\}$, and $\mathbf{y}_3' = \{y_{31}, y_{32}, \ldots, y_{3n}\}$. Then

$$\mathcal{E}\begin{bmatrix} \mathbf{y}_1 \\ \mathbf{y}_2 \\ \mathbf{y}_3 \end{bmatrix} = \begin{bmatrix} A_1 \\ A_2 \\ A_3 \end{bmatrix}\boldsymbol{\theta},$$

where

$$A_1 = \begin{bmatrix} 1 & 2 & 3 \\ 1 & 2 & 3 \\ \cdot & \cdot & \cdot \\ \cdot & \cdot & \cdot \\ \cdot & \cdot & \cdot \\ 1 & 2 & 3 \end{bmatrix}, A_2 = \begin{bmatrix} 2 & 3 & 1 \\ 2 & 3 & 1 \\ \cdot & \cdot & \cdot \\ \cdot & \cdot & \cdot \\ \cdot & \cdot & \cdot \\ 2 & 3 & 1 \end{bmatrix}, A_3 = \begin{bmatrix} 3 & 1 & 2 \\ 3 & 1 & 2 \\ \cdot & \cdot & \cdot \\ \cdot & \cdot & \cdot \\ \cdot & \cdot & \cdot \\ 3 & 1 & 2 \end{bmatrix}.$$

Then,

$$\begin{aligned} A'A &= A_1'A_1 + A_2'A_2 + A_3'A_3 \\ &= n\begin{bmatrix} 1 & 2 & 3 \\ 2 & 4 & 6 \\ 3 & 6 & 9 \end{bmatrix} + n\begin{bmatrix} 4 & 6 & 2 \\ 6 & 9 & 3 \\ 2 & 3 & 1 \end{bmatrix} + n\begin{bmatrix} 9 & 3 & 6 \\ 3 & 1 & 2 \\ 6 & 2 & 4 \end{bmatrix} \\ &= n\begin{bmatrix} 14 & 11 & 11 \\ 11 & 14 & 11 \\ 11 & 11 & 14 \end{bmatrix} = n[3I_3 + 11E_{33}]. \end{aligned}$$

Hence,

$$(A'A)^{-1} = \frac{1}{3n}[I_3 - \frac{11}{36}E_{33}], \text{ and rank } (A'A) = 3.$$

Let $Y_1 = E_{1n}\mathbf{y}_1$, $Y_2 = E_{1n}\mathbf{y}_2$ and $Y_3 = E_{1n}\mathbf{y}_3$. Also

$$\begin{aligned} A'\mathbf{y} &= A_1'\mathbf{y}_1 + A_2'\mathbf{y}_2 + A_3'\mathbf{y}_3 \\ &= \begin{bmatrix} Y_1 \\ 2Y_1 \\ 3Y_1 \end{bmatrix} + \begin{bmatrix} 2Y_2 \\ 3Y_2 \\ Y_2 \end{bmatrix} + \begin{bmatrix} 3Y_3 \\ Y_3 \\ 2Y_3 \end{bmatrix} = \begin{bmatrix} Y_1 + 2Y_2 + 3Y_3 \\ 2Y_1 + 3Y_2 + Y_3 \\ 3Y_1 + Y_2 + 2Y_3 \end{bmatrix}. \end{aligned}$$

Hence, the BLUE of $\boldsymbol{\theta}$ is given by

$$\begin{aligned} \hat{\boldsymbol{\theta}} &= (A'A)^{-1}A'\mathbf{y} \\ &= \frac{1}{3n}[I_3 - \frac{11}{36}E_{33}] \begin{bmatrix} Y_1 + 2Y_2 + 3Y_3 \\ 2Y_1 + 3Y_2 + Y_3 \\ 3Y_1 + Y_2 + 2Y_3 \end{bmatrix} \\ &= \frac{1}{3n}\begin{bmatrix} Y_1 + 2Y_2 + 3Y_3 \\ 2Y_1 + 3Y_2 + Y_3 \\ 3Y_1 + Y_2 + 2Y_3 \end{bmatrix} - \frac{11}{18}G\,E_{31}, \end{aligned}$$

where $G = Y_1 + Y_2 + Y_3$. Hence

$$\hat{\boldsymbol{\theta}} = \frac{1}{3n}\begin{bmatrix} Y_1 + 2Y_2 + 3Y_3 - \frac{11}{6}G \\ 2Y_1 + 3Y_2 + Y_3 - \frac{11}{6}G \\ 3Y_1 + Y_2 + 2Y_3 - \frac{11}{6}G \end{bmatrix}.$$

The variance-covariance matrix of $\hat{\boldsymbol\theta}$ is

$$\text{var}\,(\hat{\boldsymbol\theta}) = \frac{\sigma^2}{3n}\left[I_3 - \frac{11}{36}E_{33}\right].$$

Also

$$\begin{aligned}
SSR = \hat{\boldsymbol\theta}'\,A'\mathbf{y} &= \frac{1}{3n}\Big[(Y_1+2Y_2+3Y_3)(Y_1+2Y_2+3Y_3-\frac{11}{6}G)\\
&\quad +(2Y_1+3Y_2+Y_3)(2Y_1+3Y_2+Y_3-\frac{11}{6}G)\\
&\quad +(3y_1+Y_2+2Y_3)(3Y_1+Y_2+2Y_3-\frac{11}{6}G)\Big]\\
&= \frac{1}{3n}[(Y_1+2Y_2+3Y_3)^2+(2Y_1+3Y_2+Y_3)^2\\
&\quad +(3Y_1-Y_2+2Y_3)^2-11G^2],
\end{aligned}$$

and

$$\begin{aligned}
SSE = \sum_{i=1}^{3}\sum_{j=1}^{n} y_{ij}^2 - \frac{1}{3n}[(Y_1+2Y_2+3Y_3)^2+(2Y_1+3Y_2+Y_3)^2\\
+(3Y_1+Y_2+2Y_3)^2-11G^2].
\end{aligned}$$

Therefore, an unbiased estimator of σ^2 is given by

$$\begin{aligned}
\hat\sigma^2 &= \frac{1}{3(n-1)}\left[\sum_{i=1}^{3}\sum_{j=1}^{n} y_{ij}^2 - \frac{1}{3n}\{(Y_1+2Y_2+3Y_3)^2\right.\\
&\quad \left.+(2Y_1+3Y_2+Y_3)^2+(3Y_1+Y_2+2Y_3)^2-11G^2\}\right].\\
&= \frac{1}{3(n-1)}\left[\Sigma\,\Sigma\, y_{ij}^2 - \frac{1}{n}(Y_1^2+Y_2^2+Y_3^2)\right].
\end{aligned}$$

11. (i) Let $\mathbf{c}'\boldsymbol\theta$ be an estimable parametric function. Let $\mathbf{c}'\hat{\boldsymbol\theta}$ be the BLUE of $\mathbf{c}'\boldsymbol\theta$. To test $\mathbf{c}'\boldsymbol\theta = k$, where k is any constant, we use t test, where

$$t = \frac{(\mathbf{c}'\hat{\boldsymbol\theta}-k)}{\hat\sigma\sqrt{\mathbf{c}'(A'A)^{-1}\mathbf{c}}} \text{ with } n-r \text{ d.f.},$$

where $r = \text{rank}\,(A'A)$ and $\mathcal{E}(\mathbf{y}) = A\boldsymbol\theta$.

Consider (i) $H_0 : \theta_1 - \theta_2 = 0$. Thus $\mathbf{c}' = (1, -1)$ and $k = 0$. Also, from Exercise 6, we have

$$(A'A)^{-1} = \frac{1}{3n}\begin{bmatrix} 2 & -1 \\ -1 & 2 \end{bmatrix},\ \text{rank}\,(A'A) = 2,$$

and, $\mathbf{c}'(A'A)^{-1}\mathbf{c} = 2/n$. Then from Exercise 6, we have

$$\hat{\sigma} = \left[\frac{\Sigma\, x^2 + \Sigma\, y^2 + \Sigma\, z^2 - \frac{2}{3n}(X^2 + Y^2 + Z^2 + XY + XZ + YZ)}{3n - 2}\right]^{1/2},$$

where $X = \Sigma\, x$, $Y = \Sigma\, y$, and $Z = \Sigma\, z$. Thus for testing $\theta_1 = \theta_2$, we use

$$t = \frac{(\hat{\theta}_1 - \hat{\theta}_2)}{\hat{\sigma}\sqrt{2/n}}, \text{ with } 3n - 2 \text{ d.f.},$$

where $\hat{\theta}_1 = (2X - Y + Z)/3n$, $\hat{\theta}_2 = (-X + 2Y + Z)/3n$.

Consider (ii) $H_0 : \theta_1 = a\theta_2$, i.e. $\theta_1 - a\theta_2 = 0$.
Hence $\mathbf{c}' = [1, -a]$, $k = 0$. Also, from Exercise 7, we have

$$(A'A)^{-1} = \frac{1}{3n}\begin{bmatrix} 2 & 1 \\ 1 & 2 \end{bmatrix}, \text{ rank } (A'A) = 2$$

$$\mathbf{c}'(A'A)^{-1}\mathbf{c} = 2(a^2 - a + 1)/3n$$

$$\hat{\boldsymbol{\theta}} = \frac{1}{3n}\begin{bmatrix} 2X + Y + Z \\ X + 2Y - Z \end{bmatrix},$$

where $X = \Sigma\, x$, $Y = \Sigma\, y$, and $Z = \Sigma\, z$. From Exercise 7, we have

$$\hat{\sigma} = \left[\frac{\Sigma\, x^2 + \Sigma\, y^2 + \Sigma\, z^2 - \frac{2}{3n}(X^2 + Y^2 + Z^2 + XY + XZ + YZ)}{(3n - 2)}\right]^{1/2},$$

Hence for testing $\theta_1 = a\theta_2$, we use

$$t = \frac{(\hat{\theta}_1 - a\hat{\theta}_2)}{\hat{\sigma}\sqrt{2(a^2 - a + 1)/3n}}, \text{ with } 3n - 2 \text{ d.f.},$$

Consider (iii) $H_0 : \theta_3 = 0$. We have, from Exercise 8,

$$\hat{\theta}_3 = (2Y_1 - 2Y_2 + Y_3)/n,$$

$$\text{var}\,(\hat{\theta}_3) = \sigma^2(9/n),$$

$$\hat{\sigma} = \left[\frac{(\Sigma\Sigma\, y_{ij}^2 - \frac{1}{n}(5Y_1^2) + Y_2^2 + Y_3^2 + 4Y_1Y_2)}{3(n - 1)}\right]^{1/2}$$

where $Y_1 = \Sigma\, y_{1j}$, $Y_2 = \Sigma\, y_{2j}$, and $Y_3 = \Sigma\, y_{3j}$. Hence for testing $\theta_3 = 0$, we use

$$t = \frac{\hat{\theta}_3}{\hat{\sigma}(3/\sqrt{n})} \text{ with d.f. } 3(n - 2).$$

Consider (iv) $H_0 : \theta_1 + \theta_2 + \theta_3 = 0$. Here $\mathbf{c}' = [111]$, $k = 0$.

We have from Exercise 10,

$$(A'A)^{-1} = \frac{1}{3n}\left[I_3 - \frac{11}{36}E_{33}\right], \text{ rank } (A'A) = 3$$

$$\mathbf{c}'(A'A)^{-1}\mathbf{c} = 1/12\,n$$

$$\hat{\boldsymbol{\theta}} = \frac{1}{3n}\begin{bmatrix} Y_1 + 2Y_2 + 3Y_3 - \frac{11}{6}G \\ 2Y_1 + 3Y_2 + Y_3 - \frac{11}{6}G \\ 3Y_1 + Y_2 + 2Y_3 - \frac{11}{6}G \end{bmatrix},$$

where $Y_1 = \Sigma\, y_{1j}$, $Y_2 = \Sigma\, y_{2j}$, $Y_3 = \Sigma\, y_{3j}$. Hence $\hat{\theta}_1 + \hat{\theta}_2 + \hat{\theta}_3 = G/6n$. Also form Exercise 10, we have

$$\hat{\sigma} = \left[\frac{\Sigma\,\Sigma\, y_{ij}^2 - \frac{1}{n}(Y_1^2 + Y_2^2 + Y_3^2)}{3(n-1)}\right]^{1/2}.$$

Hence for testing $\theta_1 + \theta_2 + \theta_3 = 0$, we use

$$t = \frac{(G/6n)}{\hat{\sigma}\sqrt{1/12n}}, \text{ with d.f. } 3(n-1).$$

Consider (v) H_0: $\theta_2 = \theta_3 = 0$. In Exercise 10, we have found that

$$SSR(\theta_1, \theta_2, \theta_3) = \frac{1}{n}(Y_1^2 + Y_2^2 + Y_3^2) \text{ with d.f. 3 and}$$

$$SSE = \sum_{i=1}^{3}\sum_{j=1}^{n} y_{ij}^2 - \frac{1}{n}(Y_1^2 + Y_2^2 + Y_3^2) \text{ with d.f. } 3(n-1).$$

Now, we shall find SS due to regression when θ_1 is fitted. Then under the hypothesis $\theta_2 = \theta_3 = 0$, we have

$$\mathcal{E}\begin{bmatrix}\mathbf{y}_1 \\ \mathbf{y}_2 \\ \mathbf{y}_3\end{bmatrix} = \begin{bmatrix} E_{n1} \\ 2E_{n1} \\ 3E_{n1}\end{bmatrix}\theta_1 = A\theta_1,$$

where $A = \begin{bmatrix} E_{n1} \\ 2E_{n1} \\ 3E_{n1}\end{bmatrix}$. Hence

$$A'A = 14n,\ (A'A)^{-1} = \frac{1}{14n},\ \hat{\theta}_1 = \frac{1}{14n}(Y_1 + 2Y_2 + 3Y_3)$$

and

$$SSR(\theta_1) = \frac{1}{14n}(Y_1 + 2Y_2 + 3Y_3)^2 \text{ with d.f. 1.}$$

Therefore SS for testing $\theta_2 = \theta_3 = 0$ is given by

$$\begin{aligned} SSH &= SSR(\theta_1, \theta_2, \theta_3) - SSR(\theta_1) \\ &= \frac{1}{n}(Y_1^2 + Y_2^2 + Y_3^2) - \frac{1}{14n}(Y_1 + 2Y_2 + 3Y_3)^2 \text{ with d.f. 2.} \end{aligned}$$

Hence to test $\theta_2 = \theta_3 = 0$, we use F-statistic, where

$$F = \frac{\left[\frac{1}{n}(Y_1^2 + Y_2^2 + Y_3^2) - \frac{1}{14n}(Y_1 + 2Y_2 + 3Y_3)^2\right]/2}{[SSE]/3(n-1)}$$

with d.f. 2 and $3(n-1)$.

12. Let $x_i - \bar{x} = u_i$. Then $\Sigma\, u_i = 0$. Let $\mathbf{y}' = \{y_1, y_2, \ldots, y_n\}$, and $\boldsymbol{\theta}' = \{\alpha, \beta\}$. Then $\mathcal{E}(\mathbf{y}) = A\boldsymbol{\theta}$, where

$$A = \begin{bmatrix} 1 & u_1 \\ 1 & u_2 \\ \cdot & \cdot \\ \cdot & \cdot \\ \cdot & \cdot \\ 1 & u_n \end{bmatrix}.$$

Clearly

$$A'A = \begin{bmatrix} n & 0 \\ 0 & \Sigma\, u_i^2 \end{bmatrix}, \quad \text{rank}\,(A'A) = 2,$$

and

$$(A'A)^{-1} = \begin{bmatrix} 1/n & 0 \\ 0 & 1/\Sigma\, u_i^2 \end{bmatrix}, A'\mathbf{y} = \begin{bmatrix} n\,\bar{y} \\ \Sigma\, y_i n_i \end{bmatrix}.$$

Hence,

$$\hat{\boldsymbol{\theta}} = (A'A)^{-1}A'\mathbf{y} = \begin{bmatrix} \bar{y} \\ \Sigma\, u_i y_i / \Sigma\, u_i^2 \end{bmatrix}.$$

Thus, $\hat{\alpha} = \bar{y}$, $\hat{\beta} = \Sigma\, u_i y_i / \Sigma\, u_i^2$, and their variance-covariance matrix is

$$\sigma^2 (A'A)^{-1} = \sigma^2 \begin{bmatrix} 1/n & 0 \\ 0 & 1/\Sigma\, u_i^2 \end{bmatrix}.$$

Further,

$$\begin{aligned} SSE &= \Sigma\, y^2 - n\bar{y}^2 - \hat{\beta}\,\Sigma\, u_i y_i \\ &= \Sigma\,(y_i - \bar{y})^2 - \hat{\beta}^2\,\Sigma\, u_i^2 \end{aligned}$$

with $(n-2)$ d.f. Hence,

$$SE(\hat{\alpha}) = \sqrt{\frac{SSE}{(n-2)} \cdot \frac{1}{n}},$$

and for testing $\alpha = 0$, we use the statistic t, where

$$t = \frac{\hat{\alpha}}{SE(\hat{\alpha})} = \frac{\hat{\alpha}\sqrt{n}}{\sqrt{SSE/(n-2)}} \text{ with d.f. } (n-2).$$

Similarly,

$$SE(\hat{\beta}) = \sqrt{\frac{SSE}{(n-2)}} \cdot \frac{1}{\sqrt{\Sigma u_i^2}}.$$

Hence for testing $\beta = 0$, we use t-statistic, where

$$t = \frac{\hat{\beta}}{SE(\hat{\beta})} = \frac{\hat{\beta}\sqrt{\Sigma u_i^2}}{SSE/(n-2)} \text{ with } (n-2) \text{ d.f.}$$

13. Let $x_{ij} - \bar{x}_i = u_{ij}$, $i = 1, 2, \ldots, k$; $j = 1, 2, \ldots, n$.

$$\boldsymbol{\theta}' = (\alpha, \beta_1, \beta_2, \ldots, \beta_k), \text{ and } \mathbf{y}' = (y_1, y_2, \ldots, y_n).$$

Then, $\mathcal{E}(\mathbf{y}) = A\boldsymbol{\theta}$, where

$$A = \begin{bmatrix} 1 & u_{11} & u_{21} & \cdots & u_{k1} \\ 1 & u_{12} & u_{22} & \cdots & u_{k2} \\ \cdot & \cdot & & & \cdot \\ \cdot & \cdot & & & \cdot \\ \cdot & \cdot & & & \cdot \\ 1 & u_{1n} & u_{2n} & & u_{kn} \end{bmatrix}.$$

Therefore

$$A'A = \begin{bmatrix} n & 0 & 0 & \ldots & 0 \\ 0 & \Sigma u_{ij}^2 & \Sigma u_{1j}u_{2j} & \ldots & \Sigma u_{1j}u_{kj} \\ 0 & \Sigma u_{2j}u_{1j} & \Sigma u_{2j}^2 & \ldots & \Sigma u_{2j}u_{kj} \\ \cdot & \cdot & & & \\ \cdot & \cdot & & & \\ \cdot & \cdot & & & \\ 0 & \Sigma u_{kj}u_{1j} & \Sigma u_{kj}u_{2j} & \ldots & \Sigma u_{kj}^2 \end{bmatrix}$$

and rank $(A'A) = k + 1$.

We can write $A'A$ as $A'A = \begin{bmatrix} n & 0 \\ 0 & S \end{bmatrix}$, and hence,

$$(A'A)^{-1} = \begin{bmatrix} 1/n & 0 \\ 0 & S^{-1} \end{bmatrix},$$

where

$$S = \begin{bmatrix} \Sigma\, u_{ij}^2 & \Sigma\, u_{1j}u_{2j} & \dots & \Sigma\, u_{1j}u_{kj} \\ \Sigma\, u_{2j}u_{1j} & \Sigma\, u_{2j}^2 & \dots & \Sigma\, u_{2j}u_{kj} \\ \cdot & & & \\ \cdot & & & \\ \cdot & & & \\ \Sigma\, u_{kj}u_{1j} & \Sigma\, u_{kj}u_{2j} & \dots & \Sigma\, u_{kj}^2 \end{bmatrix}.$$

Then,

$$\hat{\boldsymbol{\theta}} = (A'A)^{-1}A'\mathbf{y}$$

$$= \begin{bmatrix} 1/n & \mathbf{0} \\ \mathbf{0} & S^{-1} \end{bmatrix} \begin{bmatrix} \Sigma\, \bar{y} \\ \Sigma\, u_{ij}y_j \\ \cdot \\ \cdot \\ \cdot \\ \Sigma\, u_{kj}y_j \end{bmatrix} = \begin{bmatrix} \bar{y} \\ S^{-1}\mathbf{z} \end{bmatrix}$$

where $\mathbf{z}' = (\Sigma\, u_{1j}y_j,\ \Sigma\, u_{2j}y_j, \dots,\ \Sigma\, u_{kj}y_j)$. Hence $\hat{\alpha} = \bar{y}$, and $\hat{\boldsymbol{\beta}} = S^{-1}\mathbf{z}$. The variance-covariance matrix of $\hat{\theta}$ is

$$\sigma^2(A'A)^{-1} = \sigma^2 \begin{bmatrix} 1/n & \mathbf{0} \\ \mathbf{0} & S^{-1} \end{bmatrix}.$$

Hence

$$\text{var}\,(\hat{\alpha}) = \sigma^2/n,\ \text{cov}\,(\hat{\alpha}, \hat{\boldsymbol{\beta}}) = 0,$$
$$\text{var}\,(\hat{\boldsymbol{\beta}}) = \sigma^2 S^{-1}.$$

To test $\beta_i = 0$, we use t-statistic, where

$$t = \frac{\hat{\beta}_i}{SE(\hat{\beta}_i)},\ \text{with } (n-k-1) \text{ d.f.}$$

The SE $(\hat{\beta}_i)$ is given by
$SE(\hat{\beta}_i) = \hat{\sigma}\sqrt{c_{ii}}$, where $S^{-1} = (c_{ij})$, and

$$\hat{\sigma}^2 = \left[\Sigma\, y^2 - (\hat{\alpha}, \hat{\boldsymbol{\beta}}) \begin{bmatrix} n\,\bar{y} \\ \mathbf{z} \end{bmatrix}\right] / (n-k-1)$$
$$= [\Sigma\, y^2 - n\bar{y}^2 - \hat{\beta}'\mathbf{z}]/(n-k-1)$$
$$= [(\Sigma\, y^2 - n\bar{y}^2) - \mathbf{z}'S^{-1}\mathbf{z}]/(n-k-1).$$

To test $\beta_i = \beta_j$, we use t-statistic, where

$$t = \frac{(\hat{\beta}_i - \hat{\beta}_j)}{SE(\hat{\beta}_i - \hat{\beta}_j)} \text{ with } (n-k-1) \text{ d.f.}$$

The SE $(\hat{\beta}_i - \hat{\beta}_j)$ is given by

$$SE\,(\hat{\beta}_i - \hat{\beta}_j) = \hat{\sigma}\sqrt{c_{ii} + c_{jj} - 2c_{ij}},$$

where $\hat{\sigma} = [(\Sigma\, y^2 - n\bar{y}^2) - \mathbf{z}'S^{-1}\mathbf{z}]/(n-k-1)$.

14. This is an example of analysis of variance for one-way classification.

Let

$$T_i = \sum_{j=1}^{n_i} y_{ij}, \quad i = 1, 2, \ldots, k, \text{ and } G = \Sigma\Sigma\, y_{ij} = \Sigma\, T_i.$$

The normal equations are obtained by minimizing $\Sigma\Sigma\,(y_{ij} - \mu - t_i)^2$ and are found to be

$$G = n\hat{\mu} + \Sigma\, n_i\hat{t}_i$$
$$T_i = n_i\,\hat{\mu} + n_i\hat{t}_i,\ i = 1, 2, \ldots, k$$

where $n = \Sigma\, n_i$. Then a set of solutions is obtained by setting $\Sigma\, n_i\hat{t}_i = 0$. Hence, we obtain

$$\hat{\mu} = G/n,\ \hat{t}_i = T_i/n_i - G/n.$$

Thus, $SSR = \Sigma\, T_i^2/n_i$. Clearly the number of independent normal equations is k. The error SS is

$$SSE = \Sigma\Sigma\, {y_{ij}}^2 - \Sigma\, T_i^2/n_i \text{ with } (n-k) \text{ d.f.}$$

and SSR has k d.f.

To test $n\mu + \Sigma\, n_i\hat{t}_i = 0$, we find $n\hat{\mu} + \Sigma\, n_i\hat{t}_i = G$. Also

$$\text{var}\,(G) = n\sigma^2.$$

Therefore,

$$SE(G) = n\hat{\sigma} = n\left[\frac{SSE}{n-k}\right]^{1/2}.$$

Hence to test $n\mu + \Sigma\, n_i t_i = 0$, we use t-statistic, where

$$t = \frac{G}{SE\,(G)} = \frac{G}{n\sqrt{SSE/(n-k)}} \text{ with } (n-k) \text{ d.f.}$$

To test $t_1 = t_2 = \ldots = t_k = 0$, we proceed as follows.

Under the hypothesis $t_1 = t_2 = \ldots = t_k = 0$, $\mathcal{E}(y_{ij}) = \mu$. We then have the following normal equation

$$G = n\overset{*}{\mu},$$

which gives $\overset{*}{\mu} = G/n$. Thus,

$$SSR(\mu) = G^2/n \text{ with 1 d.f.}$$

Hence, for testing $t_1 = t_2 = \ldots = t_k = 0$, we use F-statistic, where

$$F = \frac{[SSR - SSR(\mu)]/(k-1)}{SSE/(n-k)} \text{ with } (k-1) \text{ and } (n-k) \text{ d.f.}$$
$$= \frac{[\Sigma\, T_i^2/n_i - G^2/n]/(k-1)}{[\Sigma\Sigma\, {y_{ij}}^2 - \Sigma\, T_i^2/n_i]/(n-k)}.$$

We now consider the testing of $t_1 = t_2 = \ldots = t_k$. Write the model as

$$\mathcal{E}(y_{ij}) = \mu + t_i = \alpha + t_i', \text{ where } t_i' = t_i - \bar{t}, \bar{t} = \Sigma\, t_i/k, \alpha = \mu + \bar{t}.$$

The hypothesis $t_1 = t_2 = \ldots = t_k$ is then equivalent to the hypothesis $t_i' = 0, i = 1, 2, \ldots, k$. Thus, this case reduces to the case (ii), and hence we get the same test statistic as in (ii).

15. Let $T_{i\cdot} = \sum_{j=1}^{s} y_{ij},\ T_{\cdot j} = \sum_{i=1}^{r} y_{ij}$, and $G = \Sigma\, T_{i.} = \Sigma\Sigma\, y_{ij}$.

The normal equations are obtained by minimizing $\Sigma\Sigma\,(y_{ij} - \mu - \alpha_i - \beta_j)^2$ and are found to be as

$$\begin{aligned} G &= rs\hat{\mu} + s\Sigma\hat{\alpha}_i + r\Sigma\hat{\beta}_j \\ T_{i.} &= s\hat{\mu} + s\hat{\alpha}_i + \Sigma\hat{\beta},\ i = 1, 2, \ldots, r \\ T_{.j} &= r\hat{\mu} + \Sigma\hat{\alpha}_i + r\hat{\beta}_j,\ j = 1, 2, \ldots, s. \end{aligned}$$

Clearly, the number of independent normal equations is $r + s - 1$.

Hence,

$$SSR(\mu, \boldsymbol{\alpha}, \boldsymbol{\beta}) = \frac{\Sigma\, T_{i.}^2}{s} + \frac{\Sigma\, T_{.j}^2}{r} - \frac{G^2}{rs} \text{ with } r + s - 1 \text{ d.f.}$$

and

$$SSE = \Sigma\Sigma\, y_{i_j}^2 - \frac{\Sigma\, T_{i.}^2}{s} - \frac{\Sigma\, T_{.j}^2}{r} + \frac{G^2}{rs} \text{ with } (r-1)(s-1)\text{d.f.}$$

Solution is obtained by setting $\Sigma\hat{\alpha}_i = 0, \Sigma\hat{\beta}_j = 0$. Thus, we obtain

$$\hat{\mu} = G/rs, \hat{\alpha}_i = \frac{T_{i\cdot}}{s} - \frac{G}{rs}, \hat{\beta}_j = \frac{T_{\cdot j}}{r} - \frac{G}{rs}.$$

To test $\alpha_1 = \alpha_2 = \ldots = \alpha_r = 0$, we take $\mathcal{E}(y_{ij}) = \mu + \beta_j, i = 1, 2, \ldots, r$ and $j = 1, 2, \ldots, s$. A set of solutions for μ and β's is obtained as $\mu^* = G/rs, \beta_j^* = \dfrac{T_{\cdot j}}{r} - \dfrac{G}{rs}$ and hence

$$SSR(\mu, \boldsymbol{\beta}) = \frac{\Sigma\, T_{.j}^2}{r} \text{ with s d.f.}$$

Therefore for testing $\alpha_1 = \alpha_2 = \ldots = \alpha_r = 0$, we use F-statistic, where

$$F = \frac{[SSR(\mu, \boldsymbol{\alpha}, \boldsymbol{\beta}) - SSR(\alpha, \boldsymbol{\beta})]/(r-1)}{SSE/(r-1)(s-1)}$$

with $(r-1)$ and $(r-1)\,(s-1)$ d.f. Clearly, the above F can be written as

$$F = \frac{\left[\dfrac{\Sigma\, T_{i.}^2}{s} - \dfrac{G^2}{rs}\right]/(r-1)}{SSE/(r-1)(s-1)}.$$

Similarly, we can easily verify that the F-statistic for testing $\beta_1 = \beta_2 = \ldots = \beta_s = 0$ is given by

$$F = \frac{\left[\dfrac{\Sigma T_{.j}^2}{r} - \dfrac{G^2}{rs}\right]/(s-1)}{SSE/(r-1)(s-1)}$$

with $(s-1)$ and $(r-1)(s-1)$ d.f.

The above results can be represented by the following analysis of variance table.

Analysis of Variance Table

SOURCE	SS	d.f.
due to α's	$\dfrac{\Sigma T_{i.}^2}{s} - \dfrac{G^2}{rs}$	$r-1$
due to β's	$\dfrac{\Sigma T_{.j}^2}{r} - \dfrac{G^2}{rs}$	$s-1$
Error	*	$(r-1)(s-1)$
Total	$\Sigma\Sigma\, y^2 - \dfrac{G^2}{rs}$	$rs-1$

* obtained by subtraction

16. The normal equations for estimation of μ, t_i and β are obtained by minimizing $\Sigma_i\Sigma_j(y_{ij} - \mu - t_i - \beta x_{ij})^2$ and are found to be

(1) $Y = n\hat{\mu} + \Sigma\, n_i\hat{t}_i + \hat{\beta}X$

(2) $Y_{i.} = n_i\,\hat{\mu} + n_i\hat{t}_i + \hat{\beta}X_{i.}$, $i = 1, 2, \ldots, k$

(3) $\Sigma\Sigma\, x_{ij}y_{ij} = X\hat{\mu} + \Sigma_i\Sigma_j x_{ij}\hat{t}_i + \hat{\beta}\Sigma\Sigma\, x_{ij}^2$,

where $Y = \Sigma\Sigma\, y_{ij}$, $X = \Sigma\Sigma\, x_{ij}$, $Y_{i.} = \Sigma_j\, y_{ij}$, and $X_{i.} = \Sigma_j\, x_{ij}$.

Clearly, the number of independent equations is $1 + k - 1 + 1 = k + 1$. A set of solutions is obtained by setting $\sum_i n_i\hat{t}_i = 0$. Multiply equations (2) by $X_{i.}/n_i$ and add over the subscript i and subtract from (3). Then we get

$$\Sigma\Sigma\, x_{ij}y_{ij} - \Sigma_i Y_{i.}X_{i.}/n_i = \hat{\beta}[\Sigma\Sigma\, x_{ij}^2 - \Sigma\, X_{i.}^2/n_i].$$

Thus, we obtain

$$\hat{\beta} = \frac{\Sigma\Sigma\, x_{ij}\, y_{ij} - \Sigma_i\, X_{i.}\, Y_{i.}/n_i}{\Sigma\Sigma\, x_{ij}^2 - \Sigma_i\, X_{i.}^2/n_i}.$$

Then, from (1) and (2), we get

$$\hat{\mu} = \bar{Y}_{..} - \hat{\beta}\bar{X}_{..}$$
$$\hat{t}_i = \bar{Y}_{i.} - \bar{Y}_{..} + \hat{\beta}(\bar{X}_{..} - \bar{X}_{i.}),$$

where $\bar{X}_{..} = \Sigma\Sigma\, x_{ij}/n$, $\bar{Y}_{..} = \Sigma\Sigma\, y_{ij}/n$, $\bar{X}_{i.} = \dfrac{\Sigma_j x_{ij}}{n_i}$, and $\bar{Y}_{i.} = \sum_j y_{ij}/n_i$.

Then, we obtain

$$SSR(\mu, \mathbf{t}, \beta) = \Sigma_i\, Y_{i.}^2/n_i + \beta^2[\Sigma\Sigma\, x_{ij}^2 - \Sigma\, X_{i.}^2/n_i]$$

with (k + 1) d.f. and

$$SSE = \Sigma\Sigma\, y_{ij}^2 - \Sigma\, Y_{i.}^2/n_i - \beta^2.[\Sigma\Sigma\, x_{ij}^2 - \Sigma\, X_{i.}^2/n_i]$$

with (n − k − 1) d.f.

We now find the regression sum of squares where μ and β are fitted. In this case the normal equations are obtained by minimizing $\Sigma\Sigma\,(y_{ij} - \mu - \beta x_{ij})^2$ and are found to be

$$Y = n\overset{*}{\mu} + \beta^* X$$

and

$$\Sigma\Sigma\, x_{ij}y_{ij} = X\overset{*}{\mu} + \overset{*}{\beta}\Sigma\Sigma\, x_{ij}^2.$$

The solutions are

$$\overset{*}{\mu} = \bar{Y}_{..} - \overset{*}{\beta}\,\bar{X}_{..}$$

and

$$\beta^* = \frac{\Sigma\Sigma\, x_{ij}y_{ij} - \frac{1}{n}GX}{\Sigma\Sigma\, x_{ij}^2 - \frac{X^2}{n}}.$$

Hence,

$$\begin{aligned} SSR(\mu, \beta) &= \frac{1}{n}Y^2 + \beta^*(\Sigma\Sigma\, x_{ij}y_{ij} - \frac{1}{n}GX) \\ &= \frac{1}{n}Y^2 + \beta^{*2}(\Sigma\Sigma\, x_{ij}^2 - \frac{X^2}{n}) \end{aligned}$$

with 2 d.f.

Therefore sum of squares for testing $t_1 = t_2 = \ldots = t_k = 0$ is given by

$$\begin{aligned} SSR(\mu, \mathbf{t}, \beta) - SSR(\mu, \beta) &= \Sigma\, Y_{i.}^2/n_i \\ &+ \beta^2[\Sigma\Sigma\, x_{ij}^2 - \Sigma\, X_{i\cdot}^2/n_i] - \frac{1}{n}Y - \beta^{*2}(\Sigma\Sigma\, x_{ij}^2 - \frac{X^2}{n}) \end{aligned}$$

with (k − 1) d.f.

We can present the above results in the form of analysis of covariance table as follows. We note that

$$\begin{aligned} \Sigma\Sigma\,(y_{ij} - \bar{Y}_{..})^2 &= (\Sigma\, Y_{i.}^2/n_i - n\bar{Y}_{..}^2) + (\Sigma\Sigma\, y_{ij}^2 - \Sigma\, Y_{i.}^2/n_i) \\ \Sigma\Sigma\,(x_{ij} - \bar{X}_{..})^2 &= (\Sigma\, X_{i}^2/n_i - n\bar{X}_{..}^2) + (\Sigma\Sigma\, x_{ij}^2 - \Sigma\, X_{i.}^2/n_i) \\ \Sigma\Sigma\,(x_{ij} - \bar{X}_{..})(y_{ij} - \bar{Y}_{..}) &= (\Sigma\, X_{i.}Y_{i.}/n_i - n\bar{X}_{..}\bar{Y}_{..}) \\ &\quad + (\Sigma\Sigma\, x_{ij}\, y_{ij} - \Sigma\, X_{i.}Y_{i.}/n_i). \end{aligned}$$

Let

$$\Sigma\, Y_{i.}^2/n_i - n\bar{Y}_{..}^2 = C_{22t},\ \Sigma\Sigma\, y_{ij}^2 - \Sigma\, Y_{i.}^2/n_i = C_{22e}$$
$$\Sigma\, X_{i.}^2/n_i - n\bar{X}_{..}^2 = C_{11t},\ \Sigma\Sigma\, x_{ij}^2 - \Sigma\, X_{i.}^2/n_i = C_{11e}$$
$$\Sigma\, X_{i.}Y_{i.}/n_i - n\bar{X}_{..}\bar{Y}_{..} = C_{12t},\ \Sigma\Sigma\, x_{ij}\, y_{ij} - \Sigma\, X_{i.}Y_{i.}/n_i = C_{12e}.$$

Then one can easily verify that

$$\hat{\beta} = \frac{C_{12e}}{C_{11e}}$$
$$\beta^* = (C_{12t} + C_{12e})/(C_{11t} + C_{11e}).$$
$$SSE = C_{22e} - \frac{C_{12e}^2}{C_{11e}}.$$

The SS due to the hypothesis for testing $t_1 = t_2 = \ldots = t_k = 0$ is given by

$$SSH = C_{22t} + \frac{C_{12e}^2}{C_{11e}} - \frac{(C_{12t} + C_{12e})^2}{(C_{11t} + C_{11e})}.$$

Hence, the F statistic for testing $t_1 = t_2 = \ldots = t_k = 0$ is

$$F = \frac{SSH/(k-1)}{SSE/(n-k-1)}, \quad \text{with } k-1 \text{ and } (n-k-1) \text{ d.f.}$$

Thus, we get the following table of analysis of covariance.

Analysis of Covariance Table
(one-way classification)

SOURCE	$\Sigma\, y^2$	$\Sigma\, xy$	$\Sigma\, x^2$	due to β		d.f.
Treatments	C_{22t}	C_{12t}	C_{11t}			
Error	C_{22e}	C_{12e}	C_{11e}	C_{12e}^2/C_{11e}	$C_{22e} - \frac{C_{12e}^2}{C_{11e}} = E$	$n-k-1$
Total (Treatment + Error)	C_{22t} $+C_{22e}$	C_{12t} $+C_{12e}$	C_{11t} $+C_{11e}$	$\frac{(C_{12t} + C_{12e})^2}{(C_{12t} + C_{11e})}$	$(C_{22t} + C_{22e}) - \frac{(C_{12t} + C_{12e})^2}{(C_{11t} + C_{11e})} = T$	$n-2$
SS for testing significance of treatment differences					$T - E$	$k-1$
F for testing significance of treatment differences					$F = \frac{(T-E)/(k-1)}{E/(n-k-1)}$	

Before applying the analysis of variance, we would like to test $\beta = 0$. We shall find in this case SS due to regression when μ and $t_1, t_2, \ldots, t_k$ are fitted. For this the normal equations are obtained by minimizing $\Sigma\Sigma\, (y_{ij} - \mu - t_i)^2$

and are found to be as

$$Y = n\tilde{\mu} + \Sigma\, n_i\tilde{t}_i$$
$$Y_{i.} = n_i\tilde{\mu} + n_i\tilde{t}_i, i = 1, 2, \dots, k.$$

Clearly, only k equations are independent. A set of solutions is obtained by putting $\Sigma\, n_i\tilde{t}_i = 0$. Thus, the solutions are obtained as

$$\tilde{\mu} = \bar{Y}_{..}, \text{ and } \tilde{t}_i = (Y_{i.}/n_i) - \bar{Y}_{..}.$$

Hence, we find

$$SSR\,(\mu, \mathbf{t}) = \Sigma\, Y_{i.}^2/n_i, \text{ with k d.f.}$$

Therefore, the sum of squares for testing $\beta = 0$ is obtained as

$SSR\ (\mu, \mathbf{t}, \beta) - SSR(\mu, \mathbf{t}) = \hat{\beta}^2[\Sigma\Sigma x_{ij}^2 - \Sigma\, X_i^2/n_i] = C_{12e}^2/C_{11e}$, with 1 d.f.

Hence, for testing $\beta = 0$, we use the F statistic where

$$F = \frac{C_{12e}^2/C_{11e}}{(C_{22e} - \dfrac{C_{12e}}{C_{11e}})/(n - k - 1)},$$

with 1 and $(n - k - 1)$ d.f.

17. Let $Y_{..} = \Sigma\Sigma\, y_{ij}, Y_{i.} = \sum_{j=1}^{s} y_{ij}, Y_{.j} = \sum_{i=1}^{r} y_{ij}, X_{..} = \Sigma\Sigma\, x_{ij},$

$$X_{i.} = \sum_{j=1}^{S} x_{ij}, X_{.j} = \sum_{i=1}^{r} x_{ij}, \bar{X}_{..} = X_{..}/rs, \bar{Y}_{..} = Y_{..}/rs.$$

The normal equations for fitting all the parameters $\mu, \alpha_1, \alpha_2, \dots, \alpha_s, t_1, t_2, \dots, t_r, \beta$ are obtained by minimizing $\Sigma\Sigma\, (y_{ij} - \mu - \alpha_j - t_i - \beta x_{ij})^2$ and are found to be

(1) $Y_{..} = r\hat{\mu} + r\Sigma\hat{\alpha}_j + \Sigma\hat{t}_i + \hat{\beta}X_{..}$
(2) $Y_{.j} = r\hat{\mu} + r\hat{\alpha}_j + \Sigma\hat{t}_i + \hat{\beta}X_{.j}, j = 1, 2, \dots, s$
(3) $Y_{i.} = s\hat{\mu} + \Sigma\hat{\alpha}_j + s\hat{t}_i + \hat{\beta}X_{i.},$
(4) $\Sigma\Sigma\, x_{ij}y_{ij} = \hat{\mu}X_{..} + \Sigma\, X_{.j}\hat{\alpha}_j + \Sigma\, X_{i.}\hat{t}_i + \hat{\beta}\Sigma\Sigma\, x_{ij}^2$

The number of independent equations is equal to $1 + s - 1 + r - 1 + 1 = r + s$. To obtain one set of solutions we put $\Sigma\hat{\alpha}_j = 0$ and $\Sigma\hat{t}_i = 0$. Then we obtain

(5) $Y_{..} = rs\hat{\mu} + \hat{\beta}X_{..}$
(6) $Y_{.j} = r\hat{\mu} + r\hat{\alpha}_j + \hat{\beta}X_{.j}$
(7) $Y_{i.} = s\hat{\mu} + s\hat{t}_i + \hat{\beta}X_{i.}$
(8) $\Sigma\Sigma\ x_{ij}\, y_{ij} = \mu X_{..} + \Sigma\, X_{.j}\hat{\alpha}_j + \Sigma\, X_{i.}\hat{t}_i + \hat{\beta}\Sigma\Sigma x_{ij}^2$

Multiply (5) by $X_{..}/rs$, and add to (8). Multiply (6) by $X_{.j}/r$ and sum over j and then subtract from (8). Multiply (7) by $X_{i.}/s$ and sum over i and then subtract from (8). We then obtain

$$\Sigma\Sigma\, x_{ij}\, y_{ij} - \frac{1}{r}\Sigma\, X_{.j}Y_{.j} - \frac{1}{s}\Sigma\, X_{i.}Y_{i.} + \frac{1}{rs}X_{..}Y_{..}$$
$$= \hat{\beta}[\Sigma\Sigma\, x_{ij}^2 - \frac{1}{r}\Sigma\, X_{.j}^2 - \frac{1}{s}\Sigma\, X_{i.}^2 + \frac{1}{rs}X_{..}^2].$$

Thus, we get

$$\hat{\beta} = \frac{\Sigma\Sigma\, x_{ij}y_{ij} - \frac{1}{r}\Sigma\, X_{.j}Y_{.j} - \frac{1}{s}\Sigma\, X_{i.}Y_{i.} + \frac{1}{rs}X_{..}Y_{..}}{\Sigma\Sigma\, x_{ij}^2 - \frac{1}{r}\Sigma\, X_{.j}^2 - \frac{1}{s}\Sigma\, X_{i.}^2 + \frac{1}{rs}X_{..}^2}$$

Then from (5), (6) and (7), we obtain

$$\hat{\mu} = \bar{Y}_{..} - \hat{\beta}\bar{X}_{..}$$
$$\hat{\alpha}_j = \frac{1}{r}Y_{.j} - \hat{\mu} - \hat{\beta}X_{.j}/r$$
$$\tilde{t}_i = \frac{1}{s}Y_{i.} - \hat{\mu} - \hat{\beta}X_{i.}/s.$$

Then the sum of squares due to regression when μ, α's, t's and β are fitted is obtained as

$$\text{SSR}\,(\mu, \boldsymbol{\alpha}, \mathbf{t}, \beta) = \frac{1}{r}\Sigma\, Y_{.j}^2 + \frac{1}{s}\Sigma\, Y_{i.}^2 - \frac{1}{rs}Y_{..}^2$$
$$+ \hat{\beta}^2[\Sigma\Sigma\, x_{ij}^2 - \frac{1}{r}\Sigma\, X_{.j}^2 - \frac{1}{s}\Sigma\, X_{i.}^2 + \frac{1}{rs}X_{..}^2],$$

with (r + s) d.f. Further, the error sum of squares is obtained as

$$\text{SSE} = \Sigma\Sigma\, y_{ij}^2 - \frac{1}{r}\Sigma\, Y_{.j}^2 - \frac{1}{s}\Sigma\, Y_{i.}^2 + \frac{1}{rs}Y_{..}^2$$
$$- \hat{\beta}^2[\Sigma\Sigma\, x_{ij}^2 - \frac{1}{r}\Sigma\, X_{.j}^2 - \frac{1}{s}\Sigma\, X_{i.}^2 + \frac{1}{rs}X_{..}^2],$$
$$= E, \text{ (say)}$$

with (r − 1)(s − 1) − 1 d.f.

To test $\beta = 0$, we fit the parameters $\mu, \alpha_1, \alpha_2, \ldots, \alpha_s$ and $t_1, t_2, \ldots, t_r$. The normal equations for fitting these parameters are obtained by minimizing $\Sigma\Sigma\,(y_{ij} - \mu - \alpha_j - t_i)^2$ and are found to be as

$$Y_{..} = rs\,\overset{*}{\mu} + r\Sigma\alpha_j^* + s\Sigma\, t_i^*$$
$$Y_{.j} = r\,\overset{*}{\mu} + r\alpha_j^* + \Sigma\, t_i^*, j = 1, 2, \ldots, s$$
$$Y_{i.} = s\,\overset{*}{\mu} + \Sigma\alpha_j^* + st_i^*, i = 1, 2, \ldots, r.$$

The number of independent equations is (r + s − 1). A set of solutions is obtained by putting $\Sigma\alpha_j^* = 0$ and $\Sigma\, t_i^* = 0$ and is given by

$$\overset{*}{\mu} = \bar{Y}_{.}, \alpha_j^* = \frac{1}{r}Y_{.j} - \overset{*}{\mu}, j = 1, 2, \ldots, s,$$
$$t_i^* = \frac{1}{s}Y_{i.} - \overset{*}{\mu}, i = 1, 2, \ldots, r.$$

Thus, the sum of squares due to regression when $\mu, \alpha_1, \alpha_2, \ldots, \alpha_s, t_1, t_2, \ldots, t_r$ are fitted is given by

$$SSR(\mu, \boldsymbol{\alpha}, \mathbf{t}) = \frac{1}{r}\Sigma\, Y_{.j}^2 + \frac{1}{s}\Sigma\, Y_{i.}^2 - \frac{1}{rs}Y_{..}^2$$

with $(r + s - 1)$ d.f. Hence sum of squares for testing $\beta = 0$ is obtained as

$$\begin{aligned} &SSR(\mu, \boldsymbol{\alpha}, \mathbf{t}, \beta) - SSR(\mu, \boldsymbol{\alpha}, \mathbf{t}) \\ &\quad = \hat{\beta}^2[\Sigma\Sigma\, x_{ij}^2 - \frac{1}{r}\Sigma\, X_{.j}^2 - \frac{1}{s}\Sigma\, X_{i.}^2 + rs\, \bar{X}_{..}^2] \end{aligned}$$

with 1 d.f. Now for testing $\beta = 0$, we apply F test, where

$$F = \frac{\hat{\beta}^2[\Sigma\Sigma\, x_{ij}^2 - \frac{1}{r}\Sigma\, X_{j.}^2 - \frac{1}{s}\Sigma\, X_{i.}^2 + rs\, \bar{X}_{..}^2]}{SSE/[(r-1)(s-1)-1]}$$

with 1 and $(r - 1)(s - 1) - 1$ d.f.
Next we proceed to test $t_1 = t_2 = \ldots = t_r = 0$. We fit the parameters $\mu, \alpha_1, \alpha_2, \ldots, \alpha_s$ and β. The normal equations for fitting these parameters are obtained by minimizing $\Sigma\Sigma\, (y_{ij} - \mu - \alpha_j - \beta x_{ij})^2$ and are found to be.

$$\begin{aligned} &Y_{..} = rs\, \tilde{\mu} + \Sigma\tilde{\alpha}_j + \tilde{\beta}X_{..} \\ &Y_{.j} = r\, \tilde{\mu} + r\, \tilde{\alpha}_j + \tilde{\beta}X_{.j}, j = 1, 2, \ldots, s \\ &\Sigma\Sigma\, x_{ij}\, y_{ij} = \mu X_{..} + \Sigma\, X_{.j}\tilde{\alpha}_j + \tilde{\beta}\Sigma\Sigma\, x_{ij}^2. \end{aligned}$$

The number of independent equations is equal to $1 + (s - 1) + 1 = s + 1$ and hence a set of solutions is obtained by substituting $\Sigma\alpha_j{}^* = 0$. We then have

(9) $Y_{..} = rs\, \tilde{\mu} + \tilde{\beta}X_{..}$
(10) $Y_{.j} = r\, \tilde{\mu} + r\tilde{\alpha}_j + \tilde{\beta}X_{.j}, j = 1, 2, \ldots, s$
(11) $\Sigma\Sigma\, x_{ij}\, y_{ij} = \tilde{\mu}X_{..} + \Sigma\, X_{.j}\tilde{\alpha}_j + \tilde{\beta}\Sigma\Sigma\, x_{ij}^2$
Multiply (10) by $X_{.j}/r$ and sum over j and then subtract from (11). We get

$$[\Sigma\Sigma\, x_{ij}y_{ij} - \frac{1}{r}\Sigma\, X_{.j}Y_{.j}] = \tilde{\beta}[\Sigma\Sigma\, x_{i_j}^2 - \frac{1}{r}\Sigma\, X_{.j}^2]$$

and consequently

$$\tilde{\beta} = \frac{\Sigma\Sigma\, x_{ij}y_{ij} - \frac{1}{r}\Sigma\, X_{.j}Y_{.j}}{\Sigma\Sigma\, x_{i_j}^2 - \frac{1}{r}\Sigma\, X_{.j}^2}$$

$$\tilde{\mu} = \bar{Y}_{..} - \tilde{\beta}\bar{X}_{..}$$

$$\tilde{\alpha}_j = \frac{1}{r}Y_{.j} - \tilde{\mu} - \tilde{\beta}X_{.j}/r, j = 1, 2, \ldots, s.$$

Hence, the sum of squares due to regression when $\mu, \alpha_1, \alpha_2, \ldots, \alpha_s$ and β are fitted is

$$SSR(\mu, \boldsymbol{\alpha}, \beta) = \frac{1}{r}\Sigma\, Y_{.j}^2 + \beta^2[\Sigma\Sigma\, x_{i_j}^2 - \frac{1}{r}\Sigma\, X_{.j}^2]$$

with (s + 1) d.f. Therefore, the sum of squares for testing $t_1 = t_2 = \ldots = t_r = 0$ is given by

$$
\begin{aligned}
&SSR(\mu, \boldsymbol{\alpha}, \mathbf{t}, \beta) - SSR(\mu, \boldsymbol{\alpha}, \beta) \\
&= \frac{1}{s}\Sigma\, Y_{i.}^2 - rs\, Y_{..}^2 + \hat{\beta}^2[\Sigma\Sigma\, x_{ij}y_{ij} - \frac{1}{r}\Sigma\, X_{.j}^2 - \frac{1}{s}\Sigma\, X_{i.}^2 + rs\, X_{..}^2] \\
&\quad - \widetilde{\beta}^2[\Sigma\Sigma\, x_{ij}y_{ij} - \frac{1}{r}\Sigma\, X_{.j}^2] = T, \text{(say)}
\end{aligned}
$$

with (r − 1) d.f. Hence for testing $t_1 = t_2 = \ldots = t_r = 0$, we apply F test, where

$$
F = \frac{T/(r-1)}{E/[(r-1)(s-1)-1]},
$$

with (r − 1) and (r − 1)(s − 1) − 1 d.f. We shall now present the above test by the analysis of covariance table. We can easily verify that

$$
\begin{aligned}
\Sigma\Sigma\,(y_{ij} - \bar{Y}_{..})^2 &= [\Sigma\Sigma\, y_{ij}^2 - \frac{1}{s}\Sigma\, Y_{.i}^2 \frac{1}{r}\Sigma\, Y_{.j}^2 + rs\,\bar{Y}_{..}^2] \\
&\quad + [\frac{1}{s}\Sigma\, Y_i^2 - rs\bar{Y}_{..}^2] + [\frac{1}{r}\Sigma\, Y_{.j}^2 - rs\bar{Y}_{..}^2] \\
&= C_{22e} + C_{22t} + C_{22b} \\
\Sigma\Sigma\,(x_{ij} - \bar{X}_{..})^2 &= [\Sigma\Sigma\, x_{ij}^2 - \frac{1}{s}\Sigma\, X_i^2 - \frac{1}{r}\Sigma\, X_{.j}^2 + rs\bar{X}_{..}^2] \\
&\quad + [\frac{1}{s}\Sigma\, X_{i.}^2 - rs\bar{X}_{..}^2] + [\frac{1}{s}\Sigma\, X_i^2 - rs\bar{X}_{..}^2] \\
&\quad + [\frac{1}{r}\Sigma\, X_{.j}^2 - rs\bar{X}_{..}^2] \\
&= C_{11e} + C_{11t} + C_{11b}
\end{aligned}
$$

$$
\begin{aligned}
&\Sigma\Sigma\,(x_{ij} - \bar{X}_{..})(y_{ij} - \bar{Y}_{..}) \\
&= [\Sigma\Sigma\, x_{ij}y_{ij} - \frac{1}{s}\Sigma\, X_{i.}Y_{i.} - \frac{1}{r}\Sigma\, X_{.j}Y_{.j} + rs\,\bar{X}_{..}\bar{Y}_{..}] \\
&\quad + [\frac{1}{s}\Sigma\, X_{i.}Y_{i.} - rs\,\bar{X}_{..}\bar{Y}_{..}] + [\frac{1}{r}\Sigma\, X_{.j}Y_{.j} - rs\,\bar{X}_{..}\bar{Y}_{..}] \\
&= C_{12e} + C_{12t} + C_{12b}.
\end{aligned}
$$

Then, we see that

$$
\hat{\beta} = \frac{C_{12e}}{C_{11e}}, \text{ and } \widetilde{\beta} = (C_{12e} + C_{12t})/(C_{11e} + C_{11t}).
$$

Hence

$$
SSE = E = C_{22e} - \frac{C_{12e}^2}{C_{11e}},
$$

$$
T = [(C_{22t} + C_{22e}) - \frac{(C_{12t} + C_{12e})^2}{C_{11t} + C_{11e}}] - [C_{22e} - \frac{C_{12e}^2}{C_{11e}}].
$$

The analysis of covariance table is given below.

Analysis of Covariance Table
(Two-way classification)

SOURCE	$\Sigma\, y^2$	$\Sigma\, xy$	$\Sigma\, x^2$	due to β		d.f.
Blocks	C_{22b}	C_{12b}	C_{11b}			$s-1$
Treatment	C_{22t}	C_{12t}	C_{11t}			$r-1$
Error	C_{22e}	C_{12e}	C_{11e}	C^2_{12e}/C_{11e}	$C_{22e}-\dfrac{C^2_{12e}}{C_{11e}}=E$	$rs-r-s$
Treatment + Error	C_{22t} $+C_{22}$	C_{12t} $+C_{12e}$	C_{11t} $+C_{11e}$	$\dfrac{(C_{12t}+C_{12e})^2}{C_{11t}+C_{11e}}$	$(C_{22t}+C_{22e})-\dfrac{(C_{12t}+C_{12e})^2}{(C_{11t}+C_{11e})}=R$	$rs-s-1$
SS for testing the significance of treatment differences					$T=R-E$	$r-1$
F for testing the significance of treatment differences					$F=\dfrac{T/r-1}{E/(rs-r-s)}$	

18. We know that

$$\text{var}\,(\hat{\boldsymbol{\theta}}) = \hat{\sigma^2}(A'A)^{-1} = 200(A'A)^{-1}$$
$$= \frac{3}{10}\begin{bmatrix} 3 & 2 & 1 \\ 2 & 4 & 2 \\ 1 & 2 & 3 \end{bmatrix}.$$

Hence

$$(A'A)^{-1} = \frac{3}{2000}\begin{bmatrix} 3 & 2 & 1 \\ 2 & 4 & 2 \\ 1 & 2 & 3 \end{bmatrix}$$

$$A'A = \frac{2000}{3}\begin{bmatrix} 3 & 2 & 1 \\ 2 & 4 & 2 \\ 1 & 2 & 3 \end{bmatrix}^{-1}$$

$$= \frac{2000}{3}\cdot\frac{1}{4}\begin{bmatrix} 2 & -1 & 0 \\ -1 & 2 & -1 \\ 0 & -1 & 2 \end{bmatrix}$$

$$= \frac{500}{3}\begin{bmatrix} 2 & -1 & 0 \\ -1 & 2 & -1 \\ 0 & -1 & 2 \end{bmatrix}.$$

Now

$$A'\mathbf{y} = A'A\hat{\boldsymbol{\theta}} = \frac{500}{3}\begin{bmatrix} 2 & -1 & 0 \\ -1 & 2 & -1 \\ 0 & -1 & 2 \end{bmatrix}\begin{bmatrix} 3 \\ 5 \\ 2 \end{bmatrix}$$

$$= \frac{500}{3}\begin{bmatrix} 1 \\ 5 \\ -1 \end{bmatrix},$$

and

$$SSR(\boldsymbol{\theta}) = \hat{\boldsymbol{\theta}}A'\mathbf{y} = [3\ 5\ 2]\frac{500}{3}\begin{bmatrix} 1 \\ 5 \\ -1 \end{bmatrix}$$

$$= \frac{13000}{3}.$$

Therefore

$$F = \frac{SSR(\boldsymbol{\theta})/3}{\hat{\sigma}^2} = \frac{13000}{9(200)} = \frac{65}{9}.$$

19. Here $\mathcal{E}(\mathbf{y}) = A\boldsymbol{\theta}$, where $\boldsymbol{\theta} = \{\alpha, \beta\}$,

$$A = \begin{bmatrix} 1 & x_1 \\ 1 & x_2 \\ \cdot & \cdot \\ \cdot & \cdot \\ 1 & x_n \end{bmatrix}, \text{ and } V = \text{diag}(d_1, d_2, \ldots, d_n).$$

The BLUE of $\boldsymbol{\theta}$ is given by

$$\hat{\boldsymbol{\theta}} = (A'V^{-1}A)^{-1}A'V^{-1}\mathbf{y}.$$

We find that

$$A'V^{-1}A = \begin{bmatrix} \Sigma\,(1/d_i) & \Sigma\,(x_i/d_i) \\ \Sigma\,(x_i/d_i) & \Sigma\,(x_i^2/d_i) \end{bmatrix}.$$

Hence, we get

$$(A'V\,A) = \frac{1}{\Delta}\begin{bmatrix} \Sigma\,(x_i^2/d_i) & -\Sigma\,(x_i/d_i) \\ -\Sigma\,(x_i/d_i) & \Sigma\,(1/d_i) \end{bmatrix}$$

where $\Delta = \Sigma\,(1/d_i) - \Sigma\,(x_i^2/d_i) - (\Sigma\,(x_i/d_i))^2$. Further we have

$$A'V^{-1}\mathbf{y} = \begin{bmatrix} \Sigma\,(y_i/d_i) \\ \Sigma\,(x_iy_i/d_i) \end{bmatrix}.$$

Hence,

$$\hat{\boldsymbol{\theta}} = \frac{1}{\boldsymbol{\Delta}}\begin{bmatrix} \Sigma\,(x_i^2/d_i)\cdot\Sigma\,(y_i/d_i) - \Sigma\,(x_i/d_i)\cdot\Sigma\,(x_iy_i/d_i) \\ \Sigma\,(1/d_i)\cdot\Sigma\,(x_iy_i/d_i) - \Sigma\,(x_i/d_i)\cdot\Sigma\,(y_i/d_i) \end{bmatrix}$$

and variance-covariance matrix of $\hat{\boldsymbol{\theta}}$ is given by

$$\text{var}(\hat{\boldsymbol{\theta}}) = \frac{1}{\boldsymbol{\Delta}}\begin{bmatrix} \Sigma\,(x_i^2/d_i) & -\Sigma\,(x_i/d_i) \\ -\Sigma\,(x_i/d_i) & \Sigma\,(1/d_i) \end{bmatrix}.$$

case (i). $x_i = i,\ d_i = \sigma^2 i, i = 1, 2, \ldots, n$
Here

$$\Sigma\,(x_i/d_i) = n/\sigma^2,\ \Sigma\, x_i^2/d_i = n(n+1)/2\sigma^2,$$
$$\Sigma\, y_i/d_i = \Sigma\,(y_i/i)/\sigma^2,$$
$$\Sigma\,(1/d_i) = \Sigma\,(1/i)/\sigma^2,\ \Sigma\, x_i y_i/d_i = \Sigma\, y_i/\sigma^2$$

Hence

$$\Delta = \frac{n}{\sigma^4}\left[\frac{(n+1)}{2}\Sigma\,(1/i) - n\right]$$

$$\hat{\boldsymbol{\theta}} = \frac{1}{n\left[\dfrac{n+1}{2}\Sigma\,(1/i) - n\right]}\begin{bmatrix} \dfrac{n(n+1)}{2}\Sigma\,(y_i/i) - n\Sigma\, y_i \\ \Sigma\,(1/i)\Sigma\, y_i - n\Sigma\, y_i/i \end{bmatrix}$$

and

$$\text{var}(\hat{\boldsymbol{\theta}}) = \frac{\sigma^2}{n\left[\dfrac{n+1}{2}\Sigma\,(1/i) - n\right]}\begin{bmatrix} \dfrac{n(n+1)}{2} & -n \\ -n & \Sigma\,(1/i) \end{bmatrix}$$

case(ii). $x_i = i,\ d_i = \sigma^2 i^2,\ i = 1, 2, \ldots, n.$
Here,

$$\Sigma\,(1/d_i) = \Sigma\,(1/i^2)/\sigma^2,$$
$$\Sigma\,(x_i/d_i) = \Sigma\,(1/i)/\sigma^2,$$
$$\Sigma\,(x_i^2/d_i) = n/\sigma^2,$$
$$\Sigma\,(y_i/d_i) = \Sigma\,(y_i/i^2)/\sigma^2,$$
$$\Sigma\,(x_i y_i/d_i) = \Sigma\,(y_i/i)/\sigma^2.$$

Hence

$$\Delta = \frac{1}{\sigma^4}[n\Sigma\,(1/i^2) - (\Sigma\, 1/i)^2].$$

$$\hat{\boldsymbol{\theta}} = \frac{1}{n\Sigma\,(1/i^2) - (\Sigma\, 1/i)^2}\begin{bmatrix} n\Sigma\,(y_i/i) - (\Sigma\,\frac{1}{i})(\Sigma\frac{y_i}{i}) \\ (\Sigma\,\frac{1}{i^2})(\Sigma\frac{y_i}{i}) - (\Sigma\,\frac{1}{i})(\Sigma\,\frac{y_i}{i^2}) \end{bmatrix}$$

$$\text{var}(\hat{\boldsymbol{\theta}}) = \frac{\sigma^2}{n\Sigma\,(1/i^2) - (\Sigma\, 1/i)^2}\begin{bmatrix} n & -\Sigma\,(1/i) \\ -\Sigma\,(1/i) & \Sigma\,(1/i^2) \end{bmatrix}.$$

20. Here, we have $\mathcal{E}(\mathbf{y}) = A\boldsymbol{\theta}$, where $\boldsymbol{\theta}' = \{\theta_1, \theta_2\}$, and

$$A = \begin{bmatrix} 1 & a_1 \\ 1 & 0 \\ 1 & a_3 \end{bmatrix}$$

One can easily verify that

$$V^{-1} = \frac{5}{3}\begin{bmatrix} 2 & -1 & 0 \\ -1 & 2 & -1 \\ 0 & -1 & 2 \end{bmatrix}$$

Also, one obtains

$$A'V^{-1}A = \left(\frac{5}{3}\right)\begin{bmatrix} 2 & 0 \\ 0 & 3 \end{bmatrix},$$

$$A'V^{-1}\mathbf{y} = \left(\frac{5}{3}\right)\begin{bmatrix} y_1 + y_3 \\ \sqrt{3}\,(y_3 - y_1) \end{bmatrix},$$

and

$$(A'V^{-1}A)^{-1} = (1/10)\begin{bmatrix} 3 & 0 \\ 0 & 2 \end{bmatrix}.$$

Thus, the BLUE of $\boldsymbol{\theta}$ is given by

$$\hat{\boldsymbol{\theta}} = (1/6)\begin{bmatrix} 3(y_1 + y_3) \\ \sqrt[2]{3}(y_3 - y_1) \end{bmatrix}.$$

Hence,

$$\hat{\theta} = (y_1 + y_3)/2, \;\; \hat{\boldsymbol{\theta}}_2 = \sqrt{3}(y_3 - y_1)/3.$$

Also, the variance-covariance matrix of $\hat{\boldsymbol{\theta}}$ is given by

$$\text{var}\,(\hat{\boldsymbol{\theta}}) = (A'V^{-1}A)^{-1} = \begin{bmatrix} 3/10 & 0 \\ 0 & 1/5 \end{bmatrix}.$$

21. Here, we have $\mathcal{E}(\mathbf{y}) = A\mu$, $A = E_{n1}$. Further, one can easily verify that

$$V^{-1} = \frac{(N-1)}{N\sigma^2}\left[I_n + \frac{1}{N-n}E_{nn}\right].$$

Hence,one obtains

$$A'V^{-1}A = \frac{n(N-1)}{\sigma^2(N-n)},$$

$$(A'V^{-1}A)^{-1} = \frac{\sigma^2(N-n)}{n(N-1)},$$

and

$$A'V^{-1}\mathbf{y} = \frac{(N-1)}{\sigma^2(N-n)}\Sigma\, y.$$

Therefor, the BLUE of μ is given by

$$\hat{\mu} = (A'V^{-1}A)^{-1}A'V^{-1}y = \Sigma\, y/n = \bar{y}$$

The variance of $\hat{\mu}$ is given by

$$\text{var}\,(\hat{\mu}) = (A'V^{-1}A)^{-1} = \frac{\sigma^2(N-n)}{n(N-n)},$$

and

$$\begin{aligned}\text{SEE} &= \mathbf{y}'V^{-1}\mathbf{y} - \hat{\boldsymbol{\theta}}'A'V^{-1}\mathbf{y}\\ &= \frac{(N-1)}{N\sigma^2}\sum_1^n (y_i - \bar{y})^2.\end{aligned}$$

Now $\mathcal{E}(\text{SEE}) = (n-1)$, which gives

$$\hat{\sigma}^2 = \frac{(N-1)}{N(n-1)}\Sigma\,(y_i - \bar{y})^2.$$

Hence, the estimator of var $(\hat{\mu})$ is given by

$$\text{Est. var}\,(\hat{\mu}) = \left(\frac{N-n}{nN}\right)\frac{\Sigma\,(y_i-\bar{y})^2}{(n-1)}.$$

22. Here, we have $\mathcal{E}(\mathbf{y}) = A\boldsymbol{\theta}$, where $\boldsymbol{\theta}' = \{\hat{\mu}, \alpha\}$ and

$$A' = \begin{bmatrix} 1 & 1 & 1 & \dots & 1\\ x_1 & x_2 & x_3 & \dots & x_n\end{bmatrix},$$

where $x_i = \sqrt{3}(2i-n-1)/(n+1)$. Then, we get

$$A'V^{-1} = \frac{(n+1)(n+2)}{12\sigma^2}\begin{bmatrix} 1 & 0 & 0 & \dots & 0 & 1\\ a_1 & a_2 & a_3 & \dots & a_{n-1} & a_n\end{bmatrix},$$

where

$$\begin{aligned} a_1 &= 2x_1 - x_2,\\ a_{i+1} &= -x_i + 2x_{i+1} - x_{i+2}, i = 1, 2, \dots, n-2\\ a_n &= -x_{n-1} + 2x_n\end{aligned}$$

Substituting the values of x_i in a's, we find that

$$\begin{aligned} a_1 &= -\sqrt{3}, a_{i+1} = 0, i = 1, 2, \dots, n-2\\ a_n &= \sqrt{3}.\end{aligned}$$

Hence

$$A'V^{-1} = \frac{(n+1)(n+2)}{12\sigma^2}\begin{bmatrix} 1 & 0 & 0 & \dots & 0 & 1\\ -\sqrt{3} & 0 & 0 & \dots & 0 & \sqrt{3}\end{bmatrix},$$

$$A'V^{-1}A = \frac{(n+1)(n+2)}{6\sigma^2}\begin{bmatrix} 1 & 0\\ 0 & \dfrac{3(n-1)}{n+1}\end{bmatrix},$$

$$(A'V^{-1}A)^{-1} = \frac{2\sigma^2}{(n-1)(n+2)(n+1)}\begin{bmatrix} 3(n-1) & 0\\ 0 & (n+1)\end{bmatrix},$$

and

$$A'V^{-1}\mathbf{y} = \frac{(n+1)(n+2)}{12\sigma^2}\begin{bmatrix} 1 & 0 & \dots & 1 \\ \sqrt{3} & 0 & \dots & \sqrt{3} \end{bmatrix}\begin{bmatrix} y_1 \\ y_2 \\ \cdot \\ \cdot \\ y_n \end{bmatrix}$$

$$= \frac{(n+1)(n+2)}{12\sigma^2}\begin{bmatrix} y_1 + y_n \\ \sqrt{3}(y_n - y_1) \end{bmatrix}.$$

Therefore

$$\hat{\boldsymbol{\theta}} = (A'V^{-1}A^{-1})^{-1}A'V^{-1}\mathbf{y}$$
$$= \frac{1}{6(n-1)}\begin{bmatrix} 3(n-1)(y_1 + y_n) \\ \sqrt{3}(n+1)(y_n - y_1) \end{bmatrix}$$

and

$$\hat{\mu} = (y_1 + y_n)/2,\ \hat{\alpha} = (n+1)(y_n - y_1)/2\sqrt{3}\,(n-1).$$

Further, the variance-covariance matrix of $\hat{\boldsymbol{\theta}}$ is given by

$$\text{var}\,(\hat{\boldsymbol{\theta}}) = (A'V^{-1}A)^{-1} = \frac{2\sigma^2}{(n-1)(n+2)(n+1)}\begin{bmatrix} 3(n-1) & 0 \\ 0 & (n+1) \end{bmatrix}$$

Hence

$$\text{var}\,(\hat{\mu}) = 6\sigma^2/(n+1)(n+2)$$
$$\text{var}\,(\hat{\alpha}) = 2\sigma^2/(n-1)(n+2)$$
$$\text{cov}\,(\hat{\mu}, \hat{\alpha}) = 0.$$

23. We know that the value of F statistic for testing $\theta_1 = \theta_2 = \theta_3 = 0$ is given by

$$F = \frac{(\hat{\boldsymbol{\theta}}' A'V^{-1}\mathbf{y})/r}{SSE/(n-r)}$$

Now, we are given the value of the denominator in the above value of F. We must find the numerator. Now, from the normal equations we have

$$A'V^{-1}\mathbf{y} = (A'V^{-1}A)\hat{\boldsymbol{\theta}}.$$

Hence,

$$\boldsymbol{\theta}'A'V^{-1}\mathbf{y} = \boldsymbol{\theta}(A'V^{-1}A)\boldsymbol{\theta}.$$

We find the value of $A'V^{-1}$ A from

$$\text{var}\,(\hat{\boldsymbol{\theta}}) = (A'V^{-1}A)^{-1} = \frac{1}{20}\begin{bmatrix} 3 & 2 & 1 \\ 2 & 4 & 2 \\ 1 & 2 & 3 \end{bmatrix}.$$

We get

$$A'V^{-1}A = 20\begin{bmatrix} 3 & 2 & 1 \\ 2 & 4 & 2 \\ 1 & 2 & 3 \end{bmatrix}^{-1} = 5\begin{bmatrix} 2 & -1 & 0 \\ -1 & 2 & -1 \\ 0 & -1 & 2 \end{bmatrix}$$

Therefore

$$\hat{\boldsymbol{\theta}}'A'V^{-1}A\hat{\boldsymbol{\theta}} = \begin{bmatrix} 6 & 10 & 4 \end{bmatrix} \cdot 5 \begin{bmatrix} 2 & -1 & 0 \\ -1 & 2 & -1 \\ 0 & -1 & 2 \end{bmatrix} \begin{bmatrix} 6 \\ 10 \\ 4 \end{bmatrix} = 520.$$

Hence, since r = 3,

$$F = (520/3)/100 = 1.73.$$

24. In Exercise 20, we have obtained

$$\hat{\boldsymbol{\theta}} = (1/6)\begin{bmatrix} 3(y_1 + y_3) \\ 2\sqrt{3}(y_3 - y_1) \end{bmatrix},$$

$$A'V^{-1}\mathbf{y} = (5/3)\begin{bmatrix} (y_1 + y_3) \\ \sqrt{3}(y_3 - y_1) \end{bmatrix}.$$

Hence SS due to regression when θ_1 and θ_2 are fitted is given by

$$\begin{aligned} SSR(\theta_1, \theta_2) &= \boldsymbol{\theta}'A'V^{-1}\mathbf{y} \\ &= 5(3y_1^2 + 3y_3^2 - 2y_1y_3)/6 \end{aligned}$$

with 2 d.f. Also, the error SS is given by

$$\begin{aligned} SSE &= \mathbf{y}'V^{-1}\mathbf{y} - SSR(\theta_1, \theta_2) \\ &= \frac{5}{3}(2y_1^2 + 2y_2^2 + 2y_3^2 - 2y_1y_2 - 2y_2y_3) - SSR(\theta_1, \theta_2) \\ &= 5(y_1 - 2y_2 + y_3)^2/6, \end{aligned}$$

with 1 d.f. Now under the hypothesis $\theta_2 = 0$, we have

$$\mathcal{E}(y_i) = \theta_1, i = 1, 2, 3$$

so that $A = E_{31}$. The normal equation is then given by

$$A'V^{-1}\mathbf{y} = (A'V^{-1}A)\theta_1.$$

Now $A'V^{-1}\mathbf{y} = 5(y_1 + y_3)/3$

$$A'V^{-1}A = 10/3.$$

Hence, we obtain $\theta_1^* = (y_3 + y_1)/2$. Therefore

$$SSR(\theta_1) = \theta_1^* A'V^{-1}\mathbf{y} = 5(y_1 + y_3)^2/6$$

with 1 d.f. Hence SS for testing $\theta_2 = 0$ is given by

$$SSR(\theta_1, \theta_2) - SSR(\theta_1) = 5(y_3 - y_1)^2/3$$

with 1 d.f. Therefore, the F statistics for testing $\theta_2 = 0$ is obtained as

$$F = \frac{2(y_3 - y_1)^2}{(y_1 - 2y_2 + y_3)^2}$$

with 1 and 1 d.f.
If we wish to test the hypothesis $\theta_1 = \theta_2$, then the F statistic is given by

$$F = \frac{SSR(\theta_1, \theta_2)/2}{SSE/1}$$
$$= \frac{(3y_1^2 + 3y_3^2 - 2y_3y_1)}{2(y_1 - 2y_2 + y_3)^2}$$

with 2 and 1 d.f.

25. In Exercise 22, we have obtained

$$\hat{\boldsymbol{\theta}} = \left\{ \frac{y_1 + y_n}{2}, \quad \frac{(n+1)\,(y_n - y_1)}{2\sqrt{3}\,(n-1)} \right\}',$$

and

$$A'V^{-1}\mathbf{y} = \frac{(n+1)(n+2)}{12\sigma^2} \cdot \begin{bmatrix} (y_1 + y_n) \\ \sqrt{3}(y_n - y_1) \end{bmatrix}.$$

Hence SS due to regression when μ and α are fitted is given by

$$SSR(\mu, \alpha) = \hat{\boldsymbol{\theta}}' A'V^{-1}\mathbf{y} = \frac{(n+1)(n+2)}{12\sigma^2(n-1)}[n(y_1^2 + y_n^2) - 2y_1y_n],$$

with 2 d.f. We now find the error SS. The error SS is

$$SSE = \mathbf{y}'V^{-1}\mathbf{y} - SSR(\mu, \alpha).$$

Now

$$\mathbf{y}'V^{-1} = (y_1, y_2, \ldots, y_n) \cdot \frac{(n+1)(n+2)}{12^2}$$
$$\times \begin{bmatrix} 2 & -1 & 0 & \ldots & 0 & 0 \\ -1 & 2 & -1 & \ldots & 0 & 0 \\ 0 & -1 & 2 & \ldots & 0 & 0 \\ \cdot & \cdot & \cdot & \cdots & \cdot & \cdot \\ \cdot & \cdot & \cdot & \cdots & \cdot & \cdot \\ \cdot & \cdot & \cdot & \cdots & \cdot & \cdot \\ 0 & 0 & 0 & \ldots & -1 & 2 \end{bmatrix}$$
$$= \frac{(n+1)(n+2)}{12\sigma^2}[a_1, a_2, \ldots, a_n],$$

where

$$a_1 = 2y_1 - y_2$$
$$a_{i+1} = -ay_i + 2y_{i+1} - y_{i+2}, i = 1, 2, \ldots, n-2$$
$$a_n = 2y_n - y_{n-1}$$

Hence

$$\mathbf{y}'V^{-1}\mathbf{y} = \frac{(n+1)(n+2)}{12\sigma^2}[a_1y_1 + a_ny_n + \sum_1^{n-2} a_i + 1y_i + 1]$$
$$= \frac{(n+1)(n+2)}{12\sigma^2}\left[\sum_1^n y_iy_{i+1}\right]$$

and

$$\text{SSE} = \frac{(n+1)(n+2)}{6\sigma^2}\left[\sum_1^n {y_i}^2 - \sum_1^{n-1} y_iy_{i+1}\right]$$
$$- \frac{(n+1)(n+2)}{12\sigma^2(n-1)}[n(y_1^2 + y_n^2) - 2y_1y_n]$$
$$= \frac{(n+1)(n+2)}{12\sigma^2(n-1)}\left[2(n-1)(\sum_1^n y_i^2 - \sum_1^{n-1} y_iy_{i+1}) - n(y_1^2 + y_n^2) + 2y_1y_n\right]$$

with $(n-2)$ d.f. Therefore, the value of F statistic for testing $\mu = \alpha = 0$ is given by

$$F = \frac{[n(y_1^2 + y_n^2) - 2y_1y_n]/2}{[2(n-1)(\Sigma\, y_i^2 - \Sigma\, y_iy_{i+1}) - n(y_1^2 + y_n^2) + 2y_1y_n]/(n-2)}$$

with 2 and $(n-2)$ d.f.

We now obtain the value of F statistic for testing $\alpha = 0$.

Under $\alpha = 0$, we have

$$\mathcal{E}(\mathbf{y}) = E_{n1} \cdot \mu.$$

Hence $A = E_{n1}$. We then obtain

$$A'V^{-1} = E_{1n}V^{-1} = \frac{(n+1)(n+2)}{12\sigma^2}[1, 0, 0, \ldots, 0, 1].$$

Therefore,

$$A'V^{-1}A = E_{1n}V^{-1}E_{n1} = \frac{2(n+1)(n+2)}{12\sigma^2}$$

and

$$A'V^{-1}\mathbf{y} = \frac{(n+1)(n+2)}{12\sigma^2}(y_1 + y_n).$$

Hence, from $A'V^{-1}\mathbf{y} = (A'V^{-1}A)\overset{*}{\mu}$ we get $\overset{*}{\mu} = (y_1 + y_n)/2$.
Therefore SS due to regression when μ is fitted is given by

$$SSR(\mu) = \frac{(n+1)(n+2)}{24\sigma^2}(y_1 + y_n)^2$$

with 1 d.f. Hence SS for testing $\alpha = 0$ is given by

$$\begin{aligned} &SSR(\mu, \alpha) - SSR(\mu) \\ &= \frac{(n+1)(n+2)}{12\sigma^2(n-1)}[n(y_1^2 + y_n^2) - 2y_1y_n] - \frac{(n+1)(n+2)}{24\sigma^2}(y_1 + y_n)^2 \\ &= \frac{(n+1)(n+2)}{24\sigma^2(n-1)}(y_1 - y_n)^2 \end{aligned}$$

with 1 d.f. The F statistic for testing $\alpha = 0$ is given by

$$F = \frac{(y_1 - y_n)^2}{2[2(n-1)(\Sigma\, y_i^2 - \Sigma\, y_i y_{i+1}) - n(y_1^2 + y_n^2) + 2y_1y_n]/(n-2)}$$

with 1 and $(n-2)$ d.f.

26. We are given

$$N = \begin{bmatrix} 1 & 0 & 0 & 0 \\ 1 & 0 & 1 & 0 \\ 1 & 0 & 1 & 0 \\ 0 & 1 & 0 & 1 \\ 0 & 1 & 0 & 1 \\ 0 & 0 & 0 & 1 \end{bmatrix}.$$

Then the C matrix is given by

$$C = R - NK^{-1}N'$$

The column totals of N give $k_1 = 3$, $k_2 = 2$, $k_3 = 2$ and $k_4 = 3$, while row totals of N give $r_1 = 1$, $r_2 = 2$, $r_3 = 2$, $r_4 = 2$, $r_5 = 2$ and $r_6 = 1$. Thus

$$R = \text{diag}(1, 2, 2, 2, 2, 1) \text{ and } K = \text{diag}(3, 2, 2, 3).$$

Hence,

$$NK^{-1}N' = \begin{bmatrix} 1/3 & 1/3 & 1/3 & 0 & 0 & 0 \\ 1/3 & 5/6 & 5/6 & 0 & 0 & 0 \\ 1/3 & 5/6 & 5/6 & 0 & 0 & 0 \\ 0 & 0 & 0 & 5/6 & 5/6 & 1/3 \\ 0 & 0 & 0 & 5/6 & 5/6 & 1/3 \\ 0 & 0 & 0 & 1/3 & 1/3 & 1/3 \end{bmatrix}$$

$$C = \begin{bmatrix} 2/3 & -1/3 & -1/3 & 0 & 0 & 0 \\ -1/3 & 7/6 & -5/6 & 0 & 0 & 0 \\ -1/3 & -5/6 & 7/6 & 0 & 0 & 0 \\ 0 & 0 & 0 & 7/6 & -5/6 & -1/3 \\ 0 & 0 & 0 & -5/6 & 7/6 & -1/3 \\ 0 & 0 & 0 & -1/3 & -1/3 & 2/3 \end{bmatrix} = \begin{bmatrix} X & \mathbf{0} \\ \mathbf{0} & Y \end{bmatrix},$$

and rank (C) = rank (X) + rank (Y).

Now, since the columns in X and Y are null columns, the ranks of X and Y are 2 each. Hence rank (C) = 4. Therefore

d.f. associated with adj. treatment SS = 4,
d.f. associated with error SS = n − b − rank (C) = 2.

27. Consider the first design. We write its incidence matrix as follows.

	1	2	3	4	5	6	7	8
A	1	0	1	1	0	0	0	0
B	1	1	0	1	0	0	0	0
C	1	1	1	0	0	0	0	0
D	0	1	1	1	1	0	1	1
E	0	0	0	0	0	1	1	1
F	0	0	0	0	1	1	1	0
G	0	0	0	0	1	1	0	1

Moving from the cell (A, 1) in the direction of the arrow, we arrive at the cell (E, 6), encountering all but one treatment in the cell (D, 2). Since all treatments are not encountered, the design is not connected.

Consider the second design. We write its incidence matrix as follows.

	1	2	3	4	5	6
A	1	1	1	1	1	1
B	0	0	0	0	0	1
C	0	0	0	0	1	1
D	0	0	0	1	1	0
E	0	0	1	1	0	0
F	0	1	1	0	0	0
G	1	1	0	0	0	0

Moving from the cell (A, 1) in the direction of the arrow, we arrive at the cell (B, 1), encountering all the treatments and hence the design is connected.

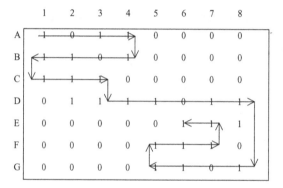

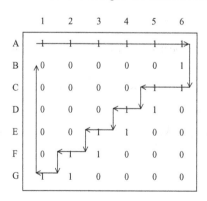

Consider the third design. We write its incidence matrix as follows.

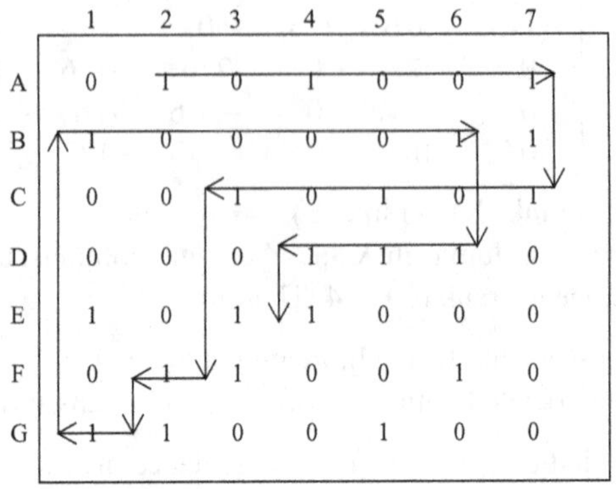

Moving from the cell (A, 2) in the direction of the arrow, we arrive at the cell (E, 4) encountering all but two treatments, namely those in the cells (F, 6) and (G, 5). Hence the design is not connected.

28. See the Hint given after Exercise 28 on page 62.

29. We have

$$N = aE_{vb}.$$

Hence

$$R = ab\, I_v, \text{ and } K = av\, I_b.$$

Therefore

$$\begin{aligned} C &= R - NK^{-1}N' \\ &= ab\, I_b - ab\, E_{vb}\left(\frac{1}{av} I_b\right) ab\, E_{bv} \\ &= ab\, I_{bv} - \frac{ab}{v} E_{vv} = ab\left(I_v - \frac{1}{v} E_{vv}\right). \end{aligned}$$

Hence, the design is connected balanced. We now show that it is orthogonal. Consider $CR^{-1}N$.

$$\begin{aligned} CR^{-1}N &= ab\, I_v - (1/v)E_{vv}(1/ab)I_v(a)E_{vb} \\ &= a\, E_{vb} - E_{vb} = \mathbf{0}. \end{aligned}$$

Hence the design is orthogonal. Since the design is connected, we can also apply the condition

$$n_{ij} = \frac{r_i k_j}{n}$$

for showing it to be orthogonal. Since $n_{ij} = a$, $r_i = ab$, $k_j = av$, and $n = abv$, the above condition is satisfied.

30. We are given $N^* = E_{vb} - N$.

The design is equi-replicate and regular. Hence

$$R = r\, I_v,\ K = k\, I_b,\ E_{1v}N = k\, E_{1b},\ N\, E_{b1} = rE_{v1}.$$

Therefore,

$$\begin{aligned} C &= R - NK^{-1}N' \\ &= r\, I_v - (1/k)NN'. \end{aligned}$$

Consider

$$\begin{aligned} E_{1v}N^* &= E_{1v}(E_{vb} - N) = v\, E_{1b} - k\, E_{1b} \\ &= (v - k)E_{1b} = k^*\, E_{1b}. \end{aligned}$$

Hence each block of N^* contains $k^* = v - k$ treatments, and the design N^* is regular. Further,

$$N^*E_{b1} = (E_{vb} - N)E_{b1} = b\, E_{v1} - r\, E_{v1} = r^*E_{v1},$$

where $r^* = b - r$. Thus, every treatment occurs $r^* = b - r$ times in the design N^*. Hence the design N^* is equi-replicate. Now,

$$\begin{aligned} C^* &= R^* - N^*K^{*\,-1}N^{*'} \\ &= r^*\, I_v - (E_{vb} - N)(1/k^*)(E_{bv} - N') \\ &= r^*\, I_v - (1/k^*)[b\, E_{vv} - r\, E_{vv} - r\, E_{vv} + NN'] \\ &= r^*\, I_v - (1/k^*)[(b - 2r)E_{vv} + NN']. \end{aligned}$$

But $NN' = -k\, C + kr\, I_v$. Consequently

$$\begin{aligned} C^* &= r^*\, I_v - (1/k^*)[(b - 2r)E_{vv} - k\, C + kr\, I_v] \\ &= (r^* - \frac{kr}{k^*})I_v + \frac{k}{k^*}C - \frac{b - 2r}{k^*}E_{vv}. \end{aligned}$$

Now suppose N is a connected balanced design. Then

$$C = a\,[I_v - (1/v)\, E_{vv}],$$

for some positive integer a and

$$\begin{aligned} C^* &= \frac{(r^*k^* - kr)}{k^*}I_v + \frac{ka}{k^*}(I_v - (1/v)E_{vv}) - \frac{b - 2r}{k^*}E_{vv} \\ &= \frac{(r^*k^* - rk + ka)}{k^*}I_v - \frac{1}{k^*v}(bv - 2rv + ka)\, E_{vv} \\ &= \frac{(bv - 2rv + ka)}{k^*v}[I_v - (1/v)E_{vv}]. \end{aligned}$$

Thus, the design N^* is a connected balanced.

31. We are given that the design N is orthogonal. Hence

$$CR^{-1}N = \mathbf{0}$$

Now $R = r\, I_v$, $K = k\, I_b$. Hence, $CR^{-1}N = \mathbf{0}$ gives

$$CN = \mathbf{0}.$$

Now, in Exercise 30, we have shown that the C-matrix of the design N^* is

given by

$$C^* = \frac{r^*k^* - rk}{k^*} I_v - \frac{b - 2r}{k^*} E_{vv} + \frac{k}{k^*} C.$$

Since $v = 2k$, and $b = 2r$, hence $r^* = b - r = r$, and $k^* = v - k = k$. Therefore, we get $r^*k^* = rk$. Hence

$$C^* = C,$$

and $R^* = r^* I_v = r I_v = R$. Therefore,

$$\begin{aligned} C^*R^{*-1}N^* &= C R^{-1}(E_{vb} - N) \\ &= r C E_{vb} - \mathbf{0} \\ &= r(C E_{v1})E_{1b} - \mathbf{0} \\ &= \mathbf{0} - \mathbf{0} = \mathbf{0} \end{aligned}$$

Hence, the design N^* is orthogonal.

32. For a connected balanced design, its C-matrix is given by

$$C = a [I_v - (1/v) E_{vv}],$$

where a is some positive constant. Now, the characteristic roots of C are given by the roots of

$$|C - \lambda I| = 0$$

i.e., by

$$|C a - \lambda I - (a/v) E_{vv}| = 0$$

i.e., by $\quad a(a - \lambda)^{v-1} = 0.$

Hence, the non-zero characteristic of C is given by $\lambda = a = \theta$, say. Hence

$$C = \theta[I_v - (1/v) E_{vv}].$$

33. Since the design is connected balanced, we have, by Exercise 32,

$$C = I_v - (\theta/v) E_{vv}.$$

But, we know that

$$\begin{aligned} C &= R - N K^{-1}N' \\ &= r I_v - (1/k) NN' \end{aligned}$$

Hence equating the above two expressions for C, we get

$$N N' = k [(r - \theta) I_v + (\theta/v) E_{vv}].$$

The characteristic roots of $N N'$ are given by the roots of

$$|N N' - \lambda I| = 0$$

i.e., by

$$(kr - \lambda)[k(r - \theta) - \lambda]^{v-1} = 0.$$

Hence the characteristic roots of $N N'$ are given by

$$\begin{aligned} \lambda &= kr, \text{ repeated once} \\ &= k(r-\theta), \text{ repeated } (v-1) \text{ times.} \end{aligned}$$

34. Since, the design is connected, rank $(C) = v - 1$. We put the side restriction $E_{1v}\mathbf{t} = 0$. Consider the matrix

$$\begin{bmatrix} C & E_{v1} \\ E_{1v} & 0 \end{bmatrix}.$$

Clearly, the rank of the above matrix is $v + 1$ and hence it is non-singular. Let

$$\begin{bmatrix} C & E_{v1} \\ E_{1v} & 0 \end{bmatrix}^{-1} = \begin{bmatrix} B_{11} & B_{12} \\ B_{21} & B_{22} \end{bmatrix}.$$

Then, we obtain

$$\begin{aligned} &C\,B_{11} + E_{v1}\,B_{21} = I_v,\ C\,B_{12} + E_{v1}\,B_{22} = 0, \\ &E_{1v}\,B_{11} = \mathbf{0},\ E_{1v}\,B_{12} = I_1. \end{aligned}$$

Solving these, one obtains

$$\begin{aligned} B_{21} &= E_{1v}/v,\ B_{22} = \mathbf{0}. \\ C\,B_{11} &= I_v - (1/v)E_{vv},\ E_{1v}B_{11} = \mathbf{0}. \end{aligned}$$

Hence,

$$C\,B_{11}C = C,$$

which shows that B_{11} is a g-inverse of C. Also, we can show that $B_{11}C\,B_{11} = B_{11}$. Hence, the solution of $\mathbf{Q} = c\hat{\mathbf{t}}$ can be taken as $\mathbf{t} = B_{11}\mathbf{Q}$. Hence,

$$\text{var}\ (\hat{t}) = B_{11}(\sigma^2 C)B_{11} = \sigma^2\ B_{11}.$$

Let the elements of B_{11} be donoted by b_{ij}. The BLUE of an elementary treatment contrast $t_i - t_j$ is given by $\hat{t}_i - \hat{t}_j$ with variance given by

$$\text{var}\ (\hat{t}_i - \hat{t}_j) = \sigma^2(b_{ii} + b_{jj} - 2b_{ij})$$

Hence, the average variance of the BLUEs of elementary treatment contrasts is,

$$\begin{aligned} &\frac{\sigma^2}{v(v-1)} \sum_{i \neq j}\sum (b_{ii} + b_{jj} - 2b_{ij}) \\ &= \frac{\sigma^2}{v(v-1)}[2(v-1)\text{tr}(B_{11}) - 2\sum_{i \neq j}\sum\ b_{ij}] \\ &= \frac{\sigma^2}{v(v-1)}[2(v-1)\text{tr}(B_{11}) - 2(\sum_i\sum_j\ b_{ij} - \text{tr}\ B_{11})] \\ &= \frac{\sigma^2}{v(v-1)}[2v\ \text{tr}\ B_{11}]. \end{aligned}$$

Since $E_{1v}B_{11} = \mathbf{0}$, the characteristic roots of C $B_{11} = I_v - (1/v)E_{vv}$ are 0 and 1 repeated $v - 1$ times. Hence, the if θ is a non-zero characteristic root of C, then θ^{-1} is a non-zero characteristic root of B_{11}. Thus, if $\theta_1, \theta_2, \ldots, \theta_{v-1}$ are the $(v - 1)$ non-zero characteristic roots of C, then $\theta_1^{-1}, \theta_2^{-1}, \ldots, \theta_{v-1}{}^{-1}$ are the non-zero characteristic roots of B_{11} and hence $\text{tr}\,(B_{11}) = \sum_{i=1}^{v-1} \theta_i^{-1}$. Hence, the average variance of BLUEs of elementary treatment contrasts becomes $\frac{2\sigma^2}{(v-1)}\sum_{i=1}^{v-1} \theta_i^{-1}$.

35. The average variance of BLUEs of elementary treatment contrasts in a connected design is

$$\frac{2\sigma^2}{(v-1)}\sum_{i=1}^{v-1}(1/\theta_i),$$

where $\theta_1, \theta_2, \ldots, \theta_{v-1}$ are the non-zero characteristic roots of the C matrix of the design. Now

$$\theta_{\min} \le \theta_i, \ i = 1, 2, \ldots, v - 1.$$

Hence

$$1/\theta_{\min} \ge 1/\theta_i, \ i = 1, 2, \ldots, \ v - 1$$

$$\frac{(v-1)}{\theta_{\min}} \ge \sum_{i=1}^{v-1}(1/\theta_i)$$

$$\frac{2\sigma^2}{\theta_{\min}} \ge \frac{2\sigma^2}{(v-1)}\sum_{i=1}^{v-1}(1/\theta_i).$$

Further,

$$\theta_{\max} \ge \theta_i, \ i = 1, 2, \ldots, \ v - 1$$

$$\frac{1}{\theta_{\max}} \le \frac{1}{\theta_i}, \ i = 1, 2, \ldots, \ v - 1.$$

Hence,

$$\frac{(v-1)}{\theta_{\max}} \le \sum_{i=1}^{v-1}(1/\theta_i)$$

$$\frac{2\sigma^2}{\theta_{\max}} \le \frac{2\sigma^2}{(v-1)}\sum_{i=1}^{v-1}(1/\theta_i).$$

Thus, we get

$$\frac{2\sigma^2}{\theta_{\max}} \le \frac{2\sigma^2}{(v-1)}\sum_{i=1}^{v-1}(1/\theta_i) \le \frac{2\sigma^2}{\theta_{\min}}.$$

36. Since, the design is connected, rank (C) = v − 1. Let the v − 1 non-zero characteristic roots of C be $\theta_1, \theta_2, \ldots, \theta_{v-1}$. Let $\boldsymbol{\xi}_1, \boldsymbol{\xi}_2, \ldots, \boldsymbol{\xi}_{(v-1)}$ be the associated orthogonal characteristic vectors of C. Then, the (v − 1) linearly independent estimable treatment constrasts are given by

$$\boldsymbol{\xi}_i{}'\hat{\mathbf{t}},\ i = 1, 2, \ldots,\ v - 1.$$

and its BLUE is given by $\boldsymbol{\xi}_i{}'\hat{\mathbf{t}}$, where $\hat{\mathbf{t}}$ is any solution of $\mathbf{Q} = C\hat{\mathbf{t}}$.
Now

$$C\boldsymbol{\xi}_i = \theta_i\boldsymbol{\xi}_i,\ i = 1, 2, \ldots,\ v - 1.$$

Hence $\boldsymbol{\xi}_i'\mathbf{Q} = \boldsymbol{\xi}_i'C\hat{\mathbf{t}} = \theta_i\boldsymbol{\xi}_i'\hat{\mathbf{t}}$ and $\boldsymbol{\xi}_i'\hat{\mathbf{t}} = \boldsymbol{\xi}_i'\mathbf{Q}/\theta_i$. Thus, $\boldsymbol{\xi}_i'\mathbf{Q}/\theta_i$ is the BLUE of $\boldsymbol{\xi}_i'\mathbf{t}$. Also

$$\begin{aligned}\text{var}\,(\boldsymbol{\xi}_i'\hat{\mathbf{t}}) &= \text{var}\,(\boldsymbol{\xi}_i'\mathbf{Q}/\theta_i) = \boldsymbol{\xi}_i'(\sigma^2C)\boldsymbol{\xi}_i/\theta_i^2\\ &= \sigma^2/\theta_i,\ i = 1, 2, \ldots,\ v - 1.\end{aligned}$$

Now $\boldsymbol{\ell}'\mathbf{t}$ is an estimable treatment contrast. Therefore the vector $\boldsymbol{\ell}$ lies in the space generated by the orthogonal vectors ξ_i, $i = 1, 2, \ldots, v - 1$. Hence,

$$\boldsymbol{\ell} = \sum_{i=1}^{v-1} a_i\,\boldsymbol{\xi}_i,$$

where a_i's are constants not all zero. Further, note that $\boldsymbol{\ell}'\boldsymbol{\ell} = \sum_1^{v-1} a_i^2$. Then the BLUE of $\boldsymbol{\ell}'\mathbf{t}$ is given by $\boldsymbol{\ell}'\hat{\mathbf{t}} = \sum a_i\boldsymbol{\xi}_i'\hat{\mathbf{t}} = \sum_{i=1}^{v-1} a_i(\boldsymbol{\xi}_i'\mathbf{Q})/\theta_i$, and its variance is given by

$$\begin{aligned}\text{var}\,(\boldsymbol{\ell}'\hat{\mathbf{t}} &= \text{var}\,(\Sigma\ a_i(\boldsymbol{\xi}_i'\mathbf{Q})/\theta_i)\\ &= \Sigma\ a_i^2\text{var}\,(\boldsymbol{\xi}_i'\mathbf{Q}/\theta_i)\\ &= \sigma^2\sum_1^{v-1} a_i^2/\theta_i.\end{aligned}$$

Now $\theta_i \leq \theta_{max}$, $i = 1, 2, \ldots, v - 1$. Hence,

$$\Sigma\ a_i^2/\theta_i \geq \Sigma\ a_i^2/\theta_{max} = \frac{\boldsymbol{\ell}'\boldsymbol{\ell}}{\theta_{max}},$$

and

$$\text{var}\,(\boldsymbol{\ell}'\hat{\mathbf{t}}) \geq \sigma^2\boldsymbol{\ell}'\boldsymbol{\ell}/\theta_{max}.$$

Similarly $\theta_i \geq \theta_{min}$, $i = 1, 2, \ldots, v - 1$, and

$$\Sigma\ a_i^2/\theta_i \leq \Sigma\ a_i^2/\theta_{min} = \boldsymbol{\ell}'\boldsymbol{\ell}'/\theta_{min}.$$

Hence

$$\text{var}\,(\boldsymbol{\ell}'\hat{\mathbf{t}}) \leq \sigma^2\boldsymbol{\ell}'\boldsymbol{\ell}/\theta_{min}.$$

Thus

$$\frac{\sigma^2 \ell' \ell'}{\theta_{max}} \leq \text{var}(\ell' \hat{\mathbf{t}}) \leq \frac{\sigma^2 \ell' \ell'}{\theta_{min}}.$$

37. The efficiency of a binary design is defined as

$$E = \frac{\bar{V}_R}{\bar{V}} = \frac{2\sigma^2}{\bar{r}\ \bar{V}},$$

where $\bar{V}_R = 2\sigma^2/\bar{r}$ the average variance of BLUEs of elementary contrasts in a randomized design, $\bar{r} = \Sigma\, r_i/v$ = average number of replications in a randomized block design, $\bar{V}$ = average variance of BLUEs of elementary treatment contrasts in the given binary design.

Now $\bar{r} = \Sigma\, r_i/v = n/v$. Further, since the given binary design is connected, we have

$$\bar{V} = \frac{2\sigma^2}{(v-1)} \Sigma\,(1/\theta_i) = \frac{2\sigma^2}{H}$$

where H is the harmonic mean of the non-zero characteristic roots $\theta_1, \theta_2, .., \theta_{v-1}$ of the C matrix of the design. Let A be the arithmetic mean of $\theta_1, \theta_2, \ldots, \theta_{v-1}$. That is

$$\begin{aligned} A &= \Sigma\, \theta_i/(v-1) \\ &= \text{tr}\, C/(v-1). \end{aligned}$$

Now $C = \text{diag}(r_1, r_2, \ldots, r_v) - NK^{-1}N'$, hence

$$\begin{aligned} \text{tr}\, C &= \Sigma\, r_i - \sum_{i=1}^{v} \sum_{j=1}^{b} n_{ij}^2/kj \\ &= n - \sum_{j=1}^{b} (k_j/k_j) = n - b, \\ A &= (n-b)/(v-1) \end{aligned}$$

and $E = \dfrac{2\sigma^2}{\bar{r}\bar{V}} = \dfrac{v\, 2\sigma^2\, H}{n\,(2\sigma^2)} = \dfrac{vH}{n}$. But $H \leq A$; and hence

$$E \leq \frac{v\,A}{n} = \frac{v(n-b)}{n(v-1)}.$$

38. Since the design is connected balanced, the non-zero characteristic roots of its C matrix are all equal. Let θ be its non-zero characteristic root. Then

$$\Sigma\, \theta = (v-1)\theta = \text{tr}\, C.$$

But in Exercise 37, we have proved that

$$\text{tr}\, C = n - b.$$

Hence $\theta = (n-b)/(v-1)$.

39. We have

$$C = R - N\,K^{-1}N'$$
$$= r\,I_v - N\,K^{-1}N'$$

since the design is equi-replicated. Next, since the design is connected balanced, we have

$$C = \theta\left[I_v - \frac{1}{v}E_{vv}\right].$$

Equating the above two expressions of C, we get

$$N\,K^{-1}N' = (r - \theta)\,I_v + \frac{1}{v}E_{vv}.$$

Hence

$$N\,K^{-1}N' = r \cdot (r - \theta)^{v-1}.$$

But for a connected balanced design, it is shown in Exercise 38 that

$$\theta = \frac{n - b}{v - 1} = \frac{vr - b}{v - 1},$$

since n = vr. Hence

$$r - \theta = (b - r)/(v - 1) \neq 0,$$

since the design is incomplete block design. Therefore,

$$|N\,K^{-1}\,N'| \neq 0,$$

and

$$\text{rank}\,(N\,K^{-1}N') = v,$$

that is,

$$\text{rank}\,N = v = \min\,(v, b).$$

Hence $b \geq v$.

40. In Exercise 33, we have proved that the incidence matrix N of an equi-replicated regular connected balanced design satisfies the relation

$$N\,N' = k[(r - \theta)I_v + (\theta/v)\,E_{vv}],$$

where θ is the non-zero characteristic root of the matrix C. Now, in Exercise 38, we have proved that

$$\theta = \frac{n - b}{v - 1}.$$

But $n = bk$, and hence $\theta = b(k - 1)/(v - 1)$. Let

$$\lambda = k\theta/v = bk(k - 1)/v(v - 1) = \frac{r(k - 1)}{(v - 1)}.$$

Then

$$k(r - \theta) = kr - \lambda v.$$

But since $\lambda(v - 1) = r(k - 1)$, we get $rk - \lambda v = r - \lambda$. Hence

$$k(r - \theta) = r - \lambda.$$

Thus,

$$N N' = (r - \lambda)I_v + \lambda E_{vv}.$$

41. Consider a connected design. The average variance of the BLUEs of elementary treatment contrasts in this design is

$$\bar{V} = \frac{\sigma^2}{(v - 1)} \sum_{i=1}^{v-1} \frac{1}{\theta_i},$$

where $\theta_1, \theta_2, \ldots, \theta_{v-1}$ are the non-zero characteristic roots of the matrix C. Further,

$$\text{tr}\, C = \sum_{i=1}^{v-1} \theta_i = n - b.$$

We minimize $\bar{V}$ subject to the restriction $\Sigma\, \theta_i = n - b$ for getting the most efficient design. So, we minimize

$$\phi = \Sigma\, \frac{1}{\theta_i} - \lambda(n - b - \Sigma\, \theta_i)$$

with respect to $\theta_1, \theta_2, \ldots, \theta_{v-1}$. Equating the partial derivative of ϕ with respect to θ_i to zero, we get

$$-\theta_i^{-2} + \lambda = 0$$

i.e. $\theta_i = 1/\sqrt{\lambda}, i = 1, 2, \ldots, v - 1$.
Hence, all θ's are equal. Thus, the most efficient design is balanced. Therefore, the criterion that a connected design may have maximum efficiency is that it must be balanced, i.e. all the non-zero characteristic roots of its C matrix must be equal. From $\Sigma\, \theta_i = n - b$, we get the equal root $\theta = (n - b)/(v - 1)$.

42. The average variance of BLUEs of elementary treatment contrasts in a connected design is given by

$$\bar{V} = \frac{2\sigma^2}{(v - 1)} \sum_{i=1}^{v-1} (1/\theta_i).$$

Now, if the connected design is balanced, then

$$\theta_i = \theta = \frac{n - b}{v - 1}, i = 1, 2, \ldots, (v - 1).$$

Hence, for a connected balanced design,

$$\bar{V} = \frac{2\sigma^2}{\theta} = \frac{2\sigma^2(v-1)}{(n-b)}.$$

Therefore the efficiency of a connected balanced design is given by

$$E = \frac{2\sigma^2}{\bar{r}\bar{V}} = \frac{2v\sigma^2(n-b)}{n \cdot 2\sigma^2(v-1)} = \frac{v(n-b)}{n(v-1)}.$$

43. Here,

$$v = b = 4,\ r_1 = r_2 = r_3 = r_4 = 3,\ k_1 = k_2 = k_3 = k_4 = 3,$$

$$N\,N' = \begin{bmatrix} 3 & 2 & 2 & 2 \\ 2 & 3 & 2 & 2 \\ 2 & 2 & 3 & 2 \\ 2 & 2 & 2 & 3 \end{bmatrix} = I_4 + 2E_{44}.$$

Hence

$$\begin{aligned} C &= R - N\,K^{-1}N' \\ &= 3I_4 - \frac{1}{3}NN' = 3I_4 - \frac{1}{3}(I_4 + 2E_{44}) \\ &= (8/3)I_4 - (2/3)E_{44}. \\ &= (8/3)[I_4 - (1/4)E_{44}]. \end{aligned}$$

Therefore, the design is connected balanced. We have proved in Exercise 42, that the efficiency factor E of a connected design is given by

$$E = \frac{v(n-b)}{n(v-1)}.$$

Substituting $v = b = 4$, and $n = 12$, we get $E = 8/9$.

44. Clearly the incidence matrix of the design is

$$N = [\overset{\leftarrow\ \ r\ \ \rightarrow}{A, A, \ldots, A}]$$

where

$$A = \operatorname{diag}(E_{k1}, E_{k1}, \ldots, E_{k1}).$$

Hence

$$NN' = r \cdot AA' = r\operatorname{diag}(E_{kk}, E_{kk}, \ldots, E_{kk}) = r\,D,$$

where $D = \operatorname{diag}(E_{kk}, E_{kk}, \ldots, E_{kk})$. Next consider $CR^{-1}N$.

$$\begin{aligned} CR^{-1}N &= \frac{1}{r}CN \\ &= \frac{1}{r}[R - NK^{-1}N']N \\ &= \frac{1}{r}[rI - \frac{1}{k}D]N \\ &= N - \frac{1}{k}DN \end{aligned}$$

Now

$$\begin{aligned} DN &= \text{diag}(E_{kk}, E_{kk}, \ldots, E_{kk})[A, A, \ldots, A] \\ &= [DA, DA, \ldots, DA], \end{aligned}$$

and

$$\begin{aligned} DA &= \text{diag}(E_{kk}, E_{kk}, \ldots, E_{kk})\text{diag}(E_{k1}, E_{k1}, \ldots, E_{k1}) \\ &= k\,\text{diag}(E_{k1}, E_{k1}, \ldots, E_{k1}) \\ &= kA. \end{aligned}$$

Hence

$$DN = k[A, A, \ldots, A] = kN,$$

and, we get

$$CR^{-1}N = N - N = \mathbf{0}.$$

Therefore, the design is orthogonal.

45. One can verify that

$$\begin{aligned} \begin{bmatrix} C & -N \\ N' & K \end{bmatrix} \begin{bmatrix} I & 0 \\ -K^{-1}N' & I \end{bmatrix} &= \begin{bmatrix} C + NK^{-1}N' & -N \\ 0 & K \end{bmatrix} \\ &= \begin{bmatrix} R & -N \\ \mathbf{0} & K \end{bmatrix} \end{aligned}$$

Taking determinants, we get

$$\begin{vmatrix} C & -N \\ N' & K \end{vmatrix} = |R| \cdot |K| = (r_1\, r_2 \ldots r_v)(k_1\, k_2 \ldots k_b).$$

46. Consider (i) $N = E_{vb}$. For this design one can easily prove that

$$C = b[I_v - (1/v)\, E_{vv}].$$

The characteristic roots of this C matrix are 0 with multiplicity 1 and b with multiplicity $(v - 1)$. Hence the design is connected balanced. The variance of the BLUE of an elementary treatment contrast is $2\sigma^2/b$.
Consider the design (ii) $C = a[I_v - (1/v)E_{vv}]$. The characteristic roots of this matrix are 0 with multiplicity 1 and a with multiplicity $v - 1$. Hence this design is connected balanced. Also the variance of the BLUE of an elementary treatment contrast in this case is $2\sigma^2/a$.

47. We know that in a one-way design the BLUE of an estimable treatment contrast $\boldsymbol{\ell}'\mathbf{t}$ is given by $\boldsymbol{\ell}'\hat{\mathbf{t}}$ where $\hat{\mathbf{t}}$ is any solution of

$$\mathbf{Q} = C\hat{\mathbf{t}}.$$

Let C^- be any g-inverse of C. Then a solution of $\mathbf{Q} = C\hat{\mathbf{t}}$ can be taken as

$$\hat{\mathbf{t}} = C^-\hat{\mathbf{Q}}.$$

The BLUE of $\boldsymbol{\ell}'\mathbf{t}$ is therefore $\boldsymbol{\ell}'C^{-}\mathbf{Q}$. Now let $C\,\mathbf{p} = \boldsymbol{\ell}$. Then $\mathbf{p} = C^{-}\boldsymbol{\ell}$, and $\mathbf{p}' = \boldsymbol{\ell}'C^{-}$. Thus,

$$\boldsymbol{\ell}'C^{-}\mathbf{Q} = \mathbf{p}'\mathbf{Q}.$$

Hence the BLUE of $\boldsymbol{\ell}'\mathbf{Q}$ can be taken as $\mathbf{p}'\mathbf{Q}$, where $C\mathbf{p} = \boldsymbol{\ell}$. Further the variance of $\mathbf{p}'\mathbf{Q}$ is given by

$$\begin{aligned}\mathrm{var}(\mathbf{p}'\mathbf{Q}) &= \mathbf{p}' \cdot \mathrm{Var}(\mathbf{Q})\mathbf{p}\\ &= \mathbf{p}'(\sigma^2 C)\mathbf{p}\\ &= \sigma^2\mathbf{p}'C\mathbf{p}\\ &= \sigma^2\mathbf{p}'\boldsymbol{\ell}.\end{aligned}$$

Let $C = aI_v - (1/v)E_{vv}$. Then, every treatment contrast is estimable, since rank $C = v - 1$. The BLUE of a treatment contrast $\boldsymbol{\ell}'\mathbf{t}$ is therefore given by

$$\mathbf{p}'\mathbf{Q},$$

where $C\mathbf{p} = \boldsymbol{\ell}$. i.e., $a\,\mathbf{p} - (a/v)E_{vv}\mathbf{p} = \boldsymbol{\ell}$.
Since rank $C = v - 1$, we put one side restriction $E_{1v}\mathbf{p} = 0$, for finding a solution of $C\mathbf{p} = \boldsymbol{\ell}$. Hence, we get

$$a\,\mathbf{p} = \boldsymbol{\ell} \quad \text{or} \quad \mathbf{p} = \boldsymbol{\ell}/a.$$

Thus, the BLUE is given by

$$\mathbf{p}'\mathbf{Q} = \boldsymbol{\ell}'\mathbf{Q}/a.$$

The variance of $\mathbf{p}'\mathbf{Q} = \boldsymbol{\ell}'\mathbf{Q}/a$ is then given by

$$\mathrm{var}(\boldsymbol{\ell}'\mathbf{Q}/a) = \sigma^2\boldsymbol{\ell}'\boldsymbol{\ell}/a.$$

As a particular case, take

$$\boldsymbol{\ell} = \{0, 0, \ldots, \overset{i}{1}, 0, 0, \ldots, \overset{j}{-1}, 0, \ldots, 0\}',$$

Then $\boldsymbol{\ell}'\mathbf{t} = t_i - t_j$, $\boldsymbol{\ell}'\mathbf{Q} = Q_i - Q_j$, and $\boldsymbol{\ell}'\boldsymbol{\ell} = 2$.

Hence the results follow.

48. Here rank $(C) = v - 1$. So we put one side contrast as $\mathbf{r}'\mathbf{t} = 0$. Hence rank of $[C \quad \mathbf{r}/\sqrt{n}] = v$. Thus the rank of the $(v+1) \times (v+1)$ matrix

$$A = \begin{bmatrix} C & \mathbf{r}/\sqrt{n} \\ \mathbf{r}'/\sqrt{n} & 0 \end{bmatrix}$$

is $v + 1$ and this matrix is non-singular. Let its inverse be

$$\begin{bmatrix} B_{11} & E_{v1}/\sqrt{r} \\ E_{1v}/\sqrt{n} & 0 \end{bmatrix}.$$

Then, upon multiplication, we see that

$$C\,B_{11} + \mathbf{r}\,E_{1v}/n = I_v,$$
$$\mathbf{r}'\,B_{11} = \mathbf{0}.$$

This gives $C\,B_{11} = I_v - \mathbf{r}\,E_{1v}/n$. Now consider the matrix $B_{11} + E_{vv}/n$. We have

$$\begin{aligned}[C + \mathbf{rr}'/n][B_{11} + E_{vv}/n] &= C\,B_{11} + \mathbf{r}\,E_{1v}/n \\ &= I_v.\end{aligned}$$

Also,

$$\begin{aligned}[B_{11} + E_{vv}/n][C + \mathbf{rr}'/n] &= B_{11}C + E_{v1}\mathbf{r}'/n \\ &= I_v.\end{aligned}$$

Since $C\,B_{11} + \mathbf{r}\,E_{1v}/n = I_v$, we see that $C + \mathbf{rr}'/n$ is non-singular and its inverse is $B_{11} + E_{vv}/n$, where B_{11} is given by $C\,B_{11} = I_v - \mathbf{r}\,E_{1v}/n$, and $\mathbf{r}'\,B_{11} = \mathbf{0}$.

Further,

$$[C + \mathbf{rr}'/n]^{-1}\mathbf{r} = [B_{11} + E_{vv}/n]\mathbf{r} = E_{vv}\mathbf{r}/n = E_{v1},$$

since $\mathbf{r}'B_{11} = \mathbf{0}$ and B_{11} is symmetric. Now we have

$$\begin{aligned}C \cdot [C + \mathbf{rr}'/n]^{-1}C &= C[B_{11} + E_{vv}/n]C \\ &= C\,B_{11}C \\ &= [I_v - \mathbf{r}E_{1v}/n]C \\ &= C\end{aligned}$$

since $E_{1v}C = \mathbf{0}$. Hence $[C + \mathbf{r}\,\mathbf{r}'/n]^{-1}$ is a g-inverse of C.

49. We know that a set of solutions of normal equations in any block design is given by

$$\begin{aligned}\hat{\mu} &= G/n \\ \hat{\alpha} &= K^{-1}\mathbf{B} - (G/n)E_{b1} - K^{-1}N'\hat{\mathbf{t}} \\ \mathbf{Q} &= C\,\hat{\mathbf{t}}.\end{aligned}$$

Now in Exercise 48 we have proved that $(C + \mathbf{r}\,\mathbf{r}'/n)^{-1}$ is a g-inverse of C and that

$$(C + \mathbf{r}\,\mathbf{r}'/n)^{-1} = B_{11} + E_{vv}/n,$$

where B_{11} is given by

$$C\,B_{11} = I_v - \mathbf{r}E_{1v}/n \text{ and } \mathbf{r}'B_{11} = \mathbf{0}.$$

Hence a solution of $\hat{\mathbf{t}}$ is given by

$$\hat{\mathbf{t}} = (C + \mathbf{r}\,\mathbf{r}'/n)^{-1}\mathbf{Q}.$$

Therefore a set of solutions of normal equations in the intrablock analysis is given by

$$\begin{aligned}\hat{\mu} &= G/n \\ \hat{\boldsymbol{\alpha}} &= K^{-1}\mathbf{B} - (G/n)E_{b1} - K^{-1}N'(C + \mathbf{r}\,\mathbf{r}'/n)^{-1}\mathbf{Q} \\ \hat{\mathbf{t}} &= (C + \mathbf{r}\,\mathbf{r}'/n)^{-1}\mathbf{Q}.\end{aligned}$$

Further, clearly

$$\begin{aligned}
\text{var}(\hat{\mathbf{t}}) &= (\mathrm{C} + \mathbf{r}\,\mathbf{r}'/\mathrm{n})^{-1} \cdot \text{var}(\mathbf{Q})(\mathrm{C} + \mathbf{r}\,\mathbf{r}'/\mathrm{n})^{-1} \\
&= \sigma^2(\mathrm{C} + \mathbf{r}\,\mathbf{r}'/\mathrm{n})^{-1}\mathrm{C}(\mathrm{C} + \mathbf{r}\,\mathbf{r}'/\mathrm{n})^{-1} \\
&= \sigma^2(\mathrm{C} + \mathbf{r}\,\mathbf{r}'/\mathrm{n})^{-1}\mathrm{C}(\mathrm{B}_{11} + \mathrm{E}_{\mathrm{vv}}/\mathrm{n}) \\
&= \sigma^2(\mathrm{C} + \mathbf{r}\mathbf{r}'/\mathrm{n})^{-1}\mathrm{C}\,\mathrm{B}_{11} \\
&= \sigma^2[\mathrm{B}_{11} + \mathrm{E}_{\mathrm{vv}}/\mathrm{n}]\mathrm{C}\,\mathrm{B}_{11} \\
&= \sigma^2\mathrm{B}_{11}\mathrm{C}\mathrm{B}_{11} \\
&= \sigma^2\mathrm{B}_{11}[\mathrm{I}_{\mathrm{v}} - \mathbf{r}\,\mathrm{E}_{1\mathrm{v}}/\mathrm{n}] \\
&= \sigma^2\mathrm{B}_{11},
\end{aligned}$$

since $\mathbf{r}'\,\mathrm{B}_{11} = \mathbf{0}$. Now

$$[\mathrm{C} + \mathbf{r}\,\mathbf{r}'/\mathrm{n}]^{-1} = \mathrm{B}_{11} + \mathrm{E}_{\mathrm{vv}}/\mathrm{n}.$$

Hence

$$\mathrm{B}_{11} = [\mathrm{C} + \mathbf{r}\,\mathbf{r}'/\mathrm{n}]^{-1} - \mathrm{E}_{\mathrm{vv}}/\mathrm{n},$$

and

$$\text{var}(\hat{\mathbf{t}}) = \sigma^2[(\mathrm{C} + \mathbf{r}\,\mathbf{r}'/\mathrm{n})^{-1} - \mathrm{E}_{\mathrm{vv}}/\mathrm{n}].$$

The intrablock estimate of $\boldsymbol{\ell}'\mathbf{t}$ is $\boldsymbol{\ell}'\hat{\mathbf{t}} = \boldsymbol{\ell}'(\mathrm{C} + \mathbf{r}\,\mathbf{r}'/\mathrm{n})^{-1}\mathbf{Q}$.
Now

$$\begin{aligned}
\text{var}(\boldsymbol{\ell}'\hat{\mathbf{t}}) &= \boldsymbol{\ell}' \cdot \text{var}(\hat{\mathbf{t}}) \cdot \boldsymbol{\ell} \\
&= \sigma^2\boldsymbol{\ell}'[(\mathrm{C} + \mathbf{r}\,\mathbf{r}'/\mathrm{n})^{-1} - \mathrm{E}_{\mathrm{vv}}/\mathrm{n}]\boldsymbol{\ell} \\
&= \sigma^2\boldsymbol{\ell}'(\mathrm{C} + \mathbf{r}\,\mathbf{r}'/\mathrm{n})^{-1}\boldsymbol{\ell},
\end{aligned}$$

since $\mathrm{E}_{1\mathrm{v}}\,\boldsymbol{\ell} = 0$.

50. Consider the following solution of normal equations in the intrablock analysis of variance of a one-way design.

$$\begin{aligned}
&\hat{\mu} = \mathrm{G}/\mathrm{n},\ \hat{\alpha} = \mathrm{K}^{-1}\mathbf{B} - (\mathrm{G}/\mathrm{n})\mathrm{E}_{\mathrm{b}1} - \mathrm{K}^{-1}\mathrm{N}'\hat{\mathbf{t}} \\
&\mathbf{Q} = \mathrm{C}\hat{\mathbf{t}}.
\end{aligned}$$

Then the sum of squares due to regression when all parameters are fitted is given by

$$\text{SSR}(\mu, \boldsymbol{\alpha}, \mathbf{t}) = \mathbf{B}'\mathrm{K}^{-1}\mathbf{B} + \hat{\mathbf{t}}'\mathbf{Q}.$$

Next, we consider another solution of normal equations in the intrablock analysis of variance of a one-way design as

$$\begin{aligned}
&\mu^* = \mathrm{G}/\mathrm{n}, \mathbf{t}^* = \mathrm{R}^{-1}\mathrm{I} - (\mathrm{G}/\mathrm{n})\mathrm{E}_{\mathrm{v}1} - \mathrm{R}^{-1}\mathrm{N}\,\boldsymbol{\alpha}^* \\
&\mathrm{P} = \mathrm{D}\boldsymbol{\alpha}^*.
\end{aligned}$$

Then the sum of squares due to regression when all parameters are fitted is

given by

$$\text{SSR}(\mu, \boldsymbol{\alpha}, \mathbf{t}) = \mu^*(G) + \boldsymbol{\alpha}'^*\mathbf{B} + t'^*I$$
$$= \mathbf{T}'R^{-1}\mathbf{T} + \boldsymbol{\alpha}'^*\mathbf{P}$$

Now since the sum of squares due to regression when all parameters are fitted remains the same no matter which solution of normal equations we take, we get equating the above two sums of squares

$$\begin{aligned}\boldsymbol{\alpha}'^*\mathbf{P} &= \mathbf{B}K^{-1}\mathbf{B} + \hat{\mathbf{t}}'\mathbf{Q} - \mathbf{T}'R^{-1}\mathbf{T}\\ &= (\mathbf{y}'\mathbf{y} - G^2/n) - [(\mathbf{y}'\mathbf{y} - G^2/n) - (\mathbf{B}'K^{-1}\mathbf{B} - G^2/n) - \mathbf{t}'\mathbf{Q}]\\ &\quad -(\mathbf{T}'K^{-1}\mathbf{T} - G^2/n)\\ &= \text{Total SS} - \text{Error SS} - \text{Unadj. Treatment SS.}\end{aligned}$$

51. Total SS $= \mathbf{y}'\mathbf{y} - G^2/n = \mathbf{y}'(I_n - n^{-1}E_{nn})\mathbf{y}$
Now in the intrablock analysis of variance, we have $\mathcal{E}(\mathbf{y}) = A\boldsymbol{\theta}$ and var $(\mathbf{y}) = \sigma^2 I_n$, where $\boldsymbol{\theta}' = (\mu, \boldsymbol{\alpha}', \mathbf{t}')$ and

$$A = \begin{bmatrix} E_{n1} \cdot \text{diag}(E_{k_1}1, E_{k_2}1, \ldots, E_{k_b}1) & \begin{matrix} A_1 \\ A_2 \\ \cdot \\ \cdot \\ \cdot \\ A_b \end{matrix} \end{bmatrix}$$

and A_j is a $k_j \times v$ matrix such that its s-th row has a unit element in the column corresponding to the treatment applied to the s-th plot in the j-th block and zero elsewhere. Then,

$$\begin{aligned}\mathcal{E}[\mathbf{y}'\mathbf{y} - G^2/n] &= \sigma^2 \text{tr}[(I_n - n^{-1}E_{nn})I_n] + \boldsymbol{\theta}'A'(I_n - n^{-1}E_{nn})A\boldsymbol{\theta}\\ &= (n-1)\sigma^2 + \boldsymbol{\theta}'[A'A - n^{-1}A'E_{nn}A]\boldsymbol{\theta}.\end{aligned}$$

Now $E_{1n}A = [n, \mathbf{k}', \mathbf{r}']$ and

$$A'A = \begin{bmatrix} n & \mathbf{k}' & \mathbf{r}' \\ \mathbf{k} & K & N' \\ \mathbf{r} & N & R \end{bmatrix}.$$

Noting $A'E_{nn}A = (E_{1n}A)'(E_{1n}A)$, we obtain

$$\begin{aligned}\mathcal{E}[\mathbf{y}'\mathbf{y} - G^2/n] = (n-1)\sigma^2 &+ \boldsymbol{\alpha}'K\boldsymbol{\alpha} + \mathbf{t}'R\mathbf{t} + 2\mathbf{t}'N\boldsymbol{\alpha}\\ &- n^{-1}(\mathbf{k}'\boldsymbol{\alpha} + \mathbf{r}'\mathbf{t})^2.\end{aligned}$$

Now, the unadjusted block SS is given by

$$\mathbf{B}'K^{-1}\mathbf{B} - n^{-1}G^2 = \mathbf{B}'(K^{-1} - n^{-1}E_{bb})\mathbf{B}.$$

Next from the normal equations, one gets

$$\begin{aligned}\mathcal{E}(\mathbf{B}) &= \mu\mathbf{k} + K\boldsymbol{\alpha} + N\mathbf{t} = L\boldsymbol{\theta},\\ \text{var}(\mathbf{B}) &= \sigma^2 K,\end{aligned}$$

where $L = [\mathbf{k}, K, N']$ and $\boldsymbol{\theta}' = [\mu, \boldsymbol{\alpha}', \mathbf{t}']$. Hence

$$\mathcal{E}[\mathbf{B}'K^{-1}\mathbf{B} - G^2/n] = \sigma^2 \text{tr}[(K^{-1} - n^{-1}E_{bb})K] + \boldsymbol{\theta}'L'[K^{-1} - n^{-1}E_{bb}]L\boldsymbol{\theta}.$$

Now

$$\begin{aligned}\boldsymbol{\theta}'L'K^{-1}L\boldsymbol{\theta} &= (\mu\mathbf{k}' + \boldsymbol{\alpha}'K + \mathbf{t}'N)K^{-1}(\mu\mathbf{k} + K\boldsymbol{\alpha} + N'\mathbf{t}) \\ &= n\mu^2 + 2\mu\boldsymbol{\alpha}'\mathbf{k} + 2\mu\mathbf{t}'\mathbf{r} + \boldsymbol{\alpha}'K\boldsymbol{\alpha} + 2\mathbf{t}'N\boldsymbol{\alpha} \\ &\quad + \mathbf{t}'NK^{-1}N'\mathbf{t} \\ \boldsymbol{\theta}'L'E_{bb}L\boldsymbol{\theta} &= (\boldsymbol{\theta}'L'E_{b1})(E_{1b}L\boldsymbol{\theta}) \\ &= (n\mu + \boldsymbol{\alpha}'\mathbf{k} + \mathbf{t}'\mathbf{r})^2.\end{aligned}$$

Hence upon substitution and after some simplification, we get

$$\mathcal{E}[\mathbf{B}'K^{-1}\mathbf{B} - G^2/n] = (b-1)\sigma^2 + \boldsymbol{\alpha}'K\boldsymbol{\alpha} + 2\mathbf{t}'N\boldsymbol{\alpha} + \mathbf{t}'NK^{-1}N'\mathbf{t} - n^{-1}(\mathbf{k}'\boldsymbol{\alpha} + \mathbf{r}'\mathbf{t})^2.$$

Clearly,

$$\mathcal{E}(\text{Intrablock Error SS}) = (n - b - v + g)\sigma^2.$$

Further, the unadjusted treatment sum of squares is given by

$$\mathbf{T}'R^{-1}\mathbf{T} - n^{-1}G^2 = \mathbf{T}'(R^{-1} - n^{-1}E_{vv})\mathbf{T}.$$

Now from the normal equations, one obtains

$$\begin{aligned}\mathcal{E}(\mathbf{T}) &= \mathbf{r} + N\boldsymbol{\alpha} + R\mathbf{t} = N\boldsymbol{\theta}, \\ \text{var}(\mathbf{T}) &= \sigma^2 R,\end{aligned}$$

where $M = [\mathbf{r}, N, \mathbf{t}]$ and $\boldsymbol{\theta}' = [\mu, \boldsymbol{\alpha}', \mathbf{t}']$. Hence

$$\mathcal{E}[\mathbf{T}'R^{-1}\mathbf{T} - n^{-1}G^2] = \sigma^2 \text{tr}[(R^{-1} - n^{-1}E_{vv})R] + \boldsymbol{\theta}'M'[R^{-1} - n^{-1}E_{vv}]M\boldsymbol{\theta}.$$

Now

$$\begin{aligned}\boldsymbol{\theta}'M'R^{-1}M\boldsymbol{\theta} &= (\mu\mathbf{r}' + \boldsymbol{\alpha}'N + \mathbf{t}'R)R^{-1}(\mu\mathbf{r} + N\boldsymbol{\alpha} + R\mathbf{t}) \\ &= n\mu^2 + 2\mu\boldsymbol{\alpha}'\mathbf{k} + 2\mu\mathbf{t}'\mathbf{r} + \mathbf{t}'R\mathbf{t} + 2\mathbf{t}'N\boldsymbol{\alpha} \\ &\quad + \boldsymbol{\alpha}'N'R^{-1}N\boldsymbol{\alpha} \\ \boldsymbol{\theta}'N'E_{vv}M\boldsymbol{\theta} &= (\boldsymbol{\theta}'M'E_{v1})(E_{1v}M\boldsymbol{\theta}) \\ &= (n\,\mu + \boldsymbol{\alpha}'\mathbf{k} + \mathbf{t}'\mathbf{r})^2.\end{aligned}$$

Hence upon substitution and after simplication, we get

$$\mathcal{E}[\mathbf{T}'R^{-1}\mathbf{T} - G^2/n] = (v-1)\sigma^2 + \mathbf{t}'R\mathbf{t} + 2\mathbf{t}'N\,\boldsymbol{\alpha} + \boldsymbol{\alpha}'N'R^{-1}N\boldsymbol{\alpha} - n^{-1}(\mathbf{k}'\boldsymbol{\alpha} + \mathbf{r}'\mathbf{t})^2.$$

Now adjusted treatment SS is given by

$$\hat{\mathbf{t}}'\mathbf{Q} = (\text{Total SS}) - (\text{Unadj. Block SS}) - (\text{Error SS}).$$

Taking expectation and substitution the expected values of total SS, unadjusted block SS and error SS, we get

$$\mathcal{E}(\hat{\mathbf{t}}'\mathbf{Q}) = (v - g)\sigma^2 + \mathbf{t}'C\,\mathbf{t}.$$

Lastly the adjusted block SS is obtained as

$$(\text{adj. block SS}) = (\text{Total SS}) - (\text{Error SS}) - (\text{Unadj. treatment SS}).$$

Taking expectation and substituting the expected values of total SS, unadjusted treatment SS and error SS, we get

$$\mathcal{E}[\text{adj. block SS}] = (b - g)\sigma^2 + \boldsymbol{\alpha}'D\boldsymbol{\alpha}.$$

52. From the normal equations of the intrablock analysis of variance one obtains

$$\text{var}\begin{bmatrix} G \\ \mathbf{B} \\ \mathbf{T} \end{bmatrix} = \sigma^2 \begin{bmatrix} n & \mathbf{k}' & \mathbf{r}' \\ \mathbf{k} & K & N' \\ \mathbf{r} & N & R \end{bmatrix}.$$

Now pre-multiply $\begin{bmatrix} G \\ \mathbf{B} \\ \mathbf{T} \end{bmatrix}$ by the matrix

$$L = \begin{bmatrix} n^{-1} & \mathbf{0} & \mathbf{0} \\ \mathbf{0} & I_b & -N'R^{-1} \\ \mathbf{0} & -N\,K^{-1} & I_v \end{bmatrix}.$$

Then, we get

$$L = \begin{bmatrix} G \\ \mathbf{B} \\ \mathbf{T} \end{bmatrix} = \begin{bmatrix} G/n \\ \mathbf{P} \\ \mathbf{Q} \end{bmatrix}.$$

Hence

$$\begin{aligned} \text{var}\begin{bmatrix} G/n \\ \mathbf{P} \\ \mathbf{Q} \end{bmatrix} &= L\,\text{var}\begin{bmatrix} G \\ \mathbf{B} \\ \mathbf{T} \end{bmatrix} L' \\ &= \sigma^2 L \begin{bmatrix} n & \mathbf{k}' & \mathbf{r}' \\ \mathbf{k} & K & N' \\ \mathbf{r} & N & R \end{bmatrix} L' \\ &= \sigma^2 \begin{bmatrix} n^{-1} & \mathbf{0} & \mathbf{0} \\ \mathbf{0} & D & -DK^{-1}N' \\ \mathbf{0} & -CR^{-1}N & C \end{bmatrix}, \end{aligned}$$

which shows that

$$\text{var}(\mathbf{P}) = \sigma^2 D, \ \text{var}(\mathbf{Q}) = \sigma^2 C$$
$$\text{cov}(\mathbf{Q}', \mathbf{P}) = -\sigma^2 C R^{-1} N, \ \text{cov}(\mathbf{P}, \mathbf{Q}) = -\sigma^2 D K^{-1} N'.$$

Now, one can easily prove that $(D K^{-1} N') = C R^{-1} N$.

53. The equations for obtaining t_s in the itnrablock analysis are

$$\mathbf{Q} = C\,\mathbf{t}$$

where $\mathbf{Q} = T - \frac{1}{k} N \mathbf{B}$, $C = R - \frac{1}{k} N N'$. The elements of $\mathbf{Q}$ are denoted by $Q_1, Q_2, \ldots, Q_v$. The elements of C are given by

$$C_{ss} = r_s(k-1)/k$$
$$C_{ss'} = -\frac{1}{k}\lambda_{ss'}, \ s \neq s'.$$
$$Q_s = r_s \frac{(k-1)}{k} t_s - \frac{\lambda\, s1}{k} t_1, \ldots, \frac{sv}{k} t_v, \ s = 1, 2, \ldots, v.$$

The normal equations in the combined inter and intrablock analysis are

$$\begin{bmatrix} W_2 G \\ w_1\,\mathbf{Q} + (w_2/k) N\,\mathbf{B} \end{bmatrix} = \begin{bmatrix} w_2 bk & w_2 \mathbf{r} \\ w_2 \mathbf{r} & w_1 C + (w_2/k) N N' \end{bmatrix} \begin{bmatrix} \mu \\ \mathbf{t} \end{bmatrix}.$$

Setting $\mathbf{r}'\mathbf{t} = 0$, we get $\hat{\mu} = G/bk$ and

$$w_1 \mathbf{Q} + w_2 \left(\frac{1}{k} N\,\mathbf{B} - \frac{G}{bk}\mathbf{r} \right) = \left[w_1 C + \frac{w_2}{k} N N' \right] \hat{\mathbf{t}}.$$

The s-th element in $w_1 \mathbf{Q} + w_2 \left(\frac{1}{k} N B - \frac{G}{bk}\mathbf{r} \right)$ is

$$w_1 Q_s + w_2 Q_s',$$

where $Q_s' = \frac{1}{k} B_{(s)} - \frac{G}{bk} r_s$, and $B_{(s)}$ = sum of blocks in which the s-th treatment occurs. One can verify that

$$Q_s' = T_s - Q_s - (G/bk)\, r_s.$$

Denote the elements of the matrix $w_1 C + (w_2/k) N N'$ by $W_{ss'}$. Then we see that

$$W_{ss} = w_1 r_s \frac{(k-1)}{1} + \frac{w_2}{k} r_s = r_s \frac{(k-1)}{k} \left[w_1 + \frac{w_2}{k-1} \right]$$
$$W_{ss'} = -\frac{w_1}{k}\lambda_{ss'} + \frac{w_2}{k}\lambda_{ss'} = -\frac{(w_1 - w_2)}{k}\lambda_{ss'}, \ s \neq s'.$$

Hence, the equations in the combined inter and intrablock analysis are given by

$$w_1 Q_s + w_2 Q_s' = r_s \frac{(k-1)}{k}\left[w_1 + \frac{w_2}{k-1}\right] t_s - \frac{(w_1 - w_2)}{k}\lambda_{s1}\, t_1 - \cdots - \frac{(w_1 - w_2)}{k}\lambda_{sv}\, t_v,$$
$$s = 1, 2, \ldots, v.$$

Comparing these equations with those in the intrablock analysis, we see that these equations are obtained from those in the itnrablock analysis by replacing $Q_s, r_s, \lambda_{ss'}$ by $P_s = w_1 Q_s + w_2 Q_s'$, $R_s = r_s\left[w_1 + \dfrac{w_2}{k-1}\right]$, and $\Lambda_{ss'} = (w_1 - w_2)\lambda_{ss'}$.

54. (i) To show var $(\mathbf{Q}) = C/w_1$:
Consider the model,

$$y_{ixj} = \mu + \alpha_j + t_i + e_{xij}, \quad x = 0, 1, \ldots, n_{ij}$$
$$i = 1, 2, \ldots, v;\ j = 1, 2, \ldots, b.$$

Summing over x and j, we get

$$T_i = r_i + \sum_J n_{ij}\alpha_j + r_i t_i + \sum_{x,j} e_{xij}.$$

Summing over x and i, we get

$$B_j = k\mu + k\alpha_j + \sum_i n_{ij} t_i + \sum_{x,i} e_{xij}.$$

Multiplying this by $\dfrac{n_{ij}}{k}$ and summing over j, we get

$$\frac{1}{k}\sum_j n_{ij} B_j = r_i\,\mu + \sum_j n_{ij}\alpha_j + \frac{1}{k}\sum_{i,j} n_{ij}^2 t_i + \sum_{x,i,j} n_{ij} e_{xij}$$
$$= r_i + \sum_j n_{ij}\alpha_j + \frac{1}{k}\sum_i n_i\, t_i + \frac{1}{k_{x,}}\sum_{i,j} n_{ij}\, e_{xij}.$$

Hence,

$$Q_i = T_i - \frac{1}{k}\sum n_{ij} B_j = r_i t_i - \frac{1}{k}\sum r_i t_i + \sum_{x,j} e_{xij} - \frac{1}{k}\sum_{x,i,j} n_{ij}\, e_{xij}.$$

We thus see that $Q_i, i = 1, 2, \ldots, v$ is independent of α_j and hence is the same function of random variables e's both in the intrablock and the combined inter and intrablock analyses. Hence var($\mathbf{Q}$) is the same in both types of analyses. But var($\mathbf{Q}$) $= \sigma_e^2$ C in the intrablock analysis. Hence in the combined inter and intrablock analysis,

$$\text{var}(\mathbf{Q}) = \sigma_e^2\, C = C/w_1.$$

(ii) To show that var $(\mathbf{Q}_1) = C_1/w_2$:

We have $\mathbf{Q}_1 = \frac{1}{k} N\, \mathbf{B} - \frac{G}{bk}\, \mathbf{r}$. Hence

$$\text{var}(\mathbf{Q}_1) = \frac{1}{k^2} N.\ \text{var}(\mathbf{B})N' + G\frac{1}{b^2k^2}\mathbf{r}.\ \text{var}(G)\,\mathbf{r}'$$
$$-\frac{2}{bk^2} N\ \text{cov}(\mathbf{B}, G)\mathbf{r}'.$$

Now, since $B_j = k\mu + k\alpha_j + \sum_i n_{ij} t_i + \sum_{x,i} e_{xij}$,

$$\text{var}(B_j) = k^2\sigma_b^2 + k\sigma_e^2 = k(\sigma_e^2 + k\sigma_e^2) = k/w_2,$$

and cov $(B_j, B_{j1}) = 0$. Hence var $(\mathbf{B}) = (k/w_2)I_b$. Also $G = E_{1b}\mathbf{B}$. Hence

$$\begin{aligned}
\text{var}(G) &= E_{1b}\ \text{var}(\mathbf{B})\ E_{b1} \\
&= (bk/w_2), \\
\text{cov}(\mathbf{B}', G) &= \text{cov}(\mathbf{B},\ E_{1b}\,\mathbf{B}) \\
&= \text{cov}(\mathbf{B},\ \mathbf{B})E_{b1} \\
&= (k/w_2)\,E_{b1}.
\end{aligned}$$

and

$$\begin{aligned}
\text{var}(\mathbf{Q}_1) &= \frac{1}{kw_2} N\,N' + \frac{1}{bkw_2}\mathbf{r}\,\mathbf{r}' - \frac{2}{bkw_2} N\,E_{b1}\,\mathbf{r}' \\
&= \left[\frac{1}{kw_2} N\,N' - \frac{1}{bkw_2}\mathbf{r}\,\mathbf{r}'\right] = C_1/w_2,
\end{aligned}$$

where $C_1 = \frac{1}{k} N\,N' - \frac{1}{bk}\mathbf{r}\,\mathbf{r}'$.

(iii) To show that var $(\mathbf{Q}_1) = C_1/w_1$:

From the normal equations in the analysis with recovery of interblock information, we get

$$\text{var}\begin{bmatrix} w_2 G \\ w_1\,\mathbf{Q} + \frac{w_2}{k} N\,B \end{bmatrix} = \begin{bmatrix} w_2\,bk & w_2\,\mathbf{r}' \\ w_2\,\mathbf{r} & w_1 C + \frac{w_2}{k} N\,N' \end{bmatrix}.$$

Now

$$\left[-\frac{1}{b\,k}\mathbf{r}, I_v\right]\begin{bmatrix} W_2\,G \\ w_1\,\mathbf{Q} + \frac{w_2}{k} N\,\mathbf{B} \end{bmatrix} = w_1\,\mathbf{Q} + w_2\,\mathbf{Q}_1,$$

where $\mathbf{Q}_1 = \frac{1}{k}\mathrm{N}\,\mathbf{B} - \frac{1}{b\,k}\mathbf{r}\,\mathbf{r}'$. Hence

$$\mathrm{var}\,(w_1\,\mathbf{Q} + w_2\mathbf{Q}) = \left[-\frac{1}{b\,k}\mathbf{r},\ I_v\right] \mathrm{var}\begin{bmatrix} w_2\,G \\ w_1\mathbf{Q} + \frac{w_2}{k}\mathrm{N}\,\mathbf{B}\end{bmatrix} \times \begin{bmatrix} -\frac{1}{b\,k}\mathbf{r}' \\ I_v \end{bmatrix}$$
$$= w_1 C + w_2 C_1,$$

where $C_1 = \frac{1}{k}\mathrm{N}\,\mathrm{N}' - \frac{1}{bk}\mathbf{r}\,\mathbf{r}'$.

(iv) To show that var $(\mathbf{Q},\ \mathbf{Q}_1) = \mathbf{0}$:

We have shown that var $(w_1\,\mathbf{Q} + w_2\,\mathbf{Q}_1) = w_1\,C + w_2\,C_1$. Hence

$$w_1^2\ \mathrm{var}\,(\mathbf{Q}) + w_2^2\ \mathrm{var}\,(\mathbf{Q}_1) + 2w_1\,w_2\ \mathrm{cov}\,(\mathbf{Q},\ \mathbf{Q}_1)$$
$$= w_1 C + w_2 C_1.$$

But var $(\mathbf{Q}) = C/w_1$, and var $(\mathbf{Q}_1) = C_1/w_2$, hence, we get

$$w_1\,C + w_2 C_1 + 2\,w_1 w_2\ \mathrm{cov}\,(\mathbf{Q},\ \mathbf{Q}_1) = w_1\,C + w_2\,C_1,$$

which gives cov $(\mathbf{Q},\ Q_1) = \mathbf{0}$.

55. For (i), see Exercise 49.

We consider (ii). The normal equations in the interblock analysis are given by

$$\begin{bmatrix} k & G \\ \mathrm{N} & \mathbf{B} \end{bmatrix} = \begin{bmatrix} b\,k^2 & k\,\mathbf{r}' \\ k\,\mathbf{r} & \mathrm{N}\,\mathrm{N}' \end{bmatrix} \begin{bmatrix} \tilde{\mu} \\ \mathbf{t} \end{bmatrix}.$$

A solution of the above equations is obtained as

$$\mu = 0, \quad \mathrm{N}\,\mathbf{B} = (\mathrm{N}\,\mathrm{N}')\tilde{\mathbf{t}}.$$

Since $(\mathrm{N}\,\mathrm{N}')$ is non-singular, we obtain

$$\tilde{\mathbf{t}} = (\mathrm{N}\,\mathrm{N}')^{-1}\mathrm{N}\,\mathbf{B}.$$

Hence, the interblock estimate of a treatment contrast $\boldsymbol{\ell}'\mathbf{t}$ is

$$\boldsymbol{\ell}'(\mathrm{N}\,\mathrm{N}')^{-1}\,\mathrm{N}\,\mathbf{B}.$$

Also, from the above normal equations, we get var $(\mathrm{N}\,\mathbf{B}) = (k/w_2)\mathrm{N}\mathrm{N}'$. Hence the variance of the interblock estimate of $\boldsymbol{\ell}'\mathbf{t}$ is given by $\boldsymbol{\ell}'(\mathrm{N}\,\mathrm{N}')^{-1}$ $(k/w_2)\mathrm{N}\,\mathrm{N}' \cdot (\mathrm{N}\,\mathrm{N}')^{-1}\boldsymbol{\ell} = (k/w_2)\boldsymbol{\ell}'(\mathrm{N}\,\mathrm{N}')^{-1}\boldsymbol{\ell}$.

We now consider (iii). The combined intra and interblock normal equations are

$$\begin{bmatrix} w_2\,G \\ w_1\,\mathbf{Q} + \frac{w_2}{k}\,\mathrm{N}\,\mathrm{B} \end{bmatrix} = \begin{bmatrix} w_2\,bk & w_2\,\mathbf{r}' \\ w_2\,\mathbf{r} & w_1\,C + \frac{w_2}{k}\mathrm{N}\,\mathrm{N}' \end{bmatrix} \begin{bmatrix} \hat{\mu} \\ \mathbf{t} \end{bmatrix}.$$

A solution of the above equations is taken as $\mu = 0$ and

$$w_1\,\mathbf{Q} + \frac{w_2}{k}\mathrm{N}\,\mathbf{B} = \left(w_1\,\mathrm{C} + \frac{w_2}{k}\mathrm{N}\,\mathrm{N}'\right)\hat{\mathbf{t}}.$$

Hence,

$$\hat{\mathbf{t}} = \left(w_1\,\mathrm{C} + \frac{w_2}{k}\mathrm{N}\,\mathrm{N}'\right)^{-1}\left(w_1\,\mathbf{Q} + \frac{w_2}{k}\mathrm{N}\,\mathbf{B}\right).$$

Thus the combined intra and interblock estimate of $\boldsymbol{\ell}'\,\mathbf{t}$ is given by

$$\boldsymbol{\ell}'\left(w_1\,\mathrm{C} + \frac{w_2}{k}\mathrm{N}\,\mathrm{N}'\right)^{-1}\left(w_1\,\mathbf{Q} + \frac{w_2}{k}\mathrm{N}\,\mathbf{B}\right).$$

Also from the above normal equations, we get

$$\mathrm{var}\left(w_1\,\mathbf{Q} + \frac{w_2}{k}\mathrm{N}\,\mathrm{B}\right) = \left(w_1\,\mathrm{C} + \frac{w_2}{k}\mathrm{N}\,\mathrm{N}'\right).$$

Hence,

$$\begin{aligned}
&\mathrm{var}\left[\boldsymbol{\ell}'\left(w_1\,\mathrm{C} + \frac{w_2}{k}\mathrm{N}\,\mathrm{N}'\right)^{-1}\left(w_1\,\mathbf{Q} + \frac{w_2}{k}\mathrm{N}\,\mathrm{B}\right)\right]\\
&= \boldsymbol{\ell}'\left(w_1\,\mathrm{C} + \frac{w_2}{k}\mathrm{N}\,\mathrm{N}'\right)^{-1}\cdot\left(w_1\,\mathrm{C} + \frac{w_2}{k}\mathrm{N}\,\mathrm{N}'\right)\\
&\quad\times\left(w_1\,\mathrm{C} + \frac{w_2}{k}\mathrm{N}\,\mathrm{N}'\right)^{-1}\cdot\boldsymbol{\ell}\\
&= \boldsymbol{\ell}'\left(w_1\,\mathrm{C} + \frac{w_2}{k}\mathrm{N}\,\mathrm{N}'\right)^{-1}\boldsymbol{\ell}.
\end{aligned}$$

If T_1 and T_2 are independent unbiased estimators of $\boldsymbol{\theta}$, with variances σ_1^2 and σ_2^2, then the linear combination of T_1 and T_2 which has the least variance is

$$(\sigma_2^2 T_1 + \sigma_1^2 T_2)/(\sigma_1^2 + \sigma_2^2).$$

Hence using this result, the linear combination of estimates (i) and (ii), having the least variance is

$$\frac{\frac{k}{w_2}\boldsymbol{\ell}'(\mathrm{NN}')^{-1}\boldsymbol{\ell}\cdot\boldsymbol{\ell}'(\mathrm{C} + \frac{\mathbf{rr}'}{n})^{-1}\mathbf{Q} + w_1\boldsymbol{\ell}'(\mathrm{C} + \frac{\mathbf{rr}'}{n})^{-1}\boldsymbol{\ell}\boldsymbol{\ell}'(\mathrm{NN}')^{-1}\mathrm{N}\,\mathrm{B}}{\boldsymbol{\ell}'\left[w_1\left(\mathrm{C} + \frac{\mathbf{rr}'}{n}\right)^{-1} + \frac{k}{w_2}(\mathrm{NN}')^{-1}\right]\boldsymbol{\ell}}$$

Now let

$$\left(\mathrm{C} + \frac{\mathbf{r}\,\mathbf{r}'}{n}\right)^{-1}\mathbf{Q} = (\mathrm{N}\,\mathrm{N}')\mathrm{N}\,\mathbf{B} = \mathbf{p},\ \text{say}.$$

Then

$$\mathbf{Q} = \left(\mathrm{C} + \frac{\mathbf{r}\,\mathbf{r}'}{n}\right)\mathbf{p}\ \text{and}\ \mathrm{N}\,\mathbf{B} = (\mathrm{N}\,\mathrm{N}')\mathbf{p},$$

and

$$E_{1v}\mathbf{Q} = E_{1v}\left(C + \frac{\mathbf{r}\,\mathbf{r}'}{n}\right)\mathbf{p} = (E_{1v}C + \mathbf{r}')\,\mathbf{p}.$$

But, since $E_{1v}\mathbf{Q} = \mathbf{0}$, and $E_{1v}C = 0$, we get $\mathbf{r}'\mathbf{p} = 0$. Hence

$$\mathbf{Q} = C\,\mathbf{p}.$$

Thus

$$\left(w_1\,\mathbf{Q} + \frac{w_2}{k}N\,\mathbf{B}\right) = \left(w_1\,C + \frac{w_2}{k}N\,N'\right)\mathbf{p},$$

and, $\mathbf{p} = \left(w_1\,C + \frac{w_2}{k}N\,N'\right)^{-1}\left(w_1\,\mathbf{Q} + \frac{w_2}{k}N\,\mathbf{B}\right)$.

We now consider the linear combination of estimates (i) and (ii) having the least variance which is

$$\frac{\frac{k}{w_2}\boldsymbol{\ell}'(NN')^{-1}\boldsymbol{\ell}\cdot\boldsymbol{\ell}'(C+\frac{\mathbf{r}\mathbf{r}'}{n})^{-1}\mathbf{Q} + w_1\boldsymbol{\ell}'(C+\frac{\mathbf{r}\mathbf{r}'}{n})^{-1}\boldsymbol{\ell}\boldsymbol{\ell}'(NN')^{-1}N\,B}{\boldsymbol{\ell}'\left[w_1\left(C+\frac{\mathbf{r}\mathbf{r}'}{n}\right)^{-1} + \frac{k}{w_2}(NN')^{-1}\right]\boldsymbol{\ell}}$$

$$= \frac{\left[\frac{k}{w_2}\boldsymbol{\ell}'(N\,N')^{-1}\boldsymbol{\ell} + w_1\boldsymbol{\ell}'\left(C+\frac{\mathbf{r}\mathbf{r}'}{n}\right)^{-1}\boldsymbol{\ell}\right]\boldsymbol{\ell}'\mathbf{p}}{\boldsymbol{\ell}'\left[w_1\left(C+\frac{\mathbf{r}\mathbf{r}'}{n}\right)^{-1} + \frac{k}{w_2}(NN')^{-1}\right]\boldsymbol{\ell}}$$

$$= \frac{\boldsymbol{\ell}'\left[w_1\left(C+\frac{\mathbf{r}\mathbf{r}'}{n}\right)^{-1} + \frac{k}{w_2}(NN')^{-1}\right]\boldsymbol{\ell}\cdot\boldsymbol{\ell}'\mathbf{p}}{\boldsymbol{\ell}'\left[w_1\left(C+\frac{\mathbf{r}\mathbf{r}'}{n}\right)^{-1} + \frac{k}{w_2}(NN')^{-1}\right]\boldsymbol{\ell}}$$

$$= \boldsymbol{\ell}'\mathbf{p} = \boldsymbol{\ell}'\cdot\left(w_1\,C + \frac{w_2}{k}N\,N'\right)^{-1}\left(w_1\,\mathbf{Q} + \frac{w_2}{k}N\,\mathbf{B}\right)$$

$$= \text{estimate (iii)}.$$

56. The combined intra and interblock estimate of $\mathbf{t}$ is given by (see Exercise 55)

$$\hat{\mathbf{t}} = \left(w_1\,C + \frac{w_2}{k}N\,N'\right)^{-1}\left(w_1\,\mathbf{Q} + \frac{w_2}{k}N\,B\right).$$

Also

$$\text{var}\,(\mathbf{t}) = \left(w_1\,C + \frac{w_2}{k}N\,N'\right)^{-1}\cdot\text{var}\left(w_1\,\mathbf{Q} + \frac{w_2}{k}N\,B\right)$$
$$\times\left(w_1\,C + \frac{w_2}{k}N\,N'\right)^{-1}$$

$$= \left(w_1\, C + \frac{w_2}{k} N\, N'\right)^{-1} \left(w_1\, C + \frac{w_2}{k} N\, N'\right)$$
$$\times \left(w_1\, C + \frac{w_2}{k} N\, N'\right)$$
$$= \left(w_1\, C + \frac{w_2}{k} N\, N'\right)^{-1}.$$

The average of the combined intra and interblock estimates of all elementary treatment contrast is given by

$$\bar{V} = \Sigma \sum_{i \neq j} \text{var}\,(\hat{t}_i - \hat{t}_j)/v(v-1)$$
$$= \frac{\Sigma \sum_{i \neq j} [\text{var}(\hat{t}_i) + \text{var}\,(\hat{t}_j) - 2\,\text{cov}\,(\hat{t}_i, \hat{t}_j)]}{v(v-1)}$$
$$= \frac{2(v-1) \sum_{i} \text{var}\,(\hat{t}_i) - \sum_{i \neq j} 2\,\text{cov}\,(\hat{t}_i, \hat{t}_j)}{v(v-1)}$$
$$= \frac{2\left[v \sum_{i=1}^{v} \text{var}\,(\hat{t}_i) - \left\{\sum_{i} \text{var}\,(\hat{t}_i) + \sum_{i \neq j} \text{cov}\,(\hat{t}_i, \hat{t}_j)\right\}\right]}{v(v-1)}$$
$$= \frac{2\left[v\, \text{tr}(w_1 C + \frac{w_2}{k} NN')^{-1} - E_{1v}(w_1 C + \frac{w_2}{k} NN')^{-1} E_{v1}\right]}{v(v-1)}$$

which proves the result.
Now

$$(w_1 C + \frac{w_2}{k} NN')E_{v1} = \frac{w_2}{k} N(E_{1v} N)'$$
$$= \frac{w_2}{k} N(kE_{1b})'$$
$$= w_2 NE_{b1} = w_2\, \mathbf{r}.$$

If the design is equireplicate then, $\mathbf{r} = r\, E_{v1}$. Hence

$$E_{v1} = w_2\, r(w_1 C + \frac{w_2}{k})NN')^{-1} E_{v1},$$

and

$$E_{1v}\, E_{v1} = w_2\, r\, E_{1v}(w_1 C + \frac{w_2}{k} NN')^{-1} E_{v1}.$$

Therefore, under this case,

$$\bar{V} = \frac{2\left[\text{tr}\,(w_1 C + \frac{w_2}{k} NN')^{-1} - \frac{1}{w_2 r}\right]}{(v-1)}.$$

57. Here $N = a\, E_{vb}$. Hence, $E_{1v}\, N = a\, v\, E_{1b}$ and $N\, E_{b1} = a\, b\, E_{v1}$. Therefore $k_j = a\, v$ and $r_i = a\, b$, for $i = 1, 2, \ldots, v, j = 1, 2, \ldots, b$. Then $n = a\, b\, v$, and $NN' = (a\, E_{vb})(a\, E_{bv}) = a^2 b\, E_{vv}$.

Thus $C = R^{-1} - NK^{-1}N' = a\,b\,I_v - \frac{a^2b}{v}E_{vv}$. The normal equations for $\mathbf{t}$ are given by $\mathbf{Q} = C\hat{\mathbf{t}}$. Hence

$$\mathbf{Q} = (a\,b\,I_v - \frac{a\,b}{v}E_{vv})\hat{\mathbf{t}}.$$

Taking $E_{1v}\,\mathbf{t} = 0$, we get a solution for $\mathbf{t}$ as

$$\hat{\mathbf{t}} = \mathbf{Q}/a\,b.$$

Clearly,

$$\mathbf{Q} = \mathbf{T} - N\,K^{-1}\mathbf{B} = \mathbf{T} - \left(\frac{G}{v}\right)E_{v1}.$$

Then sum of squares for testing $\mathbf{t} = \mathbf{0}$ is given by

$$\begin{aligned}\hat{\mathbf{t}}'\mathbf{Q} &= \mathbf{Q}'\mathbf{Q}/a\,b\\ &= [\mathbf{T}' - (G/v)E_{1v}][\mathbf{T} - (G/v)E_{v1}]/a\,b\\ &= [\mathbf{T}'\mathbf{T} - G^2/v]/a\,b.\end{aligned}$$

Clearly rank (C) = v − 1. Hence the d.f. for $\hat{\mathbf{t}}'\mathbf{Q}$ is v − 1. From the structure of the matrix C, it follows that the design b is connected balanced. Since $n_{ij} = a$, $n_{i.} = \sum_{j=1}^{b} n_{ij} = ab$, $n_{.j} = \sum_{i=1}^{v} n_{ij} = a\,v$, $n = a\,b\,v$, it follows that $n_{ij} = n_{i.}n_{.j}/n$. Hence the design is also orthogonal. The analysis of variance table is given below.

Analysis of Variance Table

SOURCE	S.S.	d.f.
Blocks	$\left[(\mathbf{B}'\mathbf{B} - \frac{G^2}{b}\right]/av$	b − 1
Treatments	$\left[T'T - \frac{G^2}{v}\right]/ab$	v − 1
Error	*	abv − b − v + 1
Total	$\mathbf{y}'\mathbf{y} - \frac{G^2}{abv}$	abv − 1

* obtained by subtractions.

Further,

$$\text{var}\,(\hat{t}_i) = \sigma^2/a\,b,\ \text{cov}\,(\hat{t}_i, \hat{t}_j) = 0.$$

Hence $\text{var}\,(\hat{t}_i - \hat{t}_j) = 2\sigma^2/a\,b$. For the design $N = E_{vb}$ the variance of $(\hat{t}_i - \hat{t}_j)$ will be obtained by taking a = 1 in the corresponding result for the design $N = aE_{vb}$. Hence for the design $N = E_{vb}$,

$$\text{var}\,(\hat{t}_j - \hat{t}_j) = 2\sigma^2/b.$$

Thus, the efficiency of the design $N = a\,E_{vb}$ in comparison with the design $N = E_{vb}$ is given by

$$E = \frac{2\sigma^2/b}{(2\sigma^2/ab)} = a.$$

Thus the design $N = aE_{vb}$ is a times efficient than the design $N = E_{vb}$.

58. For any block design, we have

$$E_{1v}\,N = (k_1, k_2, \ldots, k_b)$$

and

$$NE_{b1} = (r_1, r_2, \ldots, r_v).$$

Hence in the present case, we get

$$\begin{aligned} E_{1v}\,NN' &= (E_{1v}\,N)\,N' = k\,E_{1b}\,N' = k\,(N\,E_{1b})' \\ &= k(r\,E_{v1})' = r\,k\,E_{1v}. \end{aligned}$$

Now, clearly each row of NN' will contain r in the diagonal, $\lambda_{01}(q-1)$ times, $\lambda_{10}(p-1)$ times and $\lambda_{11}(p-1)(q-1)$ times. Hence

$$E_{1v}N\,N' = [r + (q-1)\lambda_{01} + (p-1)\lambda_{10} + (p-1)(q-1)\lambda_{11}]\,E_{1v}.$$

Therefore

$$r\,k = r + (q-1)\lambda_{01} + (p-1)\lambda_{10} + (p-1)(q-1)\lambda_{11},$$

which gives

$$(q-1)\lambda_{01} + (p-1)\lambda_{10} + (p-1)(q-1)\lambda_{11} = r(k-1).$$

Now, the normal equations for $\mathbf{t}$ are given by

$$\mathbf{Q} = C\,\hat{\mathbf{t}}, \text{ where } C = R^{-1} - NK^{-1}N'.$$

In the present case,

$$C = r\,I_v - \frac{1}{k}\,N\,N', \mathbf{Q} = \mathbf{T} - \frac{1}{k}N\,\mathbf{B}$$

Let Q_{ij} be adjusted total for the treatment t_{ij}. Then, the solution of t_{ij} is given by

$$\begin{aligned} Q_{ij} &= r\,t_{ij} - \frac{1}{k}\sum_i\sum_j \lambda_{(ij)(i'j')}t_{i'j'} \\ &= (r - \frac{r}{k})t_{ij} - \frac{1}{k}\,[\sum_{j\neq j}\lambda_{(ij)(ij')}t_{ij'} \\ &\quad + \sum_{i'\neq i}\lambda_{(ij)(i'j)}t_{i'j} + \sum_{i\neq i}\sum_{j'\neq j}\lambda_{(ij)(i'j')}t_{i'j'}] \\ &= \frac{r(k-1)}{k}t_{ij} - \frac{\lambda_{01}}{k}\sum_{j'\neq j}t_{ij'} - \frac{\lambda_{10}}{k}\sum_{i'\neq i}t_{i'j} - \frac{\lambda_{11}}{k}\sum_{i'\neq i}\sum_{j'\neq j}t_{i'j'}. \end{aligned}$$

Let $\sum_j t_{ij} = t_{i.}$, $\sum_i t_{ij} = t_{.j}$, and $\sum_i \sum_j t_{ij} = t_{..}$. Then, we get

$$\begin{aligned}
Q_{ij} &= \frac{r(k-1)}{k} t_{ij} - \frac{\lambda_{01}}{k}(t_{i.} - t_{ij}) - \frac{\lambda_{10}}{k}(t_{.j} - t_{ij}) \\
&\quad - \frac{\lambda_{11}}{k}(t_{..} - t_{i.} - t_{.j} + t_{ij}) \\
&= \frac{r(k-1) + \lambda_{01} + \lambda_{10} - \lambda_{11}}{k} t_{ij} - \frac{(\lambda_{01} - \lambda_{11})}{k} t_{i.} \\
&\quad - \frac{(\lambda_{10} - \lambda_{11})}{k} t_{.j} - \frac{\lambda_{11}}{k} t_{..}.
\end{aligned}$$

We take the side condition $t_{..} = 0$. Now let $Q_{i.} = \sum_j Q_{ij}$, $Q_{.j} = \sum_i Q_{ij}$, and $Q_{..} = \sum_i \sum_j Q_{ij}$. Adding the equations over j we get

$$\begin{aligned}
k\, Q_{i.} &= a\, t_{i.} - q(\lambda_{01} - \lambda_{11}) t_{i.} \\
&= [a - q(\lambda_{01} - \lambda_{11})] t_{i.},
\end{aligned}$$

where $a = r(k-1) + \lambda_{01} + \lambda_{10} - \lambda_{11}$. Hence

$$\hat{t}_{i.} = k\, Q_{i.} / [a - q(\lambda_{01} - \lambda_{11})].$$

Similarly adding over i, we get

$$\hat{t}_{.j} = k\, Q_{.j} / [a - p(\lambda_{10} - \lambda_{11})].$$

Thus,

$$k\, Q_{ij} = a\, t_{ij} - \frac{(\lambda_{01} - \lambda_{11}) k\, Q_{i.}}{a - q(\lambda_{01} - \lambda_{11})} - \frac{(\lambda_{10} - \lambda_{11}) k\, Q_{.j}}{a - p(\lambda_{10} - \lambda_{11})}.$$

Hence, a solution for t_{ij} is given by

$$\hat{t}_{ij} = \frac{k}{a}\left[Q_{ij} + \frac{(\lambda_{01} - \lambda_{11}) Q_{i.}}{a - q(\lambda_{01} - \lambda_{11})} + \frac{(\lambda_{10} - \lambda_{11}) Q_{.j}}{a - p(\lambda_{10} - \lambda_{11})}\right],$$
$$i = 1, 2, \ldots, p,\ j = 1, 2, \ldots, q.$$

Also

$$\begin{aligned}
\operatorname{var}(\hat{t}_{ij}) &= \frac{k\sigma^2}{a}\left[1 + \frac{(\lambda_{01} - \lambda_{11})}{a - q(\lambda_{01} - \lambda_{11})} + \frac{(\lambda_{10} - \lambda_{11})}{a - p(\lambda_{10} - \lambda_{11})}\right] \\
\operatorname{cov}(\hat{t}_{ij}, \hat{t}_{ij'}) &= \frac{k\sigma^2}{a} \cdot \frac{(\lambda_{01} - \lambda_{11})}{a - q(\lambda_{01} - \lambda_{11})} \\
\operatorname{cov}(\hat{t}_{ij}, \hat{t}_{i'j}) &= \frac{k\sigma^2}{a} \cdot \frac{(\lambda_{01} - \lambda_{11})}{a - q(\lambda_{01} - \lambda_{11})}
\end{aligned}$$

and

$$\operatorname{cov}(\hat{t}_{ij}, \hat{t}_{i'j'}) = 0.$$

Now consider $\sum_{j=1}^{q} t_{ij} - \sum_{j=1}^{q} t_{i'j}$, $i \neq i'$. Its BLUE is given by

$$\sum_{j=1}^{q} \hat{t}_{ij} - \sum_{j=1}^{q} \hat{t}_{i'j} = \hat{t}_{i.} - \hat{t}_{i'.} = \frac{k[Q_{i.} - Q_{i'.}]}{a - q(\lambda_{01} - \lambda_{11})}.$$

Also,

$$\text{var}\,(\hat{t}_{i.} - \hat{t}_{i'.}) = \text{var}\,(\hat{t}_{i.}) + \text{var}\,(\hat{t}_{i'.}) - 2\,\text{cov}\,(\hat{t}_{i.}, \hat{t}_{i'.}).$$

Now

$$\begin{aligned}
\text{var}\,(\hat{t}_{i.}) &= \text{var}\left(\sum_{j=1}^{q} \hat{t}_{ij}\right) \\
&= \sum_{j=1}^{q} \text{var}\,(\hat{t}_{ij}) + 2 \sum_{j} \sum_{j} \text{cov}\,(\hat{t}_{ij}, \hat{t}_{ij'}) \\
&= \frac{k\,q\,\sigma^2}{a} A + q(q-1) \frac{k\,\sigma^2}{a} B \\
&= \frac{k\,q\,\sigma^2}{a} [A + (q-1)B],\ i = 1, 2, \ldots, p,
\end{aligned}$$

where

$$A = 1 + B + C,\ B = \frac{\lambda_{01} - \lambda_{11}}{a - q(\lambda_{01} - \lambda_{11})},\ C = \frac{\lambda_{10} - \lambda_{11}}{a - p(\lambda_{10} - \lambda_{11})}$$

and

$$\begin{aligned}
\text{cov}\,(\hat{t}_{i.}, \hat{t}_{i'.}) &= \text{cov}\,(\sum_{j} t_{ij}, \sum_{j} t_{i'j}) \\
&= \sum_{j=1}^{q} \text{cov}\,(t_{ij}, t_{i'j}) + \sum_{j \neq j'} \text{cov}\,(t_{ij}, t_{i'j}) \\
&= q\,k\,\sigma^2 C/a + 0 = k\,q\,\sigma^2 C/a.
\end{aligned}$$

Hence

$$\begin{aligned}
\text{var}\,(\hat{t}_{i}. - \hat{t}_{i}.) &= \frac{2\,q\,k\,\sigma^2}{a}[A + (q-1)B] - \frac{2\,q\,k\,\sigma^2 C}{a} \\
&= \frac{2\,q\,k\sigma^2}{a}[A + (q-1)B - C] \\
&= \frac{2\,q\,k\,\sigma^2}{a}(1 + qB).
\end{aligned}$$

Next consider $\sum_{i=1}^{p} t_{ij} - \sum_{i'=1}^{p} t_{i'j'} = t_{.j} - t_{.j'}$. Its BLUE is

$$\hat{t}_{.j} - \hat{t}_{.j'} = \frac{k(Q_{.j} - Q_{.j'})}{a - p(\lambda_{10} - \lambda_{11})}.$$

Further

$$\text{var}(\hat{t}_{.j} - \hat{t}_{.j'}) = \text{var}(\hat{t}_{.j}) + \text{var}(\hat{t}_{.j'}) - 2\text{cov}\,\hat{t}_{.j}, \hat{t}_{.j'}).$$

As before, we can easily prove that

$$\text{var}(\hat{t}_{.j}) = \frac{k\,p\,\sigma^2}{a}[A + (p-1)C], j = 1, 2, \ldots, q$$

and

$$\text{cov}(\hat{t}_{.j}, \hat{t}_{.j'}) = k\,p\,\sigma^2\,B/a.$$

Hence

$$\text{var}(\hat{t}_{.j} - \hat{t}_{.j'}) = \frac{2\,k\,p\,\sigma^2}{a}[A + (p-1)C - B]$$
$$= \frac{(2\,k\,p\,\sigma^2}{a}[1 + pC].$$

Lastly consider $t_{ij} - t_{ij'} - t_{i'j} + t_{i'j'}$. Its BLUE is given by

$$\hat{t}_{ij} - \hat{t}_{ij'} - \hat{t}_{i'j} + \hat{t}_{i'j'} = \frac{k}{a}[Q_{ij} - Q_{ij'} - Q_{i'j} + Q_{i'j'}].$$

And its variance is given by

$$\text{var}(\hat{t}_{ij} - \hat{t}_{ij'} - \hat{t}_{i'j} + \hat{t}_{i'j'})$$
$$= \frac{4\,k\,\sigma^2}{a}A - \frac{2\,k\,\sigma^2\,B}{a} - \frac{2\,k\,\sigma^2}{a}C - \frac{2\,k\,\sigma^2}{a}C - \frac{2\,k\,\sigma^2}{a}B$$
$$= \frac{4\,k\,\sigma^2}{a}(A - B - C)$$
$$= \frac{4\,k\,\sigma^2}{a}.$$

59. Let the yield of the j-th plot receiving the i-th treatment be donoted by $y_{ij}, j = 1, 2, \ldots, n_i,\ i = 1, 2, \ldots, v$. Let the treatment effects be denoted by $t_1, t_2, \ldots, t_v$. Let $E = \frac{1}{v}\sum_1^v t'_i$, and $t_i = t'_i - \bar{t} = t'_i - \mu$, where μ has been written for $\bar{t}$. Now

$$\mathcal{E}(y_{ij}) = t'_i = \mu + t_i,\ j = 1, 2, \ldots, n_i,\ i = 1, 2, \ldots, v.$$

Testing the hypothesis $t'_1 = t'_2 = \ldots = t'_v$ is equivalent to testing the hypothesis $t_i = 0, i = 1, 2, \ldots, v$. We minimize $\sum\sum(y_{ij} - t_i)^2$. The normal equations are

$$G = n\mu + \sum_1^v n_i t_i$$

and

$$T_i = n_i\mu + n_i t_i, i = 1, 2, \ldots, v.$$

Note that the number of independent equations is v. Hence putting one side condition $\sum n_i t_i = 0$, we get a set of solutions as $\hat{\mu} = G/n$ and $\hat{t}_i = \frac{T_i}{n_i} - \frac{G}{n}$. Hence sum of squares due to regression when μ and $t_1, t_2, \ldots, t_v$ are fitted is

$$\hat{\mu}G + \sum \hat{t}_i T_i = \sum T_i^2/n_i \text{ with v d.f.}$$

The error sum of squares is given by

$$\text{SSE} = \sum_i \sum_j y_{ij}^2 - \sum_i T_i^2/n_i \text{ with } (n - v)\text{d.f.}$$

Now if the hypothesis $t_i = 0, i = 1, 2, \ldots, v$ is true, then $E(y_{ij}) = \mu$, $j = 1, 2, \ldots, n_i$, $i = 1, 2, \ldots, v$. We minimize $\sum\sum (y_{ij} - \mu)^2$. The normal equation is $G = n\mu$. Hence the solution is $\overset{*}{\mu} = G/n$. Then the sum of squares when μ is fitted is G^2/n with 1 d.f. Hence, the sum of squares for testing $t_i = 0$ is obtained as

$$\sum T_i^2/n_i - G^2/n,$$

with $(v - 1)$ d.f. Thus, the F statistic for testing $t_i = 0,\ i = 1, 2, \ldots, v$ is given by

$$F = \frac{([\sum T_i^2/n_i - G^2/n]/(v - 1)}{[\sum_i \sum_j y_{ij}^2 - \sum T_i^2/n_i]/(n - v)},$$

with $(v - 1)$ and $(n - v)$ d.f. The BLUE of an elementary treatment contrast $t_i - t_j$ is given by

$$\hat{t}_i - \hat{t}_j = (T_i/n_i) - (T_j/n_j)$$

with variance given by

$$\text{var}\,(\hat{t}_i - \hat{t}_j) = \sigma^2 \left(\frac{1}{n_i} - \frac{1}{n_j} \right).$$

The average variance of BLUEs of all elementary treatment contrasts is

$$\begin{aligned}
\bar{V} &= \sum_{i \neq j}\sum \text{var}\,(\hat{t}_i - \hat{t}_j)/v(v - 1) \\
&= \sigma^2 \sum_{i \neq j}\sum \left(\frac{1}{n_i} + \frac{1}{n_j} \right) /v(v - 1) \\
&= \sigma^2 \left[\sum_{i \neq j}\sum \frac{1}{n_i} + \sum_{i \neq j}\sum \frac{1}{n_j} \right] /v(v - 1) \\
&= 2\sigma^2 (v - 1) \sum \frac{1}{n_i} /v(v - 1) \\
&= 2\sigma^2 \sum \frac{1}{n_i} /v.
\end{aligned}$$

We minimize $\bar{V}$ with respect to n_i subject to $\sum n_i = n$. Thus we minimize

$$\phi = \frac{1}{v}\sum\frac{1}{n_i} - \lambda(n - \sum n_i)$$

$$\frac{\partial\phi}{\partial n_i} = -\frac{1}{v\,n_i^2} + \lambda = 0$$

i.e. $n_i^2 = 1/\lambda v,\ n_i = 1/\sqrt{\lambda v}$ = constant. Now $\sum n_i = n$ gives $\frac{v}{\sqrt{\lambda v}} = n$, hence $\frac{1}{\sqrt{\lambda v}} = \frac{n}{v}$. Thus $n_i = \frac{n}{v}$, $i = 1, 2, \ldots, v$. Hence the test is most sensitive when n_i are all equal.

60. Here, the incidence matrix of the design is given by

$$N = \begin{bmatrix} E_{1r_1} & 0 \\ E_{1r_1} & 2E_{1r_2} \\ E_{(v-2)r_1} & E_{(v-2)r_2} \end{bmatrix}, \text{ where } r_2 = r - r_1.$$

Hence

$$NN' = \begin{bmatrix} r_1 & r_1 & r_1E_{1(v-2)} \\ r_1 & (r_1 + 4r_2) & (r + 2r)E_{1(v-2)} \\ r_1E_{(v-2)1} & (r_1 + 2r_2)E_{(v-2)1} & rE_{(v-2)}(v-2) \end{bmatrix}.$$

Therefore the C matrix of the design is give by

$$C = \text{diag}\,[r_1, (r_1 + 2r_2), r, r, \ldots, r] - \frac{1}{v}NN'$$

$$= \begin{bmatrix} r_1 - \frac{r_1}{v} & -\frac{r_1}{v} & -\frac{r_1}{v}E_{1(v-2)} \\ -\frac{r_1}{v} & r_1 + 2r_2 - \frac{r_1 + 4r_2}{v} & -\frac{r_1 + 2r_2}{v}E_{1(v-2)} \\ -\frac{r_1}{v}E_{(v-2)1} & -\frac{r_1 + 2r_2}{v}E_{(v-2)1} & rI_{v-2} - \frac{r}{v}E_{(v-2)(v-2)} \end{bmatrix}$$

We now write the reduced normal equations for $\mathbf{t}$ as

$$Q_1 = (r_1 - \frac{r_1}{v})t_1 - \frac{r_1}{v}t_2 - \frac{r_1}{v}E_{1(v-2)}\mathbf{t}_3.$$

$$Q_2 = -\frac{r_1}{v}t_1 + (r_1 + 2r_2 - \frac{r_1 + 4r_2}{v})t_2 - \frac{r_1 + 2r_2}{v}E_{1(v-2)}\mathbf{t}_3$$

$$\mathbf{Q}_3 = -\frac{r_1}{v}E_{(v-2)1}t_1 - \frac{r_1 + 2r_2}{v}E_{(v-2)1}t_2 + [rI_{v-2} - \frac{2}{v}E_{(v-2)(v-2)}]\mathbf{t}_3.$$

Next, we put the side restriction $\sum r_i t_i = 0$. i.e.

$$r_1t_1 + (r_1 + 2r_2)t_2 + rE_{1(v-2)}\mathbf{t}_3 = 0.$$

Then, we obtain

$$vrQ_1 = r_1(vr - r_2)t_1 + r_1r_2t_2$$

$$vrQ_2 = r_1r_2t_1 + [vr(r + r_2) - r_1r_2]t_2$$

$$\mathbf{Q}_3 = r\mathbf{t}_3,$$

where $\mathbf{Q}_3 = \{Q_3, Q_4, \ldots, Q_v\}'$ and $\mathbf{t}_3 = \{t_3, t_4, \ldots, t_v\}'$. Hence, we get the solutions as

$$\hat{t}_1 = [vr(r + r_2)Q_1 - r_1r_2(Q_1 + Q_2)]/\Delta$$
$$\hat{t}_2 = [-r_1r_2(Q_1 + Q_2) + vrr_1Q_2]/\Delta$$
$$\hat{\mathbf{t}}_3 = \mathbf{Q}_3/r,$$

where $\Delta = rr_1(vr + vr_2 - 2r_2)$
Hence sum of squares due to treatments is

$$\sum_{1}^{v} \hat{t}_i Q_i = \hat{t}_1 Q_1 + \hat{t}_2 Q_2 + \sum_{i=3}^{v} Q_i^2/r \quad \text{with}(v-1)\text{ d.f.}$$

The Block SS and Error SS are obtained as usual and the analysis of variance will be completed as usual. Further, we obtain

$$\begin{aligned}
\text{var}(\hat{t}_1) &= \sigma^2 \cdot \text{Coefficient of } Q_1 \text{ in the equation of } \hat{t}_1 \\
&= \sigma^2[vr(r + r_2) - r_1r_2]/\Delta, \\
\text{var}(\hat{t}_2) &= \sigma^2 r_1(vr - r_2)/\Delta, \\
\text{var}(\hat{t}_i) &= \sigma^2/r, i = 3, 4, \ldots, v \\
\text{cov}(\hat{t}_1, \hat{t}_2) &= -\sigma^2 r_1r_2/\Delta, \\
\text{cov}(\hat{t}_i, \hat{t}_j) &= 0, i \neq j = 1, 2, \ldots, v \text{ except } i = 1, \\
&\qquad\qquad j = 2, \text{ and } i = 2, j = 1.
\end{aligned}$$

Hence

$$\begin{aligned}
\text{var}(\hat{t}_1 - \hat{t}_2) &= \sigma^2 2vr^2/\Delta, \\
\text{var}(\hat{t}_1 - \hat{t}_j) &= \sigma^2\left[\frac{1}{r} + \frac{vr(r + r_2) - r_1r_2}{\Delta}\right], \\
\text{var}(\hat{t}_2 - \hat{t}_j) &= \sigma^2\left[\frac{1}{r} + \frac{r_1(vr - r_2)}{\Delta}\right], \quad j = 3, 4, \ldots, v.
\end{aligned}$$

Let $\bar{V}$ be equal to the average variance of the BLUEs of all elementary treatment contrasts. Then

$$\begin{aligned}
\bar{V} &= \frac{1}{v(v-1)}\left[\frac{2 \cdot 2vr^2\sigma^2}{\Delta} + 2(v-2)\sigma^2\left\{\frac{1}{r} + \frac{vr(r+r_2) - r_1r_2}{\Delta}\right\}\right. \\
&\quad \left. + 2(v-2)\sigma^2\left\{\frac{1}{r} + \frac{r_1(vr - r_2)}{\Delta}\right\} + (v-2)(v-3)\frac{2\sigma^2}{r}\right] \\
&= \frac{2\sigma^2}{rv(v-1)}\left[(v-1)(v-2) + \frac{2r[vr^2(v-1) - r_1r_2]}{\Delta}\right] \\
&= \frac{2\sigma^2}{r}\left[1 + \frac{rr_2^2}{\Delta} + \frac{rr_1r_2(2v-3)}{\Delta v(v-1)}\right].
\end{aligned}$$

Hence, the efficiency relative to the original randomized block design is

$$E = \frac{1}{1 + \frac{r\,r_2^2}{\Delta} + \frac{r\,r_1\,r_2(2v-3)}{\Delta v(v-1)}}$$

The loss in efficiency is given by

$$1 - E = L/(1+L),$$

where $L = \frac{r\,r_2^2}{\Delta} + \frac{r\,r_1 r_2(2v-3)}{\Delta v(v-1)}$.

61. Here the incidence matrix of the modified design is given by

$$N = \begin{bmatrix} 2I_2 & E_{2(r-2)} \\ E_{(v-2)2} & E_{(v-2)(r-2)} \end{bmatrix}.$$

Hence,

$$NN' = \begin{bmatrix} 4I_2 + (r-2)E_{22} & rE_{2(v-2)} \\ rE_{(v-2)2} & rE_{(v-2)(v-2)} \end{bmatrix}.$$

Thus,

$$C = rI_v - \frac{1}{v}N\,N'$$

$$= \begin{bmatrix} \left(r - \frac{4}{v}\right)I_2 - \frac{(r-2)}{v}E_{22} & -\frac{r}{v}E_{2(v-2)} \\ -\frac{r}{v}E_{(v-2)2} & rI_{v-2} - \frac{r}{v}E_{(v-2)(v-2)} \end{bmatrix}.$$

Then the reduced normal equations $\mathbf{Q} = C\hat{\mathbf{t}}$ give

$$\mathbf{Q}_1 = \left[\left(r - \frac{4}{v}\right)I_2 - \frac{r-2}{v}E_{22}\right]\hat{\mathbf{t}}_1 - \frac{r}{v}E_{2(v-2)}\hat{\mathbf{t}}_2$$

$$\mathbf{Q}_2 = -\frac{r}{v}E_{(v-2)2}\hat{\mathbf{t}}_1 + \left[rI_{v-2} - \frac{r}{v}E_{(v-2)(v-2)}\right]\hat{\mathbf{t}}_2,$$

where $\mathbf{Q}_1' = \{Q_1, Q_2\}$, $\mathbf{Q}_2 = \{Q_3, Q_4, \ldots, Q_v\}$, $\hat{\mathbf{t}}_1 = \{t_1, t_2\}$ and $\mathbf{t}_2' = \{t_3, t_4, \ldots, t_v\}$. We put one side restriction $E_{1v}\hat{\mathbf{t}} = 0$, i.e. $E_{12}\mathbf{t}_1 + E_{1(v-2)}\hat{\mathbf{t}}_2 = 0$. Then, we obtain the following normal equations

$$\mathbf{Q}_1 = \left[\left(r - \frac{4}{v}\right)I_2 + \frac{2}{v}E_{22}\right]\hat{\mathbf{t}}_1$$

$$\mathbf{Q}_2 = r\hat{\mathbf{t}}_2.$$

Hence, we obtain

$$\hat{t}_1 = [(vr-2)Q_1 - 2Q_2]/r(vr-4)$$

$$\hat{t}_2 = [-2Q_1 + (vr-2)Q_2]/r(vr-4)$$

$$\hat{\mathbf{t}}_2 = \mathbf{Q}_2/r.$$

Thus the sum of squares due to treatments is obtained as

$$SST = \hat{t}_1 Q_1 + \hat{t}_2 Q_2 + \sum_{i=3}^{v} Q_i^2/r$$

with (v − 1) d.f. The block SS and the errorr SS are obtained as usual and the analysis of variance table is completed. Now

$$\begin{aligned}
\text{var}(\hat{t}_1) &= \sigma^2 \cdot \text{coefficient of } Q_1 \text{ in the equation of } \hat{t}_1 \\
&= \sigma^2(vr-2)/r(vr-4) \\
\text{var}(\hat{t}_2) &= \sigma^2(vr-2)/r(vr-4). \\
\text{var}(\hat{t}_i) &= \sigma^2/r, i = 3, 4, \ldots, v. \\
\text{cov}(\hat{t}_1, \hat{t}_2) &= \sigma^2 \cdot \text{coefficient of } Q_2 \text{ in the equation of } \hat{t}_1 \\
&= -2\sigma^2/r(vr-4). \\
\text{cov}(\hat{t}_i, \hat{t}_j) &= 0, i \neq j = 1, 2, \ldots, v \text{ except } i = 1, j = 2 \text{ and} \\
&\qquad\qquad\qquad\qquad\qquad\qquad i = 2, j = 1.
\end{aligned}$$

Then, we get

$$\begin{aligned}
\text{var}(\hat{t}_1 - \hat{t}_2) &= \frac{2\sigma^2}{r}\left[1 + \frac{4}{vr-4}\right] \\
\text{var}(\hat{t}_1 - \hat{t}_j) &= \frac{2\sigma^2}{r}\left[1 + \frac{1}{vr-4}\right], j = 3, 4, \ldots, v \\
\text{var}(\hat{t}_2 - \hat{t}_j) &= \frac{2\sigma^2}{r}\left[1 + \frac{1}{vr-4}\right], j = 3, 4, \ldots, b \\
\text{var}(\hat{t}_i - \hat{t}_j) &= \frac{2\sigma^2}{r}, i \neq j = 3, 4, \ldots, v.
\end{aligned}$$

Hence the average variance of the BLUEs of all elementary treatment contrasts is given by

$$\begin{aligned}
\bar{V} &= \frac{2\sigma^2}{rv(v-1)}\left[2 + \frac{8}{vr-4} + 4(v-2)\right. \\
&\qquad \left. \times \left\{1 + \frac{1}{vr-4}\right\} + (v-2)(v-3)\right] \\
&= \frac{2\sigma^2}{r}\left[1 + \frac{4}{(v-1)(vr-4)}\right].
\end{aligned}$$

Hence the efficiency of the modified design is given by

$$E = \frac{2\sigma^2/r}{\dfrac{2\sigma^2}{r}[1+\Delta]} = \frac{1}{1+\Delta},$$

where $\Delta = 4/(v-1)(vr-4)$. The loss of efficiency is given by

$$1 - E = \Delta/(1+\Delta) = 4/[4 + (v-1)(vr-4)]$$

62. Here the incidence matrix of the design is

$$N = \begin{bmatrix} 2E_{12} & E_{1(r-2)} \\ \mathbf{0} & E_{1(r-2)} \\ E_{(v-2)2} & E_{(v-2)(r-2)} \end{bmatrix}.$$

Hence

$$NN' = \begin{bmatrix} r+2 & r-2 & (r+2)E_{1(v-2)} \\ r-2 & r-2 & (r-2)E_{1(v-2)} \\ (r+2)E_{(v-2)1} & (r-2)E_{(v-2)1} & rE_{(v-2)(v-2)} \end{bmatrix}.$$

Therefore

$$C = \begin{bmatrix} \frac{(r+2)(k-1)}{k} & -\frac{(r-2)}{k} & -\frac{(r+2)}{k}E_{1(v-2)} \\ -\frac{(r-2)}{k} & \frac{(r-2)(k-1)}{k} & -\frac{(r-2)}{k}E_{1(v-2)} \\ -\frac{(r+2)}{k}E_{(v-2)1} & -\frac{(r-2)}{k}E_{(v-2)1} & r(I_{v-2}-\frac{1}{k}E_{(v-2)(v-2)}) \end{bmatrix}.$$

Hence, the normal equations for $\mathbf{t}$ are

$$\begin{aligned} kQ_1 &= (r+2)(k-1)\hat{t}_1 - (r-2)\hat{t}_2 - (r+2)E_{1(v-2)}\hat{\mathbf{t}}_3 \\ kQ_2 &= -(r-2)\hat{t}_1 + (r-2)(k-1)\hat{t}_2 - (r-2)E_{1(v-2)}\hat{\mathbf{t}}_3 \\ k\mathbf{Q}_3 &= -(r+2)E_{(v-2)1}\hat{t}_1 - (r-2)E_{(v-2)1}\hat{t}_2 \\ &\quad + r((I_{v-2}) - \frac{1}{k}E_{(v-2)(v-2)})\hat{\mathbf{t}}_3 \end{aligned}$$

where $\mathbf{t}' = (t_3, t_4, \ldots, t_v)$ and $\mathbf{Q}_3 = (Q_3, Q_4, \ldots, Q_v)$.
We now set $\mathbf{r}'\mathbf{t} = 0$, i.e. $(r+2)t_1 + (r-2)t_2 + rE_{1(v-2)}\mathbf{t}_3 = 0$. Hence

$$E_{1(v-2)}\mathbf{t}_3 = -\frac{1}{r}[(r+2)t_1 - (r-2)t_2].$$

Substituting this in the above equations, we get

$$\begin{bmatrix} rkQ_1 \\ rkQ_2 \end{bmatrix} = \begin{bmatrix} (r+2)(rk+2) & 2(r-2) \\ 2(r-2) & (r-2)(rk-2) \end{bmatrix} \begin{bmatrix} \hat{t}_1 \\ \hat{t}_2 \end{bmatrix}$$

$$\mathbf{Q}_3 = r\mathbf{t}_3.$$

Hence we get

$$\begin{aligned} \hat{t}_1 &= k(r-2)[(rk-2)Q_1 - 2Q_2]/\Delta \\ \hat{t}_2 &= k[(r+2)(rk+2)Q_2 - 2(r-2)Q_1]/\Delta \\ \hat{\mathbf{t}}_3 &= \mathbf{Q}_3/r \end{aligned}$$

where $\Delta = (r-2)(r^2k^2 + 2rk^2 - 8)$. Let $B_1 + B_2 = G_1$, $B_3 + B_4 + \cdots + B_r = G_2$, and $G = G_1 + G_2$. Then from

$$\mathbf{Q} = \mathbf{T} - \frac{1}{k}\mathbf{NB},$$

we get

$$Q_1 + T_1 - \frac{1}{k}(G_1 + G)$$
$$Q_2 = T_2 - \frac{1}{k}G_2$$
$$Q_i = T_i - \frac{1}{k}G, i = 3, 4, \ldots, v.$$

Hence the sum of squares for testing $\mathbf{t} = \mathbf{0}$ is

$$\hat{t}_1 Q_1 + \hat{t}_2 Q_2 + \sum_{i=3}^{v} Q_i$$
$$= \frac{k}{\Delta}\left[(r-2)(rk-2)Q_1^2 + (r+2)(rk+2)Q_2^2 - 4(r-2)Q_1Q_2\right]$$
$$+ \sum_{3}^{v} Q_i^2/r.$$

Further

$$\begin{aligned}
\text{var}(\hat{t}_1) &= \sigma^2.\text{coefficient of } Q_1 \text{ in the equation of } t_1\\
&= \sigma^2 k(r-2)(rk-2)/\Delta,\\
\text{var}(\hat{t}_2) &= \sigma^2 k(r+2)(rk+2)/\Delta,\\
\text{var}(\hat{t}_i) &= \sigma^2/r, \ i = 3, 4, \ldots, v\\
\text{cov}(\hat{t}_1, \hat{t}_2) &= -2\sigma^2 k(r-2)/\Delta\\
\text{cov}(\hat{t}_1, \hat{t}_i) &= 0, i = 3, 4 \ldots, v\\
\text{cov}(\hat{t}_2, \hat{t}_i) &= 0, i = 3, 4, \ldots, v.
\end{aligned}$$

Hence,

$$\begin{aligned}
\text{var}(\hat{t}_1 - \hat{t}_2) &= 2k\sigma^2 r(rk+2)/\Delta\\
\text{var}(\hat{t}_1 - \hat{t}_i) &= \sigma^2\left[\frac{k(r-2)(rk-2)}{\Delta} + \frac{1}{r}\right], \ i = 3, 4, \ldots, v\\
\text{var}(\hat{t}_2 - \hat{t}_i) &= \sigma^2\left[\frac{k(r+2)(rk+2)}{\Delta} + \frac{1}{r}\right], \ i = 3, 4, \ldots, v,\\
\text{var}(\hat{t}_i - \hat{t}_j) &= 2\sigma^2/r, \ i \neq j = 3, 4, \ldots, v.
\end{aligned}$$

63. Let the new varieties be denoted by $t_1, t_2, \ldots, t_v$ and the standard variety by t_0. Then, the incidence matrix of the design is seen to be

$$N = \begin{bmatrix} gE_{vv} \\ E_{1v} \end{bmatrix}.$$

Also

$$R = \begin{bmatrix} g_v I_v & \mathbf{0} \\ \mathbf{0} & v\, I_1 \end{bmatrix}, K = (g_v + 1)I_v.$$

Hence, the C matrix of the design is obtained as

$$C = R - NK^{-1}N'$$
$$= \begin{bmatrix} g_v I_v - \dfrac{g_v^2}{gv+1} E_{vv} & -\dfrac{gv}{gv+1} E_{v1} \\ -\dfrac{gv}{gv+1} E_{1v} & vI_1 - \dfrac{v}{gv+1} E_{11} \end{bmatrix}.$$

For solving the normal equations $\mathbf{Q} = C\ \mathbf{t}$, we impose the restrictions $\sum_i r_i t_i = 0$, which gives

$$g\, E_{1v}\mathbf{t} = -t_0,$$

where $\mathbf{t}' = (t_1, t_2, \ldots, t_v)$. Further, let $\mathbf{Q}' = (Q_1, Q_2, \ldots, Q_v)$ and Q_0 be the adjusted treatment total for t_0. Then we have

$$\mathbf{Q} = \mathbf{T} - \frac{g}{gv+1} G, \mathbf{Q} = \mathbf{T} - \frac{1}{gv+1} G,$$

where $\mathbf{T}' = (T_1, T_2, \ldots, T_v) =$ vector of totals of the new varieties and $T_0 =$ total of the standard variety t_0 , and $G =$ grand total. Thus, we have the following equations

$$\begin{bmatrix} \mathbf{Q} \\ Q_0 \end{bmatrix} = \begin{bmatrix} g_v I_v - \dfrac{g_v^2}{gv+1} E_{vv} & -\dfrac{gv}{gv+1} E_{v1} \\ -\dfrac{gv}{gv+1} E_{1v} & v\, I_1 - \dfrac{v}{gv+1} E_{11} \end{bmatrix} \begin{bmatrix} \mathbf{t} \\ t_0 \end{bmatrix}$$

The above equations give

$$\mathbf{Q} = (g_v I_v - \frac{g_v^2}{gv+1} E_{vv})\mathbf{t} - \frac{gv}{gv+1} t_0\, E_{v1},$$

$$Q_0 = -\frac{gv}{gv+1} E_{1v}\mathbf{t} + \frac{g_v^2}{gv+1} t_0.$$

Now taking into account the fact that $g\, E_{1v}\mathbf{t} = -t_0$, we see that

$$\mathbf{Q} = g\, v\, \mathbf{t} \quad \text{and} \quad Q_0 = v\, t_0$$

Hence

$$\mathbf{t} = \mathbf{Q}/gv \quad \text{and } t_0 = Q_0/v.$$

Thus, the SS for testing the significance of varieties is

$$(\mathbf{Q}'\mathbf{Q}/gv) + (Q_0^2/v)$$

with d.f. v.

The block SS, total SS and error SS are obtained as usual and the analysis of variance table is completed. Further,

$$\begin{aligned} \text{var}\,(\hat{t}_i) &= \sigma^2/gv, i = 1, 2, \ldots, v \\ \text{var}\,(\hat{t}_0) &= \sigma^2/v, \\ \text{cov}\,(\hat{t}_i, \hat{t}_j) &= 0 \quad i \neq j = 1, 2, \ldots, v \\ \text{cov}\,(\hat{t}_i, \hat{t}_0) &= 0 \quad i = 1, 2, \ldots, v. \end{aligned}$$

Hence,

$$\begin{aligned} \text{var}(\hat{t}_i - \hat{t}_j) &= 2\sigma^2/gv, \; i \neq j = 1, 2, \ldots, v \\ \text{var}(\hat{t}_i - \hat{t}_0) &= \sigma^2(g+1)/gv, \; i = 1, 2, \ldots, v. \end{aligned}$$

64. For a SBIBD, N, N′, NN′, and N′N are all non-singular. Further,

$$NN' = (r-\lambda)I_v + \lambda E_{vv}.$$

Hence

$$\begin{aligned} (N\,N')^{-1} &= \frac{1}{(r-\lambda)}[I_v - \frac{\lambda}{r-\lambda+\lambda v}E_{vv}] \\ &= \frac{1}{(r-\lambda)}[I_v - \frac{\lambda}{r^2}E_{vv}]. \end{aligned}$$

Hence,

$$\begin{aligned} (N')^{-1} \cdot N^{-1} &= \frac{1}{(r-\lambda)}[I_v - \frac{\lambda}{r^2}E_{vv}] \\ (N')^{-1} &= \frac{1}{(r-\lambda)}[N - \frac{\lambda}{r^2}E_{vv}N] \\ &= \frac{1}{r-\lambda}\left[N - \frac{\lambda}{r}E_{vv}\right]. \end{aligned}$$

Premultiply by N′, we get

$$\begin{aligned} I_v &= \frac{1}{(r-\lambda)}[N'N - \frac{\lambda}{r}N'E_{vv}] \\ &= \frac{1}{(r-\lambda)}[N'N - \lambda E_{vv}], \end{aligned}$$

which gives

$$N'N = (r-\lambda)I_v + \lambda E_{vv} = NN'.$$

65. (i) For a BIBD, we have

$$NN' = (r-\lambda)I_v + \lambda E_{vv}.$$

The characteristic roots of N N′ are given by the roots of the equation

$$|NN' - \theta I_v| = 0,$$

i.e. by the roots of

$$|(r - \lambda - \theta)I_v + \lambda E_{vv}| = 0.$$

Now,

$$|(r-\lambda-\theta)I_v+\lambda E_{vv}| = (r-\lambda-\theta)^{v-1}\cdot(r-\lambda-\theta+\lambda v)$$
$$= (r-\lambda-\theta)^{v-1}(rk-\theta).$$

Hence, we have

$$(r-\lambda-\theta)^{v-1}(rk-\theta)=0.$$

Thus, we obtain the characteristic roots of NN' as

$$\theta = rk, \text{ repeated once}$$
$$\theta = r-\lambda, \text{ repeated } (v-1) \text{ times.}$$

The non-zero characteristic roots of $N'N$ and NN' are same. Hence the characteristic roots of $N'N$ are $r\,k$, and $r-\lambda$.
(ii) We know that

$$|NN'| = rk(r-\lambda)^{v-1} \neq 0.$$

Hence,

$$\text{rank }(NN') = \text{rank }(N'N) = v \leq b.$$

Since, the design is non-symmetrical, $v < b$. Hence, $N'N$ is singular. Therefore $|N'N| = 0$.
(iii) We have

$$\text{tr}(N'N) = \text{tr}(NN')$$
$$= \text{tr}[(r-\lambda)I_v+\lambda E_{vv}]$$
$$= vr.$$

66. We have $bk = vr$. But $b = 4(r-\lambda)$, hence we get

$$4k(r-\lambda) = vr,$$

i.e. $$4k\left[(r-\frac{(r(k-1)}{v-1}\right] = vr,$$

i.e. $$(v-2k)^2 = v$$

i.e. $$v-2k = \pm\sqrt{v}, \text{ which gives } 2k = v \pm \sqrt{v}.$$

67. For an affine resolvable BIBD, $v = nk$, $b = nr$, we know that $b = v+r-1$ and k^2/v is a positive integer. Hence we get $\dfrac{k^2}{v} = \dfrac{k}{n} = g$, where g is a positive integer. Now

$$\lambda = \frac{(k-1)}{v-1} = \frac{r(k-1)}{b-r} = \frac{k-1}{n-1} = \frac{ng-1}{n-1} = g + \frac{g-1}{n-1}.$$

Hence, $(g-1)/(n-1) = t$ is 0 or a positive integer. Thus, $g = (n-1)t+1$, and $\lambda = (n-1)t+1+t = nt+1$. Further, $k = ng = n(nt-t+1)$. Hence

$v = nk = n^2(nt - t + 1)$, also

$$r = \frac{\lambda(v-1)}{k-1} = \frac{(nt+1)(n-1)(n^2t+n+1)}{(n-1)(nt+1)} = n^2t + n + 1.$$

Then, $b = nr = n(n^2t + n + 1)$.

68. Let N_0 be the incidence matrix of the new design. Then

$$N_0 = \begin{bmatrix} N \\ E_{1b} \end{bmatrix},$$

where N is the incidence matrix of the original BIBD. Then

$$N_0N_0' = \begin{bmatrix} NN' & NE_{b1} \\ E_{1b}N' & bE_{11} \end{bmatrix} = \begin{bmatrix} (r-\lambda)I_v + \lambda E_{vv} & rE_{v1} \\ rE_{1v} & bE_{11} \end{bmatrix}$$

Also,

$$K_0 = (k+1)I_b,\ R_0 = \begin{bmatrix} rI_v & 0 \\ 0 & bI_1 \end{bmatrix}.$$

Therefore, C matrix of the new design denoted by C_0 is seen to be

$$C_0 = R_0 - N_0K_0^{-1}N_{0'}$$
$$= \begin{bmatrix} rI_v - \frac{1}{k+1}\{(r-\lambda)I_v + \lambda E_{vv}\} & -\frac{r}{k+1}E_{v1} \\ -\frac{r}{k+1}E_{1v} & \left(b - \frac{b}{k+1}\right) \end{bmatrix}$$
$$= \begin{bmatrix} \frac{rk+\lambda}{k+1}I_v - \frac{\lambda}{k+1}E_{vv} & -\frac{r}{k+1}E_{v1} \\ -\frac{r}{k+1}E_{1v} & \frac{bk}{k+1} \end{bmatrix}.$$

Let the treatments of the original BIBD be donoted by $t_1, t_2, \ldots, t_v$ and $\mathbf{t}' = [t_1, t_2, \ldots, t_v]$. Let t_0 be the control treatments. Let $Q_i = T_i - B_{(i)}/(k+1)$, $i = 1, 2, \ldots, v$, $Q_0 = T_0 - G/(k+1)$ and $\mathbf{Q}' = [Q_1, Q_2, \ldots, Q_v]$. Then, the reduced normal equations for treatments are given by

$$\begin{bmatrix} \mathbf{Q} \\ Q_0 \end{bmatrix} = \begin{bmatrix} \frac{rk+\lambda}{k+1}I_v - \frac{\lambda}{k+1}E_{vv} & -\frac{r}{k+1}E_{v1} \\ -\frac{r}{r+1}E_{1v} & \frac{bk}{k+1} \end{bmatrix} \begin{bmatrix} \mathbf{t} \\ t_0 \end{bmatrix}.$$

Thus, we get

$$\mathbf{Q} = \left[\frac{rk+\lambda}{k+1}I_v - \frac{\lambda}{k+1}E_{v1}\right]\mathbf{t} - \frac{r}{k+1}E_{v1}t_0$$
$$Q_0 = -\frac{r}{k+1}E_{1v}\mathbf{t} + \frac{b}{k+1}t_0.$$

As usual, we take $\Sigma\, r_i t_i = 0$, i.e. $r(t_1 + t_2 + \cdots + t_v) + bt_0 = 0$. Hence, $rE_{1v}\mathbf{t} = -bt_0$, and $Q_0 = \left(\frac{b}{k+1} + \frac{b}{k+1}\right) t_0 = bt_0$. Thus, $\hat{t}_0 = Q_0/b$, and

$$\begin{aligned}
\mathbf{Q} &= \left[\frac{rk+\lambda}{k+1} I_v - \frac{\lambda}{k+1} E_{vv}\right]\mathbf{t} + \frac{r}{k+1} E_{v1} \cdot \frac{r}{b} \cdot E_{1v}\mathbf{t}. \\
&= \left[\frac{rk+\lambda}{k+1} I_v - \frac{\lambda}{k+1} E_{vv} + \frac{r^2}{b(k+1)} E_{vv}\right]\mathbf{t} \\
&= \left[pI_v + q \cdot E_{vv}\right]\mathbf{t},
\end{aligned}$$

where $p = (rk+\lambda)/(k+1)$, and $q = (r^2 - b\lambda)/b(k+1)$. One can easily show that

$$q = \frac{r^2 - b\lambda}{b(k+1)} = \frac{r^2 - \frac{vr}{k}\lambda}{b(k+1)} = \frac{r(rk - v\lambda)}{bk(k+1)} = \frac{r - \lambda}{v(k+1)}.$$

Now

$$[pI_v + qE_{vv}]^{-1} = \frac{1}{p} I_v - \frac{q}{p(p+vq)} E_{vv},$$

and

$$p + vq = \frac{rk+\lambda}{k+1} + \frac{r-\lambda}{k+1} = r.$$

Hence

$$\begin{aligned}
[pI_v + qE_{vv}]^{-1} &= \frac{(k+1)}{(rk+\lambda)} I_v - \frac{(r-\lambda)(k+1)}{v(k+1) \cdot r(rk+\lambda)} E_{vv} \\
&= \frac{k+1}{rk+\lambda} I_v - \frac{(r-\lambda)}{vr(rk+\lambda)} E_{vv},
\end{aligned}$$

$$\hat{\mathbf{t}} = \left[\frac{k+1}{rk+\lambda} I_V - \frac{(r-\lambda)}{vr(rk+\lambda)} E_{vv}\right]\mathbf{Q}.$$

Therefore,

$$\hat{t}_i = \frac{k+1}{rk+\lambda} Q_i - \frac{(r-\lambda)}{vr(rk+\lambda)} \sum_1^v Q_i,$$

and the adjusted treatment SS is

$$\hat{\mathbf{t}}'\mathbf{Q} + \hat{t}_0 Q_0 = \frac{k+1}{rk+\lambda} \sum_1^v Q_i^2 - \frac{(r-\lambda)}{vr(rk+\lambda)} \left(\sum_1^v Q_i\right)^2 + \frac{Q_0^2}{b}.$$

Now $\sum_1^v Q_i + Q_0 = 0$, hence $\sum_1^v Q_i = -Q_0$, and

$$\begin{aligned}
\hat{\mathbf{t}}'\mathbf{Q} + \hat{t}_0 Q_0 &= \frac{\mu k + 1}{rk+\lambda} \sum_1^v Q_i^2 + \left[\frac{1}{b} - \frac{r-\lambda}{bk(rk+\lambda)}\right] Q_0^2 \\
&= \frac{k+1}{rk+\lambda} \sum_1^v Q_i^2 + \frac{\lambda}{r} Q_0^2
\end{aligned}$$

with v d.f. The other sum of squares are obtained as usual. Further,

$$\hat{t}_i - \hat{t}_j = \frac{(k+1)}{(rk+\lambda)}(Q_i - Q_j),\ i \neq j = 1, 2, \ldots, v$$

$$\begin{aligned}\hat{t}_i - \hat{t}_0 &= \frac{k+1}{rk+\lambda}Q_i - \frac{(r-\lambda)}{vr(rk+\lambda)}\sum_1^v Q_i - \frac{Q_0}{b} \\ &= \frac{k+1}{rk+\lambda}Q_i + \left[\frac{r-\lambda}{bk(\gamma k+\lambda)} - \frac{1}{b}\right]Q_0 \\ &= \frac{k+1}{rk+\lambda}Q_i - \frac{\lambda}{r}Q_0,\ i = 1, 2, \ldots, v.\end{aligned}$$

Also

$$\begin{aligned}\text{var}\,(\hat{t}_i) &= \sigma^2\left[\frac{k+1}{rk+\lambda} - \frac{r-\lambda}{vr(rk+\lambda)}\right],\ i = 1, 2, \ldots, v \\ \text{cov}\,(\hat{t}_i, \hat{t}_j) &= \sigma^2\left[-\frac{(r-\lambda)}{vr(rk+\lambda)}\right],\ i \neq j = 1, 2, \ldots, v \\ \text{var}\,(\hat{t}_0) &= \sigma^2/b \\ \text{cov}\,(\hat{t}_i, \hat{t}_0) &= 0,\ i = 1, 2, \ldots, v.\end{aligned}$$

Hence,

$$\text{var}\,(\hat{t}_i - \hat{t}_j) = 2\sigma^2\left[\frac{k+1}{rk+\lambda} - \frac{r-\lambda}{vr(rk+\lambda)}\right] + \frac{2\sigma^2(r-\lambda)}{vr(rk+\lambda)}$$

$$2\sigma^2(k+1)/(rk+\lambda), \quad i \neq j = 1, 2, \ldots, v.$$

and therefore

$$\begin{aligned}\text{var}\,(\hat{t}_i - \hat{t}_0) &= \sigma^2\left[\frac{k+1}{rk+\lambda} - \frac{r-\lambda}{vr(rk+\lambda)}\frac{1}{b}\right] \\ &= \sigma^2\left[\frac{k+1}{rk+\lambda} + \frac{\lambda}{r}\right],\ i = 1, 2, \ldots, v.\end{aligned}$$

69. One can easily prove that $E_{1b}N'N = rkE_{1b}$. Also

$$\begin{aligned}N'NN'N &= N'[(r-\lambda)I_v + \lambda E_{vv}]N \\ &= (r-\lambda)N'N + \lambda(E_{1v}N)'(E_{1v}N) \\ &= (r-\lambda)N'N + \lambda k^2 E_{bb}.\end{aligned}$$

Let the blocks be denoted by $B_1, B_2, \ldots, B_b$ and let ℓ_j denote the number of common treatments between B_1 and B_j, $j = 1, 2, 3, \ldots, b$. Then, clearly $\ell_1 = k$. Then the elements of 1st row of N'N are $\ell_1, \ell_2, \ldots, \ell_b$. Then from the above two relation, we find that

$$\sum_1^b \ell_j = rk, \quad \sum_1^b \ell_j^2 = k(r-\lambda) + \lambda k^2.$$

Hence $\sum_{b}^{2} \ell_j = k(r-1)$, and $\sum_{b}^{2} \ell_j^2 = k(r-\lambda-k+\lambda k)$. Define $\bar{\ell} = \sum_{b}^{2} \ell_j/(b-1)$. Then,

$$\sum_{2}^{b}(\ell_j - \bar{\ell})^2 = k(r-\lambda-k+\lambda k) - \frac{(k^2(r-1)^2}{b-1}$$

$$= \frac{k}{(b-1)}[(b-1)(r-\lambda-k+\lambda k) - k(r-1)^2]$$

$$= \frac{k}{(b-1)}\left[\frac{(b-1)}{(v-1)}\{(v-1)(r-k)+r(k-1)^2\} - k(r-1)^2\right]$$

$$= \frac{k^2}{(b-1)(v-1)}[(b-1)(b-v+1+rk-2r)$$
$$-(r-1)^2(v-1)]$$

$$= \frac{k^2}{(b-1)(v-1)}[(b-1)(b-v) - 2r(b-v) + r(\gamma - k)]$$

$$= \frac{k^2}{(b-1)(v-1)}\left[(b-1)(b-v) - 2r(b-v) + \frac{r^2}{b}(b-v)\right]$$

$$= \frac{k^2(b-v)(b-r)^2}{b(b-1)(v-1)}.$$

Thus the necessary and sufficient condition that there will be the same number of treatments common between any two blocks of a BIBD is that $b = v$ or $b = r$. But $b = r$ gives a RBD, hence is inadmissible. Hence, $b = v$.

70. Let N be the incidence matrix of the BIBD and let N be partitioned as $N = [N_{11}\ N_{12}]$, so that N_{11} is the $v \times 2$ incidence matrix of the first two blocks between which it is assumed that there are $\times$ treatments in common. Now consider N_0 as

$$N_0 = \begin{bmatrix} N_{11} & N_{12} \\ I_2 & \mathbf{0} \end{bmatrix}.$$

Then

$$N_0 N_0' = \begin{bmatrix} NN' & N_{11} \\ N_{11'} & I_2 \end{bmatrix},$$

and

$$|N_0 N_0'| = |N N'| \cdot |I_2 - N_{11}'(NN')^{-1}N_{11}|.$$

Now for a BIBD

$$N N' = (r-\lambda)I_v + \lambda E_{vv}$$

and

$$(N N')^{-1} = \frac{1}{(r-\lambda)}I_v - \frac{\lambda}{(r-\lambda)\, rk}E_{vv}$$
$$|N N'| = rk(r-\lambda)^{v-1}.$$

Thus,

$$N_{11}'(NN')^{-1}N_{11} = \frac{1}{(r-\lambda)}N_{11}'N_{11} - \frac{\lambda}{(r-\lambda)rk}N_{11}'E_{vv}N_{11}.$$

Now $N_{11}'N_{11} = [\begin{smallmatrix} k & x \\ x & k \end{smallmatrix}]$, and $E_{1v}N_{11} = kE_{12}$. Hence,

$$\left| I_2 - \frac{1}{(r-\lambda)}N_{11}'\,N_{11} - \frac{\lambda}{(r-\lambda)rk}N_{11}'\,E_{vv}N_{11} \right|$$
$$= \left| (I_2 - \frac{1}{(r-\lambda)}N_{11}'\,N_{11} - \frac{k}{r(r-\lambda)}E_{22} \right|$$
$$= \frac{[(r-k)^2(r-\lambda)^2 - (rx-\lambda k)^2]}{r^2(r-\lambda)^2},$$

and

$$|N_0\,N_0'| = \frac{k(r-\lambda)^{v-3}}{r} \cdot [(r-k)^2(r-\lambda)^2 - (rx-\lambda k)^2].$$

Now N_0 is $(v+2) \times b$ matrix of real elements and $b > v+2$. Hence, it is well known that $|N_0\,N_{0'}| \geq 0$. Hence, we get

$$(r-k)^2(r-\lambda)^2 - (rx-\lambda k)^2 \geq 0$$
$$\text{i.e. } (rx-\lambda k)^2 \leq (r-k)^2(r-\lambda)^2$$
$$\text{i.e. } |rx-\lambda k| \leq (r-k)(r-\lambda)$$
$$\text{i.e. } -(r-\lambda-k) \leq x \leq \frac{2\lambda k + r(r-\lambda-k)}{r}.$$

71. Let the blocks be denoted by $B_1, B_2, \ldots, B_b$ and let ℓ_j denote the number of common treatments between B_1 and $B_j, j = 1, 2, \ldots, b$. Then clearly $\ell_1 = k$. Suppose $\ell_2 = x$. Then, from Exercise 69, we have

$$\sum_3^b \ell_j = k(r-1) - x,$$

and

$$\sum_3^b \ell_j^2 = k(r-\lambda-k+\lambda k) - x^2.$$

Now $\sum_3^b \ell_j^2 - \frac{(\sum \ell_j)^2}{b-2} \geq 0$, from which we get

$$k(r-\lambda-k+\lambda k) - x^2 - \frac{[k(r-1)-x]^2}{(b-2)} \geq 0$$

i.e.

$$k(r-\lambda-k+\lambda k) - \frac{k^2(r-1)^2}{(b-1)} \geq \frac{b-1}{b-2}\left[x - \frac{k(r-1)}{b-1}\right]^2.$$

In Exercise 69, we have shown that

$$k(r-\lambda-k+\lambda k)-\frac{k^2(r-1)^2}{b-v}=\frac{k^2(b-r)^2(b-v)}{b(b-1)(v-1)}.$$

Hence, we get

$$\left[x-\frac{k(x-1)}{b-1}\right]^2 \le \frac{(b-2)k^2(b-r)^2(b-v)}{b(b-1)^2(v-1)} \le \frac{k^2(b-r)^2T^2}{(b-1)^2},$$

since $T=[(b-2)(b-v)/b(v-1)]$. Thus, we get

$$\left|x-\frac{k(r-1)}{b-1}\right| \le \frac{k(b-r)T}{(b-1)}$$

from which we get the required result.

72. Clearly $M = 2N - E_{vb}$. Hence,

$$\begin{aligned} NM' &= (2N-E_{vb})(2N'-E_{bv}) \\ &= 4NN' - 2E_{vb}N' - 2nE_{bv} + bE_{vv} \\ &= 4(r-\lambda)I_v + 4\lambda E_{vv} - 2E_{v1}(NE_{b1})' - 2(NE_{b1})E_{1v} + bE_{vv} \\ &= 4(r-\lambda)I_v + 4\lambda E_{vv} - 2E_{v1}(rE_{v1})' - 2(rE_{v1})E_{1v} + bE_{vv} \\ &= 4(r-\lambda) + [b-4(r-\lambda)]E_{vv}. \end{aligned}$$

73. Clearly the elements of N are 0 and 1, and N is $v_1v_2 \times b_1b_2$ matrix. Further

$$\begin{aligned} E_{(1)(v_1v_2)}N &= (E_{1v_1} \times (E_{1v_2})(N_1 \times N_2 + \overset{*}{N}_1 \times \overset{*}{N}_2) \\ &= E_{1v_1}N_1 \times E_{1v_2}N_2 + E_{1v1}\overset{*}{N}_1 \times E_{1v_2}\overset{*}{N}_2 \end{aligned}$$

Now $E_{1v_1}N_1 = k_1E_{1b_1}, E_{1v_1}\overset{*}{N}_1 = (v_1-k_1)E_{1b_1}$, and similarly $E_{1v_2}N_2 = k_2E_{1b_2}$, and

$$E_{1v_2}\overset{*}{N}_2 = (v_2-k_2)E_{1b_2}.$$

Hence

$$\begin{aligned} E_1(v_1v_2)N &= k_1E_{1b_1} \times k_2E_{1b_2} + (v_1-k_1)E_{1b_1} \times (v_2-k_2)E_{1b_2} \\ &= k_1k_2E_{1(b_1b_2)} + (v_1-k_1)(v_2-k_2)E_{1(b_1b_2)} \\ &= [k_1k_2 + (v_1-k_1)(v_2-k_2)]E_{1(b_1b_2)}. \end{aligned}$$

Similarly, it can be proved that

$$N\,E_{(b_1b_2)1} = [r_1r_2 + (b_1-r_1)(b_2-r_2)]E_{(v_1v_2)1}.$$

Writing $b = b_1b_2$, $v = v_1v_2$, $k = k_1k_2 + (v_1 - k_1)(v_2 - k_2)$, and $r = r_1r_2 + (b_1 - r_1)(b_2 - r_2)$, we get

$$E_{1v}N = k\,E_{1b},\ N\,E_{b1} = rE_{v1}.$$

Further

$$\begin{aligned} N\,N' &= (N_1 \times N_2 + \overset{*}{N}_1 \times \overset{*}{N}_2)(N_1 \times \overset{*}{N}_1 \times \overset{*}{N}_2)' \\ &= (N_1 \times N_2 + \overset{*}{N}_1 \times \overset{*}{N}_2)(N_1' \times N_2' + \overset{*}{N}{}_1' \times \overset{*}{N}{}_2') \\ &= N_1N_1' \times N_2N_2' + \overset{*}{N}_1\,N_1' \times \overset{*}{N}_2\,N_2' + N_1\,\overset{*}{N}{}_1' \times N_2\,\overset{*}{N}{}_2' \\ &\quad + \overset{*}{N}_1\overset{*}{N}{}_1' \times \overset{*}{N}_2\overset{*}{N}{}_2'. \end{aligned}$$

Now

$$\begin{aligned} &N_1N_1' \times N_2N_2' \\ &\quad = [(r_1 - \lambda_1)I_{v_1} + \lambda_1E_{v_1v_1}] \times [(r_2 - \lambda_2)I_{v_2} + \lambda_2E_{v_2v_2}] \\ &\quad = (r_1 - \lambda_1)(r_2 - \lambda_2)I_v + \lambda_1(r_2 - \lambda_2)E_{v_1v_1} \times I_{v_2} \\ &\qquad + \lambda_2(r_1 - \lambda_1)I_{v_1} \times E_{v_2v_2} + \lambda_1\lambda_2E_{vv} \end{aligned}$$

$$\begin{aligned} \overset{*}{N}_1\,N_1' \times \overset{*}{N}_2\,N_2' &= (E_{v_1b_1} - N_1) \times N_1' \times (E_{v_2b_2} - N_2)N_2' \\ &= [r_1E_{v_1v_1} - (r_1 - \lambda_1)I_{v_1} - \lambda_1E_{v_1v_1}] \\ &\qquad \times [r_2E_{v_2v_2} - (r_2 - \lambda_2)I_{v_2} - \lambda_2E_{v_2v_2}] \\ &= (r_1 - \lambda_1)(r_2 - \lambda_2)[(E_{v_1v_1} - I_{v_1}) \times (E_{v_2v_2} - I_{v_2})] \\ &= (r_1 - \lambda_1)(r_2 - \lambda_2)[E_{vv} - I_{v_1} \times E_{v_2v_2} - E_{v_1v_1} \times I_{v_2} + I_v] \end{aligned}$$

$$\begin{aligned} N_1\,\overset{*}{N}{}_1' \times N_2\,\overset{*}{N}{}_2' &= N_1(E_{v_1b_1} - N_1)' \times N_2(E_{v_2b_2} - N_2)' \\ &= N_1(E_{b_1v_1} - N_1') \times N_2(E_{b_2v_2} - N_2') \\ &= [r_1E_{v_1v_1} - (r_1 - \lambda_1)I_{v_1} - \lambda_1E_{v_1v_1}] \\ &\qquad \times [r_2E_{v_2v_2} - (r_2 - \lambda_2)I_{v_2} - \lambda_2E_{v_2v_2}] \\ &= (r_1 - \lambda_1)(r_2 - \lambda_2)[(E_{v_2v_2} - I_{v_2}) \times (E_{v_2v_2} - I_{v_2})] \\ &= (r_1 - \lambda_1)(r_2 - \lambda_2)(E_{vv} - I_{v_1} \times E_{v_2v_2} - E_{v_1v_1} \times I_{v_2} + I_v) \end{aligned}$$

$$\begin{aligned} &\overset{*}{N}_1\overset{*}{N}{}_1' \times \overset{*}{N}_2\overset{*}{N}{}_2' \\ &\quad = (E_{v_1b_1} - N_1)(E_{b_1v_1} - N_1') \times (E_{v_2b_2} - N_2)(E_{b_2v_2} - N_2') \\ &\quad = [b_1E_{v_1v_1} - r_1E_{v_1v_1} - r_1E_{v_1v_1} + (r_1 - \lambda_1)I_{v_1} + \lambda_1E_{v_1v_1}] \\ &\qquad \times [(b_2 - 2r_2 + \lambda_2)E_{v_2v_2} + (\gamma_2 - \lambda_2)I_{v_2}] \\ &\quad = (b_1 - 2r_1 + \lambda_1)(b_2 - 2r_2 + \lambda_2)E_{vv} \\ &\qquad + (r_1 - \lambda_1)(b_2 - 2r_2 + \lambda_2)I_{v_1} \times E_{v_2v_2} \\ &\qquad + (r_2 - \lambda_2)(b_1 - 2r_1 + \lambda_1)E_{v_1v_1} \times I_{v_2} \\ &\qquad + (r_1 - \lambda_1(r_2 - \lambda_2)I_v. \end{aligned}$$

Hence

$$\begin{aligned}
NN' &= 4(r_1-\lambda_1)(r_2-\lambda_2)I_v \\
&= \frac{b}{4}I_v + [\lambda_1\lambda_2 + 2(r_1-\lambda_1)(r_2-\lambda_2) \\
&\quad + (b_1 - 2r_1 + \lambda_1)(b_2 - 2r_2 + \lambda_2)]E_{vv} \\
&\quad + (r_1-\lambda_1)(b_2 - 4r_2 + 4\lambda_2)I_{v_1} \times E_{v_1v_1} \\
&\quad + (r_2-\lambda_2)(b_1 - 4r_1 + 4\lambda_1)E_{v_1v_1} \times I_{v_2} \\
&= \frac{b}{4}I_v + [\lambda_1\lambda_2 + 3(r_1-\lambda_1)(r_2-\lambda_2) \\
&\quad + (b_1-r_1)(b_2-r_2) - (r_1-\lambda_1)(b_2-r_2) \\
&\quad - (r_2-\lambda_2)(b_1-r_1)]E_{vv}.
\end{aligned}$$

The term in the bracket of the second term is

$$\begin{aligned}
& r - r_1r_2 + \lambda_1\lambda_2 + \frac{3b_1b_2}{16} - \frac{b_1}{4}(b_2-r_2) - \frac{b_2}{4}(b_1-r_1) \\
&= r - r_1(r_2-\lambda_2) - \lambda_2(r_1-\lambda_1) + \frac{3b}{16} - \frac{b}{4} - \frac{b}{4} + \frac{b_1r_2}{4} + \frac{b_2r_1}{4} \\
&= r - \frac{r_1b_2}{4} - \frac{\lambda_2b_1}{4} + \frac{3b}{16} - \frac{b}{24} + \frac{b_1r_2}{4} + \frac{b_2r_1}{4} \\
&= r + \frac{b_1(r_2-\lambda_2)}{4} + \frac{3b}{16} - \frac{b}{2} \\
&= r + \frac{b}{16} + \frac{3b}{16} - \frac{b}{2} \\
&= r + \frac{b}{4} - \frac{b}{2} \\
&= (r - \frac{b}{4}).
\end{aligned}$$

Hence

$$\begin{aligned}
NN' &= \frac{b}{4}I_v + (r - \frac{b}{4})E_{vv} \\
&= (r-\lambda)I_v + \lambda E_{vv},
\end{aligned}$$

where $\lambda = r - \frac{b}{4}$. Thus N is the incidence matrix of a BIBD, with parameters

$$\begin{aligned}
& v = v_1v_2,\ b = b_1b_2,\ r = r_1r_2 + (b_1-r_1)(b_2-r_2) \\
& k = k_1k_2 + (v_1-k_1)(v_2-k_2),\ \lambda = r - \frac{b}{4}.
\end{aligned}$$

74. Consider the BIBD $(\overset{*}{v}, \overset{*}{b}, \overset{*}{r}, \overset{*}{k}, \overset{*}{\lambda})$ and let $\overset{*}{N}$ be the incidence matrix. We interchange the role of treatments and blocks and get a new design. The incidence matrix of this new design is $N = \overset{*}{N}'$. This operation is know as dualization. Let v, b, respectively denote the number of treatments and the

number of blocks in the new design. Then

$$\begin{aligned} v &= \text{no. of treatments in the new design} \\ &= \text{no. of blocks in the BIBD } = \overset{*}{b} \\ b &= \text{no. of blocks in the new design} \\ &= \text{no. of treatments in the BIBD } = \overset{*}{v} \end{aligned}$$

Now every treatment in the BIBD occurs exactly in $\overset{*}{r}$ blocks, hence every block in the new design contains exactly $\overset{*}{r}$ plots. Hence,

$$k = \text{ size of block in the new design } = \overset{*}{r}.$$

Now every block in the BIBD contains $\overset{*}{k}$ treatments, hence every treatment in the new design will occur exactly in $\overset{*}{k}$ blocks. Thus

$$r = \overset{*}{k}$$

Further every pair of treatments in the BIBD occurs exactly $\overset{*}{\lambda}$ blocks. Hence, every pair of blocks in the new design has $\overset{*}{\lambda}$ treatments in common. Thus $\mu = \overset{*}{\lambda}$ and the new design is a linked block design with parameters

$$v = \overset{*}{b},\ b = \overset{*}{v},\ r = \overset{*}{k},\ k = \overset{*}{r},\ \mu = \overset{*}{\lambda}.$$

$$N = \overset{*}{N}'.$$

Now the C-matrix of the new design is defined as

$$\begin{aligned} C &= \text{diag}\,(r_1, r_2, \ldots, r_v) - N\,\text{diag}\left(\frac{1}{k_1}, \ldots, \frac{1}{k_b}\right) N' \\ &= rI_v - \frac{1}{k} NN' \\ &= rI_v - \frac{1}{k}\, \overset{*}{N}{}'\overset{*}{N} \end{aligned}$$

The D matrix of the new design is given by

$$\begin{aligned} D &= \text{diag}\,(k_1, k_2, \ldots, k_b) - N'\,\text{diag}\left(\frac{1}{r_1}, \cdots, \frac{1}{r_v}\right) N \\ &= kI_b - \frac{1}{r} N'N \\ &= kI_b - \frac{1}{r}\, \overset{*}{N}\overset{*}{N}{}' \\ &= kI_b - \frac{1}{r}[(\overset{*}{r} - \overset{*}{\lambda})\, \overset{*}{I}_v - \overset{*}{\lambda}\overset{**}{E}_{vv}] \\ &= kI_b - \frac{1}{r}[(k - \mu)I_b + \mu E_{bb}] \\ &= \frac{k(r-1) + \mu}{r} I_b - \frac{\mu}{r} E_{bb}. \end{aligned}$$

The characteristic roots of a matrix a $I_n + dE_{nn}$ are given by a + nd and a with multiplicities 1 and (n – 1), respectively. Hence, the characteristic roots of D are

$$\frac{k(r-1)+\mu}{r} - b\frac{\mu}{r} = \frac{k(r-1)-\mu(b-1)}{r}$$

$$= \frac{k(r-1)-\overset{*}{\lambda}(\overset{*}{v}-1)}{r} = \frac{\overset{*}{r}(\overset{*}{k}-1)-\overset{*}{\lambda}(\overset{*}{v}-1)}{r}$$

$$= 0, \text{ repeated once,}$$

and $\frac{k(r-1)+\mu}{r}$, with multiplicity (b – 1). Hence, clearly the rank of D is (b – 1).

75. In Exercise 51, we have derived the expected values of the sum of squares which occur in the intrablock analysis of variance in the case of a one-way design. The expected values of sums of squares of the intrablock analysis of variance table of a BIBD are obtained as a particular case of the general results derived in Exercise 51, by taking

$$K = kI_b,\ R = rI_v,$$
$$\mathbf{k}' = k\,E_{1b}$$
$$\mathbf{r}' = rE_{1v},\ n = b\,k.$$

Hence, we derive from Exercise 51,

(i) $$\mathcal{E}\left[\mathbf{y}'\mathbf{y} - \frac{G^2}{bk}\right] = (b\,k-1)\sigma^2 + k\boldsymbol{\alpha}'\boldsymbol{\alpha} + r\,\mathbf{t}'\mathbf{t} + 2\mathbf{t}'N\boldsymbol{\alpha} - (k\,\Sigma\,\alpha + r\Sigma\,t)^2/b\,k$$

(ii) $$\mathcal{E}\left[\frac{1}{k}\mathbf{B}'\mathbf{B} - \frac{G^2}{bk}\right] = (b-1)\sigma^2 + k\boldsymbol{\alpha}'\boldsymbol{\alpha} + 2\mathbf{t}'N\boldsymbol{\alpha} + \frac{1}{k}\mathbf{t}'NN'\mathbf{t} - (k\,\Sigma\,\alpha + r\Sigma\,t)^2/b\,k$$

(iii) $$\mathcal{E}\left[\frac{1}{r}\mathbf{T}'\mathbf{T} - \frac{G^2}{bk}\right] = (v-1)\sigma^2 + r\mathbf{t}'\mathbf{t} + 2\mathbf{t}'N\boldsymbol{\alpha} + \frac{1}{r}\boldsymbol{\alpha}'N'N\boldsymbol{\alpha} - (k\,\Sigma\,\alpha + r\,\Sigma\,t)^2/b\,k$$

(iv) $\mathcal{E}(\hat{\mathbf{t}}'\mathbf{Q}) = (v-1)\sigma^2 + \mathbf{t}'C\mathbf{t}$

(v) $\mathcal{E}$ (adj. block SS) $= (b-1)\sigma^2 + \boldsymbol{\alpha}'D\boldsymbol{\alpha}$

(vi) $\mathcal{E}$ (Intrablock Error SS) $= (b\,k - b - v + 1)\sigma^2$.

76. We denote the parameters of the new design by putting asterisk over the symbols. Clearly when the role of blocks and treatments are interchanged, we get the new design with parameters $\overset{*}{v}, \overset{*}{b}, \overset{*}{r}, \overset{*}{k}$, where

$$\overset{*}{v} = b = v(v-1)/2,\ \overset{*}{b} = v,\ \overset{*}{r} = k = 2,\ \overset{*}{k} = r = v-1.$$

In the old design, given one block, we divide the other blocks in two groups, the first group having blocks which have one treatment in common with the given block and the second group having blocks which have no treatment in common with the given block. For example consider one block as (α, θ), where α and θ are two treatments. Thus

	First Group	Second Group
$(\alpha, \theta),$	☐	☐

The number of blocks in the first group is equal to the sum of

(i)   number of blocks in which one treatment is α and the other is different from $\theta = v - 2$

(ii)   number of blocks in which one treatment is θ and the other is different from $\alpha = v - 2$

Thus, the number of blocks in the first group is

$$n_1 = 2(v - 2).$$

The number n_2 of blocks in the second group is equal to the number of blocks obtained by taking a pair of treatments from the set of treatments except α and θ, which is clearly $(v - 2)(v - 3)/2$. Thus

$$n_2 = (v - 2)(v - 3)/2.$$

So, when the role of blocks and treatments is interchanged, in the new design, given a treatment, the other treatments can be divided into two groups, the first group containing treatment, which occur once with the given treatment in a block, while the second group containing treatments, which do not occur with the given treatment in a block, and

$$n_1 = 2(v - 2), \quad n_2 = (v - 2)(v - 3)/2$$
$$\lambda_1 = 1, \quad \lambda_2 = 0.$$

Clearly, $n_1 + n_2 = (v - 2)(v + 1)/2 = \overset{*}{v} - 1$, and

$$n_1\lambda_1 + n_2\lambda_2 = 2(v - 2) = \overset{*}{r}\,(\overset{*}{k} - 1).$$

Now, let us find the parameters of

$$P_1 = \begin{bmatrix} p_{11}^{\;1} & p_{12}^{\;1} \\ p_{21}^{\;1} & p_{22}^{\;1} \end{bmatrix}, \quad P_2 = \begin{bmatrix} p_{11}^{\;2} & p_{12}^{\;2} \\ p_{21}^{\;2} & p_{22}^{\;2} \end{bmatrix}.$$

Consider a pair of blocks, which have one treatment in common, say (α, θ) and (α, δ), and their first associate groups, the number of blocks common between the first associate groups of (α, θ) and (α, δ) are the blocks which have one treatment α and the other not equal to θ or δ and the block (α, θ).

Clearly this number is $(v-3)+1=v-2$. Thus,

$$p^1_{11}=(v-2).$$

Now consider the first associate group of (α, θ) and the second associate group of (α, δ). The blocks common between these two groups are the blocks which have one treatment θ and the other treatment different from α and δ. Clearly this number is $v-3$. Thus,

$$p^1_{12}=v-3.$$

Consider the second associate group of (α, θ) and the first associate group of (α, δ). The blocks which are common between these two groups are the blocks which have one treatment δ and the other different from α and θ. Clearly this number is $v-3$. Thus,

$$p^1_{21}=v-3.$$

Lastly, consider the second associate groups of (α, θ) and (α, δ). The blocks common between these two groups are the blocks which contain a pair of treatments taken from $(v-3)$ treatments from the set except α, θ and δ. Clearly this number is $(v-3)(v-4)/2$. Hence,

$$p^1_{22}=v-4.$$

We can verify that

$$p^1_{11}+p^1_{12}=2v-5=n_1-1,$$
$$p^1_{21}+p^1_{22}=(v-3)(v-2)/2=n_2.$$

Let us now find P_2. Consider a pair of blocks (α, θ) and (γ, δ) and their first associates. Clearly the blocks common between these two groups are (α, γ), (α, δ), (θ, γ), (θ, δ). Hence

$$p^2_{11}=4.$$

Next consider the first associate group of (α, θ) and the second associate group of (α, δ). The blocks common between these two groups are

(i) the blocks which contain one treatment α and the other treatment different from θ, γ, δ. Hence their number is $(v-4)$; and

(ii) the blocks which contain one treatment θ and the other treatment different from α, γ, δ. Hence their number is $(v-4)$.

Hence,

$$p^2_{12}=2(v-4).$$

Similarly, one can easily verify that $p^2_{21}=2(v-4)$.

Now consider the second associate groups of (α, θ) and (γ, δ). The blocks common between these two groups are those which contain a pair of treatments from the set of $(v-4)$ treatments, obtained by deleting $\alpha, \theta, \gamma, \delta$

from the complete set. Hence their number is $(v-4)(v-5)/2$. Thus

$$p_{22}^2 = (v-4)(v-5)/2.$$

We can verify that

$$p_{11}^2 + p_{12}^2 = 2(v-2) = n_1$$
$$p_{21}^2 + p_{22}^2 = (v-1)(v-4)/2 = n_2 - 1.$$

Thus, interchanging the role of blocks and treatments, we get a 2-associate PBIBD with parameters

$$\overset{*}{v} = v(v-1)/2,\ \overset{*}{b} = v,\ \overset{*}{r} = 2,\ \overset{*}{k} = v-1$$
$$n_1 = 2(v-2),\ n_2 = (v-2)(v-3)/2,$$
$$\lambda_1 = 1, \lambda_2 = 0$$
$$P_1 = \begin{bmatrix} v-2 & v-3 \\ v-3 & \dfrac{(v-3)(v-4)}{2} \end{bmatrix},$$
$$P_2 = \begin{bmatrix} 4 & 2(v-4) \\ 2(v-4) & \dfrac{(v-4)(v-5)}{2} \end{bmatrix}.$$

77. Here clearly $v = mk, b = mr$. We define a pair of treatments to be first associates if they belong to the same set and second associates if they belong to different sets. Then, clearly

$$n_1 = k-1, n_2 = k(m-1).$$

Further two treatments belonging to the set occur together in a block r times. Hence $\lambda_1 = r$. While two treatments belonging to two different sets do not occur together in a block. Hence $\lambda_2 = 0$. We can easily verify that

$$n_1 + n_2 = mk - 1 = v - 1$$
$$n_1\lambda_1 + n_2\lambda_2 = r(k-1).$$

Consider now a pair of first associates α and β, and their first asssociates. The treatments common to their first associates are clearly the $(k-2)$ other treatments belonging to their set. Hence

$$p_{11}^1 = k-2.$$

The treatments common between the first associates of α and the second associates of β. Clearly no treatments are common. Hence,

$$p_{12}^1 = 0.$$

Similarly $p_{21}^1 = 0$. Consider the second associate classes of α and β. Clearly the treatments common between these two classes are the treatments of the remaining $(m-1)$ sets. Hence,

$$p_{22}^1 = k(m-1).$$

We easily verify that

$$p_{11}^1 + p_{12}^1 = k - 2 = n_1 - 1$$
$$p_{21}^1 + p_{22}^1 = k(m - 1) = n_2.$$

Now consider a pair of second associates, α and δ, say. They belong to different sets, and clearly there are no treatments common between their first associates, and hence

$$p_{11}^2 = 0.$$

The treatments common between the first associate class of α and the second associate class of δ are the $(k - 1)$ treatments of the first associate class of α, and hence

$$p_{12}^2 = k - 1.$$

Similarly, $p_{21}^2 = k - 1$. Now consider the second associate classes of α and δ. The treatments common to the second associate classes of α and δ are the treatments of the other $(m - 2)$ sets, and hence

$$p_{22}^2 = (m - 2)k.$$

We may easily verify that

$$p_{11}^2 + p_{12}^2 = k - 1 = n_1,$$

and

$$p_{21}^2 + p_{22}^2 = mk - k - 1 = n_2 - 1.$$

Thus the design is a PBIBD with two associate classes with the parameters

$$v = mk, b = mr, r, k.$$
$$n_1 = k - 1, n_2 = k(m - 1),$$
$$\lambda_1 = r, \lambda_2 = 0$$
$$P_1 = \begin{bmatrix} k-2 & 0 \\ 0 & k(m-1) \end{bmatrix},$$
$$P_2 = \begin{bmatrix} 0 & k-1 \\ k-1 & k(m-2) \end{bmatrix}.$$

Consider a pair of treatments α and β belonging to two different sets, then it is not possible to find a chain of treatments $\alpha, \theta_1, \theta_2, \ldots, \theta_i = \beta$, such that consecutive treatments will occur together in a block. Hence, the design is not connected. Now $v = m\,k$, $b = m\,r$. So if $r \leq k$, then $b \geq v$. Hence by taking r greater than, equal to or less than k, we can make b greater than, equal to or less than v.

78. (i) Note that B_i matrices are symmetrical. Consider the sum of elements in the αth row of B_i, which is

$$b_{\alpha 0}^i + b_{\alpha 1}^i + \ldots + b_{\alpha v}^i, \quad \alpha = 1, 2, \ldots, v.$$

Since, there are n_i ith associates of α_1, there will be n_i elements in the above sum which are 1 and the rest of the elements zero. Hence, the above sum is n_i.

Hence the sum of elements in each row is equal to n_i. Thus $B_i E_{v1} = n_i E_{v1}$. Taking traspose, we get $E_{1v} B_i = n_i E_{1v}$.

(ii) Consider $\sum_{i=0}^{m} B_i$. The (α, β) th element in $\sum_{i=0}^{m} B_i$ is given by

$$b^0_{\alpha\beta} + b^1_{\alpha\beta} + \ldots + b^m_{\alpha\beta}.$$

Now since α and β are either 0th, 1st, ..., or the mth associates, only one of the elements $b^0_{\alpha\beta}, b^1_{\alpha\beta}, \ldots, b^m_{\alpha\beta}$ is 1, hence the (α, β) th element of B_i is equal 1 for $\alpha, \beta = 1, 2, \ldots, v$. Therefore

$$\sum_{i=0}^{m} B_i = E_{vv}.$$

(iii) Consider $\sum_{i=0}^{m} c_i B_i$. The (α, β) th element in $\Sigma\, c_i B_i$ is

$$c_0 b^0_{\alpha\beta} + c_1 b^1_{\alpha\beta} + \ldots + c_m b^m_{\alpha\beta} = c_i.$$

If α and β are i-th associates, then $\Sigma\, c_i B_i = \mathbf{0}$ gives $c_i = 0$. Selecting α and β as 0th, 1st, 2nd, ..., mth associates, we get $c_0 = c_1 = \ldots = c_m = 0$. This implies that B's are linearly independent

(iv) Consider $B_j B_k$. The (α, β) th element in $B_j B_k$ is given by

$$\sum_{i=0}^{v} b^j_{\alpha i} b^k_{i\beta}.$$

Now, $b^j_{\alpha i} b^k_{i\beta}$ = 1, if i occurs in the j-th associate class of α and the kth associate class of β.
= 0, otherwise.

Hence,

$\Sigma\, b^j_{\alpha i} b^k_{i\beta}$ = no. of common treatments between the jth associate class of α and the kth associate class of β.
= p^i_{jk}, if α and β are ith associates
= 0, otherwise.

Therefore,

$$B_j B_k = \sum_{i=0}^{m} p^i_{jk} B_i.$$

Further,

$$B_jB_k = B_j'B_k' = (B_kB_j)'$$
$$= (\Sigma\, p^i_{kj}B_i)' = \Sigma\, p^i_{kj}B_i$$
$$= B_kB_j.$$

(v) We have

$$B_iB_jB_k = B_i(B_jB_k) = \sum_u p^u_{jk}(B_iB_u) = \sum_{u,t} p^u_{jk}p^t_{iu}B_t.$$

Further,

$$B_iB_jB_k = (B_jB_j)B_k = \sum_u p^u_{ij}B_uB_k = \sum_{u,t} p^u_{ij}\, p^t_{uk}B_t.$$

Thus,

$$\sum_t \left[\sum_u p^u_{jk}\, p^t_{iu} - \sum_u p^u_{ij}\, p^t_{uk}\right] B_t = 0.$$

In view of (iii), i.e. the linear independence of B_i-matrices, we get

$$\sum_u p^u_{jk}\, p^t_{iu} = \sum_u p^u_{ij}\, p^t_{uk}.$$

79. (i) Consider $\sum_{i=0}^{m} P_i$. The (α, β)th element of $\sum_{i=0}^{m} P_i$ is given by

$$p^\beta_{\alpha 0} + p^\beta_{\alpha 1} + \ldots + p^\beta_{\alpha m} = \sum_{j=0}^{m} p^\beta_{\alpha j} = n_\alpha.$$

Hence the result follows.

(ii) Consider $\sum_{i=0}^{m} c_iP_i = 0_{(m+1)\times(m+1)}$.

The (α, β)th element in $\sum_{i=0}^{m} c_iP_i$ is clearly

$$c_0\, p^\beta_{\alpha 0} + c_1\, p^\beta_{\alpha 1} + \ldots + c_m\, p^\beta_{\alpha m}.$$

Hence $\Sigma\, c_iP_i = 0$, gives

$$c_0p^\beta_{\alpha 0} + c_1p^\beta_{\alpha 1} + \ldots + c_mp^\beta_{\alpha m} = 0$$

for every α and β. Selecting $\beta = 0$ and $\alpha = 0, 1, 2, \ldots, m$ in succession, we get

$$c_0 = c_1 = \ldots = c_m = 0.$$

Hence the result follows.

(iii) The (α, β)th element of P_jP_k is given by

$$a_{\alpha\beta} = \sum_{i=0}^{m} p^i_{\alpha j}\, p^\beta_{ik}.$$

Now by the result (v) proved in Exercise 78, we get

$$\sum_{i=0}^{m} p^{i}_{\alpha j}\, p^{\beta}_{ik} = \sum_{i=0}^{m} p^{i}_{jk}\, p^{\beta}_{\alpha i}.$$

Hence,

$$a_{\alpha\beta} = \sum_{i=0}^{m} p^{i}_{jk}\, p^{\beta}_{\alpha i}.$$

Now, (α, β)th element in $\sum_{i=0}^{m} p^{i}_{jk} P_i$ is given by

$$p^{0}_{jk} p^{\beta}_{\alpha 0} + p^{1}_{jk} p^{\beta}_{\alpha 1} + \ldots + p^{m}_{jk} p^{\beta}_{\alpha m} = \sum_{i=0}^{m} p^{i}_{jk} p^{\beta}_{\alpha i} = a_{\alpha\beta}.$$

Hence the result follows.

$$\text{Also, } P_j P_k = \sum_{i=0}^{m} p^{i}_{jk} P_i = \sum_{i=0}^{m} p^{i}_{kj} P_i = P_k P_j.$$

(iv) From (ii) and (iii), we see that the P_i-matrices are linearly independent and combine in the same way as B_i-matrices. Thus they form the basis of a vector space. Hence, they provide regular representation in $(m + 1) \times (m + 1)$ matrices of the algebra given by the B_i-matrices which are $v \times v$ matrices.

80. Consider NN'. The diagonal element of NN' are all r.

Now the diagonal element in the αth row and αth column of $\sum_{i=0}^{m} \lambda_i B_i$ is given by

$$\lambda_0 b^{0}_{\alpha\alpha} + \lambda_1 b^{1}_{\alpha\alpha} + \ldots + \lambda_m b^{m}_{\alpha\alpha} = \lambda_0 = r$$

since $b^{1}_{\alpha\alpha} = b^{2}_{\alpha\alpha} = \ldots = b^{m}_{\alpha\alpha} = 0$, $b^{0}_{\alpha\alpha} = 1$, and $\lambda_0 = r$.

Thus, the diagonal elements of $\sum \lambda_i B_i$ are all r. Now consider the (α, β)th element of NN'. Clearly this is equal to

$$\sum_{j=1}^{b} n_{\alpha j} n_{j\beta} = \text{no. of times the pair of treatments } \alpha \text{ and } \beta \text{ occur together in the block}$$
$$= \lambda_i, \text{ if } \alpha \text{ and } \beta \text{ are ith associates.}$$

The (α, β)th element of $\sum \lambda_i B_i$ is given by

$$\lambda_0 b^{0}_{\alpha\beta} + \lambda_1 b^{1}_{\alpha\beta} + \ldots + \lambda_m b^{m}_{\alpha\beta} = \lambda_i,$$

if α and β are ith assoicates. Hence, the (α, β)th elements of NN' and $\Sigma\, \lambda_i B_i$ are equal and

$$NN' = \sum_{i=0}^{m} \lambda_i B_i.$$

81. (i) In Exercise 78 (ii), we have proved that $\sum_{i=0}^{m} B_i = E_{vv}$

Hence $\sum_{i=0}^{m} E_{1v}B_i = vE_{1v}$. But $E_{1v}B_i = n_iE_{1v}$. Hence, we get

$$\sum_{i=0}^{m} n_iE_{1v} = vE_{1v}.$$

This gives $\sum_{i=0}^{m} n_i = v$.

(ii) In Exercise 80, we have proved that $NN' = \sum_{i=0}^{m} \lambda_iB_i$.

Hence $E_{1v}NN' = \sum_{i=0}^{m} \lambda_iE_{1v}B_i$ gives

$$r\,k\,E_{1v} = \left(\sum_{i=0}^{m} n_i\,\lambda_i\right)E_{1v}.$$

Therefore, we get $\sum_{i=0}^{m} n_i\lambda_i = r\,k$.

(iii) In Exercise 78 (iv), we have proved that $B_jB_k = \sum_{t=0}^{m} p_{jk}^tB_t$.

Hence $B_j\sum_{k=0}^{m} B_k = \sum_{t=0}^{m}\left(\sum_{k=0}^{m} p_{jk}^t\right)B_t$ and

$$B_j \cdot E_{vv} = \sum_{t=0}^{m}\left(\sum_{k=0}^{m} p_{jk}^t\right)B_t.$$

But $B_jE_{v1} = n_jE_{v1}$. Hence $B_jE_{vv} = n_jE_{vv}$. Thus, we get

$$n_jE_{vv} = \sum_{t=0}^{m}\left(\sum_{k=0}^{m} p_{jk}^t\right)B_t.$$

Let α and β be ith associates, then the (α, β)th element in $\sum_{t=0}^{m}\left(\sum_{k=0}^{m} p_{jk}^t\right)B_t$ is equal to

$$\sum_{k=0}^{m}\sum_{t=0}^{m} p_{jk}^tb_{\alpha\beta}^t = \sum_{k=0}^{m} p_{jk}^i,$$

since $b_{\alpha\beta}^i = 0$ for all $t \neq i$ and $b_{\alpha\beta}^i = 1$. Thus, we get

$$n_j = \sum_{k=0}^{m} p_{jk}^i.$$

(iv) In Exercise 78 (v), we have proved that

$$\sum_{u=0}^{m} p_{jk}^{u}\, p_{iu}^{t} = \sum_{u=0}^{m} p_{jk}^{u} p_{uk}^{t}.$$

Put $t = 0$, in the above equation. Since $p_{iu}^{0} = n_i$, if $i = u$ and 0, otherwise, we see that the left hand side of the above equation is equal to $n_i p_{jk}^{i}$. Further $p_{uk}^{0} = n_k$ if $u = k$ and 0, otherwise, we see that the right hand side of the above equation becomes $n_k p_{ij}^{k}$. Thus we get

$$n_i p_{jk}^{i} = n_k p_{ji}^{k}.$$

Writing $n_i\, p_{jk}^{i}$ as $n_i\, p_{kj}^{i}$ and applying the above result we get

$$n_i p_{jk}^{i} = n_j p_{ki}^{j} = n_j p_{ik}^{j}.$$

82. The minimal polynomial of the square matrix A_n is the monic scalar polynomial,

$$M(x) = x^m + a_1 x^{m-1} + \ldots + a_m$$

of least degree such that

$$M(A) = A^m + a_1 A^{m-1} + \ldots + a_m I_n = 0_{n\times n}.$$

It can be easily verified that the distinct characteristic roots of A are the solutions of its minimal polynomial. Now let f(x) and g(x) be the minimal polynomials of B and P respectively. Then,

$$\begin{aligned} f(B) &= B^t + a_1 B^{t-1} + \ldots + a_t \\ &= \sum_{i=0}^{m} d_i B_i, \end{aligned}$$

Since the multiplication of B_i-matrices is closed in the set of linear functions of $B_0, B_1, \ldots, B_m$.
Next consider the representation of f(B) in P_i matrices,

$$f(P) = \sum_{i=0}^{m} d_i P_i.$$

Since f(x) is the minimal polynomial of B, we have $f(B) = 0_{vxv}$, which implies that $d_0 = d_1 = d_2 = \ldots = d_m = 0$, for B_i-matrices are linearly independent. Thus, we get

$$f(P) = \mathbf{0}_{(m+1)\times(m+1)}.$$

Since g(x) is the minimal polynomial of P, it follows that g(x) is polynomial of lowest degree such that $g(P) = 0_{(m+1)\times(m+1)}$. Hence g(x) divides f(x). Similarly, we can show that f(x) divides g(x). Thus, we have f(x) = g(x). Hence the distinct characteristic roots of B and P, being the roots of f(x) = g(x) = 0, are the same.

Selecting $c_i = \lambda_i$ in the definition of B, we note that the distinct characteristic roots of $N N'$ and $\sum_{i=0}^{m} \lambda_i P_i$ are the same.

Now let $A = \sum_{i=0}^{m} \lambda_i P_i$. Then the (α, β)th element of A is given by $a_{\alpha\beta} = \sum_{t=0}^{m} \lambda t p^{\beta}_{\alpha t}$. We see that $\sum_{\alpha=0}^{m} a_{\alpha\beta} = \sum_t \lambda_t \sum_\alpha p^{\beta}_{\alpha t} = \sum_t \lambda_t n_t = rk$, for every B. Now the characteristic roots of A are given by

$$|A - \theta\, I| = 0.$$

Note that

$$|A - \theta\, I| = \begin{vmatrix} a_{00} - \theta & a_{01} & a_{02} & \dots & a_{0m} \\ a_{10} & a_{11} - \theta & a_{12} & \dots & a_{1m} \\ \cdot & & & & \\ \cdot & & & & \\ \cdot & & & & \\ a_{m0} & a_{m1} & a_{m2} & & a_{mm} - \theta \end{vmatrix}.$$

Adding the rows to the first row and taking $(r\,k - \theta)$ as a factor and subtracting the first column from the remaining columns, we get

$$\begin{aligned} |A - \theta\, I| &= (r\,k - \theta) \\ &\times \begin{vmatrix} a_{11} - a_{10} - \theta & a_{12} - a_{10} & \dots & a_{1m} - a_{10} \\ a_{21} - a_{20} & a_{22} - a_{20} - \theta & \dots & a_{2m} - a_{20} \\ \cdot & & & \\ \cdot & & & \\ \cdot & & & \\ a_{m1} - a_{m0} & a_{m2} - a_{m0} & \dots & a_{mm} - a_{m0} - \theta \end{vmatrix} \\ &= (r\,k - \theta)|L - \theta\, I| \end{aligned}$$

where $L = (\ell_{ij})$, $\ell_{ii} = a_{ii} - a_{i0}$, and $\ell_{ij} = a_{ij} - a_{i0}$, $i \neq j = 1, 2, \dots, m$. Hence the distinct characteristic roots of $A = \Sigma\, \lambda_i P_i$ are r k and the distinct characteristic roots of L. Now

$$\begin{aligned} \ell_{ii} = a_{ii} - a_{i0} &= \sum_t \lambda_t p^{i}_{it} - \sum_t \lambda_t p^{0}_{it} \\ &= \lambda_0 p^{i}_{i0} + \sum_{t=1}^{m} \lambda_t p^{i}_{it} - n_i \lambda_i. \end{aligned}$$

Since $p^i_{i0} = 1$ and $\lambda_0 = r$, we see that

$$\ell_{ii} = r - n_i\,\lambda_i + \sum_{t=1}^{m} \lambda_t p^i_{it},\; i = 1, 2, \ldots, m.$$

Also for $i \neq j = 1, 2, \ldots, m$, we have, since $a_{i0} = n_i\,\lambda_i$,

$$\begin{aligned}\ell_{ij} = a_{ij} - a_{i0} &= \sum_{t=0}^{m} \lambda_t p^j_{it} - n_i\lambda_i,\\ &= \lambda_0 p^j_{i0} + \sum_{t=1}^{m} \lambda_t p^j_{it} - n_i\lambda_i,\\ &= \sum_{t=1}^{m} \lambda_t p^j_{it} - n_i\,\lambda_i,\end{aligned}$$

for $p^j_{i0} = 0$ for $i \neq j$.

Thus, we see that the distinct characteristic roots of $\sum_{i=0}^{m} \lambda_t P_t$ are rk and the distinct characteristic roots of the matrix $L = (\ell_{ij})$, where

$$\ell_{ii} = r - n_i\lambda_i + \sum_{t=1}^{m} \lambda_t p^i_{it},\; i = 1, 2, \ldots, m$$

$$\ell_{ij} = \sum_{t=1}^{m} \lambda_t p^j_{it} - n_i\lambda_i,\; i \neq j = 1, 2, \ldots, m.$$

Hence, the distinct characteristic roots of $N\,N'$ are $r\,k$ and the distinct characteristic roots of the matrix $L = (\ell_{ij})$, where ℓ_{ij} are defined as above.

83. We know that the rank of a matrix is equal to the number of its non-zero charateristic roots. Since NN' has one characteristic root 0 with multiplicity u, the number of non-zero characteristic roots of NN' is equal to $v - u$. Hence

$$\text{rank}\,(NN') = v - u = \text{rank}\,(N'N).$$

Hence it follows that $b \geq v - u$.

When the design is resolvable, N consists of r sets of (b/r) columns each, such that 1 occurs only once in each row of the set. By adding the 1st, 2nd, ..., $\left(\frac{b}{r} - r\right)$th columns to the (b/r)th column of a set, we get a column consisting of 1 only. As there are r sets in N, we have

$$\text{rank}(N) \leq b - (r - 1).$$

But $\text{rank}(N) = \text{rank}(NN') = v - u$. Hence

$$v - u \leq b - (r - 1).$$

Thus, we get $b \geq v - u + (r - 1)$.

84. The distinct characteristic roots of NN' are given by rk and the distinct characteristic roots of $L = (\ell_{ij})$, (see Exercise 82)

$$\ell_{11} = r - n_1\lambda_1 + \lambda_1 p_{11}^1 + \lambda_2 p_{12}^1 = r - \lambda_1 - (\lambda_1 - \lambda_2)p_{12}^1$$
$$\ell_{22} = r - n_2\lambda_2 + \lambda_1 p_{21}^2 + \lambda_2 p_{22}^2 = r - \lambda_2 - (\lambda_1 - \lambda_2)p_{12}^2$$
$$\ell_{12} = -n_1\lambda_1 + \lambda_1 p_{11}^2 + \lambda_2 p_{12}^2 = -(\lambda_1 - \lambda_2)p_{12}^2$$
$$\ell_{21} = -n_2\lambda_2 + \lambda_1 p_{21}^1 + \lambda_2 p_{22}^1 = (\lambda_1 - \lambda_2)p_{12}^1.$$

The distinct characteristic roots of L are given by

$$|L - \theta I_2| = 0.$$

Now

$$\begin{aligned}
&|L - \theta I_2| \\
&= \begin{vmatrix} r - \lambda_1 - (\lambda_1 - \lambda_2)p_{12}^1 - \theta & -(\lambda_1 - \lambda_2)p_{12}^2 \\ (\lambda_1 - \lambda_2)p_{12}^1 & r - \lambda_2 + (\lambda_1 - \lambda_2)p_{12}^2 - \theta \end{vmatrix} \\
&= \begin{vmatrix} r - \lambda_1 - \theta & r - \lambda_2 - \theta \\ (\lambda_1 - \lambda_2)p_{12}^1 & r - \lambda_2 + (\lambda_1 - \lambda_2)p_{12}^2 - \theta \end{vmatrix} \\
&= \begin{vmatrix} r - \lambda_1 - \theta & \lambda_1 - \lambda_2 \\ (\lambda_1 - \lambda_2)p_{12}^1 & r - \lambda_2 - \theta + (\lambda_1 - \lambda_2)p \end{vmatrix}
\end{aligned}$$

where $p = p_{12}^2 - p_{12}^1$. Thus

$$\begin{aligned}
|L - \theta I_2| &= (r - \lambda_1 - \theta)(r - \lambda_2 - \theta) \\
&\quad + (r - \lambda_1 - \theta)(\lambda_1 - \lambda_2)p - (\lambda_1 - \lambda_2)^2 p_{12}^1 \\
&= \theta^2 - \theta(2r - \lambda_1 - \lambda_2) + (r - \lambda_1)(r - \lambda_2) \\
&\quad + (r - \lambda_1)(\lambda_1 - \lambda_2)p - \theta p(\lambda_1 - \lambda_2) - (\lambda_1 - \lambda_2)^2 p_{12}^1 \\
&= \theta^2 - \theta[A + p(\lambda_1 - \lambda_2)] + Q + (r - \lambda_1)(\lambda_1 - \lambda_2)p \\
&= 0
\end{aligned}$$

where

$$A = 2r - \lambda_1 - \lambda_2, \quad Q = (r - \lambda_1)(r - \lambda_2) - (\lambda_1 - \lambda_2)^2 p_{12}^1.$$

Hence the roots of $|L - \theta I_2| = 0$ are given by

$$\begin{aligned}
\theta_i = \frac{1}{2}\{&A + p(\lambda_1 - \lambda_2) + (-1)^i[A^2 + 2pA(\lambda_1 - \lambda_2) \\
&+ p^2(\lambda_1 - \lambda_2)^2 - 4Q - 4(r - \lambda_1)(\lambda_1 - \lambda_2)p]^{1/2}\},\ i = 1, 2
\end{aligned}$$

Now, we can easily verify that

$$A^2 - 4Q = (\lambda_1 - \lambda_2)^2(1 + 4p_{12}^1)$$
$$2pA(\lambda_1 - \lambda_2) - 4(r - \lambda_1)(\lambda_1 - \lambda_2)p = 2r(\lambda_1 - \lambda_2)^2.$$

Hence

$$A^2 + 2p\,A(\lambda_1 - \lambda_2) + p^2(\lambda_1 - \lambda_2)^2 - 4Q - 4(r - \lambda_1)(\lambda_1 - \lambda_2)p$$
$$= (\lambda_1 - \lambda_2)^2(p^2 + 1 + 4p_{12}^1 + 2p)$$
$$= (\lambda_1 - \lambda_2)^2(p^2 + 2\beta + 1) = (\lambda_1 - \lambda_2)^2\Delta.$$

where $\Delta = p^2 + 2\beta + 1$, and $\beta = 2p_{12}^1 + p$. Therefore,

$$\theta_i = r - \frac{1}{2}(\lambda_1 + \lambda_2) + \frac{1}{2}p(\lambda_1 - \lambda_2) + \frac{1}{2}(-1)^i(\lambda_1 - \lambda_2)\sqrt{\Delta},$$
$$i = 1, 2.$$

We now determine the multiplicities of θ_1 and θ_2. Denote by α_i the multiplicity of θ_i, $i = 1, 2$. Clearly

$$1 + \alpha_1 + \alpha_2 = v,$$

and hence, $\alpha_1 + \alpha_2 = v - 1$. Further, since the trace of a matrix is equal to the sum of its characteristic roots, we get

$$\mathrm{tr}(NN') = rk + \alpha_1\theta_1 + \alpha_2\theta_2.$$

But $\mathrm{tr}(NN') = vr$. Hence

$$\alpha_1\theta_1 + \alpha_2\theta_2 = r(v - k).$$

Then α_1 and α_2 are found by solving the equations

$$\alpha_1 + \alpha_2 = v - 1$$
$$\alpha_1\theta_1 + \alpha_2\theta_2 = r(v - k).$$

Multiplying the first by θ_2 and subtracting from the second we get

$$\alpha_1 = \frac{r(v - k) - \theta_2(v - 1)}{\theta_1 - \theta_2}$$

Now from the values of θ_1 and θ_2, we find

$$\theta_1 - \theta_2 = -(\lambda_1 - \lambda_2)\sqrt{\Delta}.$$

Also,

$$r(v - k) - \theta_2(v - 1)$$
$$= r(v - 1) - r(k - 1) - \theta_2(v - 1)$$
$$= (v - 1)(r - \theta_2) - n_1\lambda_1 - n_2\lambda_2$$
$$= (n_1 + n_2)\left[\frac{1}{2}(\lambda_1 + \lambda_2) - \frac{1}{2}(\lambda_1 - \lambda_2)p - \frac{1}{2}(\lambda_1 - \lambda_2)\sqrt{\Delta}\right]$$
$$-n_1\lambda_1 - n_2\lambda_2$$
$$= \frac{-1}{2}(n_1 - n_2)(\lambda_1 - \lambda_2) - \frac{1}{2}(n_1 + n_2)(\lambda_1 - \lambda_2)p$$
$$-\frac{1}{2}(n_1 + n_2)(\lambda_1 - \lambda_2)\sqrt{\Delta}.$$

Hence,

$$\alpha_1 = \frac{n_1 + n_2}{2} + \frac{[(n_1 - n_2) + (n_1 + n_2)p]}{2\sqrt{\Delta}}$$

and

$$\begin{aligned}\alpha_2 &= (v - 1) - \alpha_1 = n_1 + n_2 - \alpha_1\\ &= \frac{n_1 + n_2}{2} - \frac{[(n_1 - n_2) + (n_1 + n_2)p]}{2\sqrt{\Delta}}.\end{aligned}$$

Hence the result.

85. The parameters of a Group Divisible design are

$$\begin{aligned}&v = m\,n, b, r, k\\ &n_1 = n - 1, n_2 = n(m - 1) = v - n,\\ &\lambda_1, \lambda_2.\\ &P_1 = \begin{bmatrix} n-2 & 0 \\ 0 & n(m-1) \end{bmatrix}, P_2 = \begin{bmatrix} 0 & n-1 \\ n-1 & n(m-2) \end{bmatrix}.\end{aligned}$$

Hence, we get

$$p = p^2_{12} - p^1_{12} = n - 1, \beta = p^2_{12} + p^1_{12} = n - 1,$$

and therefore $\Delta = (p + 1)^2 = n^2$. Hence, the characteristic roots of NN' of a GD design are

$$\begin{aligned}\theta_0 &= r\,k\\ \theta_i &= r - \frac{1}{2}(\lambda_1 + \lambda_2) + \frac{1}{2}(\lambda_1 - \lambda_2)[p + (-1)^i\sqrt{\Delta}]\\ &= r - \frac{1}{2}(\lambda_1 + \lambda_2) + \frac{1}{2}(\lambda_1 - \lambda_2)[n - 1 + (-1)^i n], i = 1, 2.\end{aligned}$$

Thus,

$$\begin{aligned}\theta_1 &= r - \frac{1}{2}(\lambda_1 + \lambda_2) + \frac{1}{2}(\lambda_1 - \lambda_2)(-1) = r - \lambda_1\\ \theta_2 &= r - \frac{1}{2}(\lambda_1 + \lambda_2) + \frac{1}{2}(\lambda_1 - \lambda_2)(2n - 1)\\ &= r - \lambda_1 + n(\lambda_1 - \lambda_2)\\ &= r + \lambda_1(n - 1) - n\lambda_2.\end{aligned}$$

Now since $n_1\lambda_1 + n_2\lambda_2 = r(k - 1)$, we get

$$\begin{aligned}&r + (n - 1)\lambda_1 + (v - n)\lambda_2 = rk\\ &r + (n - 1)\lambda_1 - n\lambda_2 = rk - v\lambda_2.\end{aligned}$$

Hence $\theta_2 = rk - v\lambda_2$. The multiplicities are given by

$$\alpha_i = \frac{n_1 + n_2}{2} - (-1)^i\left[\frac{(n_1 - n_2) + p(n_1 + n_2)}{2\sqrt{\Delta}}\right]$$

$$= \frac{v-1}{2} - (-1)^i \left[\frac{(2n-v-1)+(n-1)(v-1)}{2n}\right]$$
$$= \frac{v-1}{2} - (-1)^i \frac{(v-2m+1)}{2}$$
$$= \frac{1}{2}[v-1-(-1)^i(v-2m+1)], i = 1, 2,$$

and therefore

$$\alpha_1 = (v-m) = m(n-1), \quad \alpha_2 = (m-1).$$

86. The parameters of an L_i design are

$$v = s^2, b, r, k$$
$$n_1 = i(s-1), n_2 = (s-i+1)(s-1)$$
$$\lambda_1, \lambda_2$$
$$P_1 = \begin{bmatrix} (i-1)(i-2)+s-2 & (s-i+1)(i-1) \\ (s-i+1)(i-1) & (s-i+1)(s-1) \end{bmatrix}$$
$$P_2 = \begin{bmatrix} i(i-1) & i(s-i) \\ i(s-1) & (s-i)(s-i-1)+s-2 \end{bmatrix}.$$

Here

$$\begin{aligned} p &= i(s-i)-(s-i+1)(i-1) \\ &= s-2i+1 \\ &= i(s-i)+(s-i+1)(i-1) \\ &= 2i(s-i)-p. \end{aligned}$$

Hence

$$\begin{aligned} \Delta = p^2+2p+1 &= p^2+4i(s-i)-2p+1 \\ &= (p-1)^2+4i(s-i) \\ &= (s-2i)^2+4i(s-i) = s^2. \end{aligned}$$

Therefore, the distinct characteristic roots of NN' of L_i design are

$$\theta_0 = rk, \text{ with multiplicity } 1$$
$$\theta_1 = r - \frac{1}{2}(\lambda_1+\lambda_2) + \frac{1}{2}(\lambda_1-\lambda_2)[p-\sqrt{\Delta}], \text{ with multiplicity } \alpha_1$$
$$\theta_2 = r - \frac{1}{2}(\lambda_1+\lambda_2) + \frac{1}{2}(\lambda_1-\lambda_2)[p+\sqrt{\Delta}], \text{ with multiplicity } \alpha_2,$$

where

$$\alpha_1 = \frac{n_1+n_2}{2} + \frac{(n_1-n_2)+p(n_1+n_2)}{2\sqrt{\Delta}}$$
$$\alpha_2 = \frac{n_1+n_2}{2} - \frac{(n_1-n_2)+p(n_1+n_2)}{2\sqrt{\Delta}}.$$

Thus

$$\begin{aligned}
\theta_1 &= r - \frac{1}{2}(\lambda_1 + \lambda_2) + \frac{1}{2}(\lambda_1 - \lambda_2)(s - 2i + 1 - s) \\
&= r - \frac{1}{2}(\lambda_1 + \lambda_2) + \frac{1}{2}(\lambda_1 - \lambda_2)(-2i + 1) \\
&= r - \lambda_2 - i(\lambda_1 - \lambda_2) \\
&= r - i\lambda_1 + \lambda_2(i - 1).
\end{aligned}$$

$$\begin{aligned}
\theta_2 &= r - \frac{1}{2}(\lambda_1 + \lambda_2) + \frac{1}{2}(\lambda_1 - \lambda_2)(2s - 2i + 1) \\
&= r - \lambda_2 + (s - i)(\lambda_1 - \lambda_2) \\
&= r + \lambda_1(s - i) - \lambda_2(s - i + 1).
\end{aligned}$$

Also,

$$\begin{aligned}
\alpha_1 &= \frac{v-1}{2} + \frac{(n_1 - n_2) + (s - 2i + 1)(v - 1)}{2s} \\
&= \frac{v-1}{2} - \frac{(s-1)(s - 2i + 1) + (s - 2i + 1)(v - 1)}{2s} \\
&= \frac{(v-1)}{2} + \frac{(s - 2i + 1)(-s + 1 + v - 1)}{2s} \\
&= \frac{(s^2-1)}{2} + \frac{(s-1)(s - 2i + 1)}{2} \\
&= (s-1)(s - i + 1)
\end{aligned}$$

and

$$\begin{aligned}
\alpha_2 &= (v - 1) - \alpha_1 \\
&= (s^2 - 1) - (s - 1)(s - i + 1) \\
&= i(s - 1).
\end{aligned}$$

87. The parameters of a triangular design are

$$\begin{aligned}
&v = n(n-1)/2, b, r, k \\
&n_1 = 2(n-2), n_2 = (n-2)(n-3)/2 \\
&\lambda_1, \lambda_2, \\
&P_1 = \begin{bmatrix} n-2 & n-3 \\ n-3 & \dfrac{(n-3)(n-4)}{2} \end{bmatrix}, \\
&P_2 = \begin{bmatrix} 4 & 2n-8 \\ 2n-8 & \dfrac{(n-4)(n-5)}{2} \end{bmatrix}.
\end{aligned}$$

Hence

$$p = n - 5,\ \beta = 3n - 11,\ \Delta = (n - 2).$$

The characteristic roots are $\theta_0 = rk$ with multiplicity 1 and

$$\begin{aligned}\theta_i &= r - \frac{1}{2}(\lambda_1 + \lambda_2) + \frac{1}{2}(\lambda_1 - \lambda_2)[p + (-1)^i\sqrt{\Delta}] \\ &= r - \frac{1}{2}(\lambda_1 + \lambda_2) + \frac{1}{2}(\lambda_1 - \lambda_2)[n - s + (-1)^i(n-2)], \\ & \qquad\qquad i = 1, 2\end{aligned}$$

Hence

$$\begin{aligned}\theta_1 &= r - 2\lambda_1 + \lambda_2 \\ \theta_2 &= r + \lambda_1(n-4) - \lambda_2(n-3).\end{aligned}$$

The multiplicities of θ_1 and θ_2 are given by α_1 and α_2, where

$$\begin{aligned}\alpha_i &= \frac{n_1 + n_2}{2} - (-1)^i \frac{[(n_1 - n_2) + p(n_1 + n_2)]}{2\sqrt{\Delta}} \\ &= \frac{v-1}{2} - (-1)^i \frac{[(n_1 - n_2) + p(n-5)]}{2(n-2)} \\ &= \frac{(n-2)(n+1)}{4} - (-1)^i \frac{(n^2 - 5n + 2)}{4} \\ \alpha_1 &= \frac{1}{2}n(n-3), \alpha_2 = (n-1).\end{aligned}$$

88. Since $k > r$, we have $b < v$. Hence NN' will be singular and therefore NN' will have zero as a characteristic root. Now in Exercise 82 we have proved that the distinct characteristic roots of NN' other than rk are given by the characteristic roots of the matrix $L = (\ell_{ij})$, where

$$\begin{aligned}\ell_{ij} &= \sum_{t=1}^{m} \lambda_t p_{it}^j - n_i\lambda_i, \;\; i \neq j = 1, 2, \ldots, m \\ \ell_{ii} &= r + \sum_{t=1}^{m} \lambda_t p_{it}^i - n_i\lambda_i, \;\; i = 1, 2, \ldots, m.\end{aligned}$$

Hence, the matrix L will have zero as a characteristic root. Hence $|L| = 0$.

For a 2-associate class PBIBD, where $k > r$, we have

$$\begin{vmatrix} r + \sum_{t=1}^{2} \lambda_t p_{1t}^1 - n_1\lambda_1 & \sum_{t=1}^{2} \lambda_t p_{1t}^2 - n_1\lambda_1 \\ \sum \lambda_t p_{2t}^1 - n_2\lambda_2 & r + \sum \lambda_t p_{2t}^2 - n_2\lambda_2 \end{vmatrix} = 0.$$

Now

$$\begin{aligned}r + \lambda_1 p_{11}^1 + \lambda_2 p_{12}^1 - n_1\lambda_1 &= r - \lambda_1(n_1 - p_{11}^1) + \lambda_2 p_{12}^1 \\ &= r - \lambda_1 - p_{12}^1(\lambda_1 - \lambda_2),\end{aligned}$$

and

$$\lambda_1 p_{11}^2 + \lambda_2 p_{12}^2 - n_1\lambda_1 = -p_{12}^2(\lambda_1 - \lambda_2),$$
$$\lambda_1 p_{21}^1 + \lambda_2 p_{22}^1 - n_2\lambda_2 = p_{12}^1(\lambda_1 - \lambda_2),$$
$$r + \lambda_1 p_{21}^2 + \lambda_2 p_{22}^2 - n_2\lambda_2 = r - \lambda_2 + p_{12}^2(\lambda_1 - \lambda_2).$$

Thus, we get

$$\begin{vmatrix} r - \lambda_1 - p_{12}^1(\lambda_1 - \lambda_2) & -p_{12}^2(\lambda_1 - \lambda_2) \\ p_{12}^1(\lambda_1 - \lambda_2) & r - \lambda_2 + p_{12}^2(\lambda_1 - \lambda_2) \end{vmatrix} = 0.$$

Adding the second row to the first row, we get

$$\begin{vmatrix} r - \lambda_1 & r - \lambda_2 \\ p_{12}^1(\lambda_1 - \lambda_2) & r - \lambda_2 + p_{12}^2(\lambda_1 - \lambda_2) \end{vmatrix} = 0$$

i.e. $(r - \lambda_1)(r - \lambda_2) + (\lambda_1 - \lambda_2)[p_{12}^2(r - \lambda_1) - p_{12}^1(r - \lambda_2)] = 0.$

89. Let the blocks of the given BIBD be denoted by $B_1, B_2, \ldots, B_b$. Let B_1 have d disjoint blocks $B_2, B_3, \ldots, B_{d+1}$ and the block $B_i (i = d + 2, d + 3, \ldots, b)$ have x_i treatments in common with B_1. Then, we have

$$\sum_{i=d+2}^{b} x_i = k(r - 1),$$

$$\sum_{i=d+2}^{b} x_i(x_i - 1) = k(k - 1)(r - 1).$$

Then, we get

$$\Sigma\, x_i^2 = k(r - \lambda - k + k\lambda)$$

Now, define $\bar{x} = \Sigma\, x_i/(b - d - 1) = \dfrac{k(r - 1)}{b - d - 1}$,

$$\Sigma\, (x_i - \bar{x})^2 = k(r - \lambda - k + k\lambda) - \frac{k^2(\gamma - 1)^2}{b - d - 1}.$$

Since $\Sigma\, (x_i - \bar{x})^2 \geq 0$, we get

$$d \leq b - 1 - \frac{k(r - 1)^2}{(r - \lambda - k + k\lambda)}.$$

If $d = b - 1 - \dfrac{k(\gamma - 1)^2}{(r - \lambda - k + k\lambda)}$, $\Sigma\, (x_1 - \bar{x})^2 = 0$,
and hence

$$x_i = \bar{x} = \frac{k(r - 1)}{b - d - 1} = \frac{(r - \lambda - k + k\lambda)}{(r - 1)},$$

since $d = b - 1 - [k(\gamma - 1)^2/(r - \lambda - k + \lambda\lambda)]$.

90. When each treatment of a BIBD $(\overset{*}{v}, \overset{*}{b}, \overset{*}{r}, \overset{*}{k}, \overset{*}{\lambda})$ is replaced by a group of n treatments, clearly we get a design with

$$v = n\overset{*}{v},\ b = \overset{*}{b},\ r = \overset{*}{r},\ k = n\overset{*}{k}.$$

Now, in the new design with respect to any treatment, we can divide the remaining treatments into two classes, (i) those which are in the same group as the given one and (ii) those which are not in the same group as the given treatment. Then we have $n_1 = n - 1$, and $n_2 = n(\overset{*}{v} - 1)$.

A treatment in the given BIBD occurs in $\overset{*}{r}$ blocks, so when it is replaced by a group of n treatments, then any pair in this group will obviously occur together in $\overset{*}{r}$ blocks. Hence $\lambda_1 = \overset{*}{r}$.

Consider a pair of treatments α and β of the given BIBD. They occur together in $\overset{*}{\lambda}$ blocks. Now denote the groups of n treatments, which replace α and β by $[\alpha]$ and $[\beta]$. Then these groups $[\alpha]$ and $[\beta]$ occur together in $\overset{*}{\lambda}$ blocks. Hence any pair of treatments, one belonging to $[\alpha]$ and the other belonging to $[\beta]$ occur together in $\overset{*}{\lambda}$ blocks. Thus $\lambda_2 = \overset{*}{\lambda}$ as is evident, two treatments in the same group are 1st associates, while two treatments in different groups are 2nd associates. Consider a pair of treatments belonging to the group $[\alpha]$. Clearly, the number of treatments common to the 1st associates of them is $p^1_{11} = n - 2$. The number of treatments common to the first associates of one and the second associates of the second is $p^1_{12} = 0$. Also, the number of treatments common between their second associates is $p^1_{22} = n(\overset{*}{v} - 1)$.

Now, consider a pair of treatments one belonging to the group $[\alpha]$ and the other to $[\beta]$. The number of treatments common to their 1st associates is $p^2_{11} = 0$. The number of treatments common to the first associates of the first and the second associates of the second $p^2_{12} = n - 1$. The number of treatments common between their second associates are the treatments of the remaining $(\overset{*}{v} - 2)$ groups, hence $p^2_{22} = n(\overset{*}{v} - 2)$. Thus, we get P_1 and P_2.

91. Let the blocks of the given singular group divisible design be denoted by $B_1, B_2, \ldots, B_b$, and let B_1 have $B_2, B_3, \ldots, B_{d+1}$ disjoint blocks and let $B_i (i = d + 2, \ldots, b)$ have x_i treatments common with B_1. Then clearly

$$\sum_{i=d+2}^{b} x_i = k(r - 1).$$

Now, we consider the given singular group divisible design as obtained from the BIBD $(\overset{*}{v}, \overset{*}{b}, \overset{*}{r}, \overset{*}{k}, \overset{*}{\lambda})$ by replacing each of its treatments by a group of n treatments. Hence, the block B_1 contains $\overset{*}{k}$ groups of n treatments each. Hence, considering pairs of treatments obtained from B_1, we get

$$\begin{aligned}\Sigma\, x_i(x_i - 1) &= \overset{*}{k} n(n - 1)(\lambda_1 - 1) + [k(k - 1) - \overset{*}{k} n(n - 1)](\lambda_2 - 1)\\ &= k[(n - 1)(\lambda_1 - 1) + (k - n)(\lambda_2 - 1)].\end{aligned}$$

Thus,

$$\Sigma\, x_i^2 = k[(n-1)(\lambda_1 - 1) + (k-n)(\lambda_2 - 1) + k(\gamma - 1)$$
$$= k[n(\lambda_1 - 1) + (k-n)(\lambda_2 - 1)],$$

since $\gamma = \lambda_1$ in a singular group divisible design. Define $\bar{x} = \Sigma\, x_i/(b-d-1)$, we get

$$\Sigma\,(x_i - \bar{x})^2 = k[n(\lambda_1 - 1) + (k-n)(\lambda_2 - 1)] - \frac{k^2(\lambda_1 - 1)^2}{b-d-1}$$

Since $\Sigma\,(x_i - \bar{x})^2 \geq 0$, we get

$$d \leq b - 1 - \left[\frac{k(\lambda_1 - 1)^2}{n(\lambda_1 - 1) + (k-n)(\lambda_2 - 1)}\right].$$

If $d = b - 1 - \dfrac{k(\lambda_1 - 1)^2}{n(\lambda_1 - 1) + (k-n)(\lambda_2 - 1)}$, then $\Sigma\,(x_i - \bar{x})^2 = 0$, hence

$$x_i = \bar{x} = \frac{k(\lambda_1 - 1)}{b-d-1}$$
$$= \frac{n(\lambda_1 - 1) + (k-n)(\lambda_2 - 1)}{(\lambda_1 - 1)}$$
$$= n + [(k-n)(\lambda_2 - 1)/(\lambda_1 - 1)].$$

92. Let the blocks of a singular group divisible design be denoted by $B_1, B_2, \ldots, B_b$. Let B_1 have x_i treatments in common with the block B_i, $i = 2, 3, \ldots, b$. Then, as in Exercise 91, we can prove that

$$\sum_{i=2}^{b} x_i = k(r-1)$$
$$\Sigma\, x_i^2 = k[n(\lambda_1 - 1) + (k-n)(\lambda_2 - 1)]$$

Define $\bar{x} = \Sigma\, x_i/(b-1) = \dfrac{k(r-1)}{b-1}$. Then, we obtain

$$\Sigma\,(x_i - \bar{x})^2 = \frac{k}{(b-1)}[n(\lambda_1 - 1)(b-1) + (k-n)(\lambda_2 - 1)(b-1)$$
$$- k(r-1)^2]$$
$$= \frac{k}{(b-1)}[n(b-1)(\lambda_1 - \lambda_2) + k\{(\lambda_2 - 1)(b-1)$$
$$- (\lambda_1 - 1)(r-1)\}].$$

Note $r = \lambda_1$. Now for the singular group divisible design, $n_1 = n - 1$, $n_2 = v - n$, and $r = \lambda_1$, hence $n_1\lambda_1 + n_2\lambda_2 = r(k-1)$ gives

$$\lambda_2 = \frac{\lambda_1(k-n)}{(v-n)},$$

and $\lambda_1 - \lambda_2 = \lambda_1(v-k)/(v-n)$. Also,

$$\begin{aligned}(\lambda_2-1)(b-1)-(\lambda_1-1)(r-1) &= \frac{\lambda_1(k-n)}{(v-n)}(b-1)-\lambda_1(r-1)-b+r\\ &= \frac{\lambda_1\{(v-k)-n(b-r)\}}{(v-n)}-(b-r)\\ &= \frac{(b-r)(k-nr)}{(v-n)}-(b-r)\\ &= (b-r)(k-nr-v+n)/(v-n).\end{aligned}$$

Hence

$$\begin{aligned}\Sigma\,(x_i-\bar{x})^2 &= \frac{k}{(b-1)}\left[\frac{\lambda_1 n(b-1)(v-k)}{(v-n)} + \frac{k(b-r)(k-nr-v+n)}{(v-n)}\right]\\ &= \frac{kr(v-k)(nb-nr-v+k)}{(b-1)(v-n)}\\ &= \frac{k(v-k)(b-r)(nr-k)}{(v-n)}.\end{aligned}$$

Now $v = mn$, hence $nr - k = k(b-m)/m$ and

$$\Sigma\,(x_i-\bar{x})^2 = \frac{(k^2(v-k)(b-r)(b-m)}{m(v-n)}.$$

Thus, if all x_i are equal, then $\Sigma\,(x_i-\bar{x})^2 = 0$, which gives $b = m$. Conversely if $b = m$, $\Sigma\,(x_i-\bar{x})^2 = 0$, which gives all x_i are equal. Since $\bar{x} = k(-1)/(b-1) = k(-1)/(m-1)$, the second condition follows.

93. Consider a resolvable singular group divisible design with $b = tr$, $v = mn = tk$. Let B_{ij} denote the jth block in the ith replication, $j = 1, 2, \ldots, t$, $i = 1, 2, \ldots, r$ and x_{ij} denote the number of common treatments between B_{11} and B_{ij}, $i = 2, 3, \ldots, r$, $j = 1, 2, \ldots, t$. Then as in Exercise 91, we have

$$\sum_{i=2}^{r}\sum_{j=1}^{t} x_{ij} = k(r-1)$$

$$\sum_{i=2}^{r}\sum_{j=1}^{t} x_{ij}^2 = k[n(\lambda_1-1)+(k-n)(\lambda_2-1)].$$

Define $\bar{x} = \Sigma\Sigma\; x_{ij}/t(r-1) = k/t = k^2/v$. Then

$$\begin{aligned}\Sigma\Sigma\;(x_{ij}-\bar{x})^2 &= k[n(\lambda_1-1)+(k-n)(\lambda_2-1)] - \frac{k^2(r-1)}{t}\\ &= \frac{k}{t}[nt(\lambda_1-\lambda_2)+k\{t(\lambda_2-1)-r+1\}].\end{aligned}$$

Now we have shown in Exercise 92, that

$$\lambda_2 = \frac{\lambda_1(k-n)}{v-n}, \lambda_1 - \lambda_2 = \frac{\lambda_1(v-k)}{(v-n)}$$

and remembering that $r = \lambda_1$, and $v = mn$, we get

$$\Sigma\Sigma\ (x_{ij} - \bar{x})^2 = \frac{k}{t(v-n)}[ntr(v-k) + nk(t-1)(1-m-r)]$$

$$= \frac{nk^2(t-1)(b-m-r+1)}{t(v-n)}.$$

Since, $\Sigma\Sigma\ (x_{ij} - \bar{x})^2 \geq 0$, we get $b \geq m + r - 1$.

If the resolvable singular group divisible design is affine, then all x_{ij} are equal to $\bar{x}$, and hence $\Sigma\Sigma\ (x_{ij} - \bar{x})^2 = 0$, which gives $b = m + r - 1$. Conversely if $b = m + r - 1$, then $\Sigma\Sigma\ (x_{ij} - \bar{x})^2 = 0$ and hence all x_{ij} are equal, and the design is affine resolvable. Since $\bar{x} = k^2/v$, the second condition follows.

94. The characteristic roots of NN' of a group divisible design are (see Exercise 85) given by

$$\theta_0 = rk, \text{ with multiplicity } 1$$
$$\theta_1 = r - \lambda_1, \text{ with multiplicity } v - m$$
$$\theta_2 = r + \lambda_1(n-1) - n\lambda_2, \text{ with multiplicity } m - 1.$$

Now for a singular group divisible design, $\theta_1 = r - \lambda_1 = 0$. Hence,

$$\begin{aligned}\text{rank}(NN') &= \text{no. of non-zero characteristic roots}\\ &= 1 + m - 1\\ &= m.\end{aligned}$$

Hence,

$$m = \text{rank}(NN') = \text{rank}(N'N) \leq b,$$

and $b \geq m$. Further if the singular group divisible is resolvable, then

$$m = \text{rank}\ (NN') = \text{rank}\ (N'N) \leq b - (r-1).$$

Hence, we get $b \geq m + r - 1$.

95. The characteristic roots of NN' of a GD design (see Exercise 85) are given

$$\theta_0 = rk, \text{ with multiplicity } 1,$$
$$\theta_1 = r - \lambda_1, \text{ with multiplicity } v - m$$
$$\theta_2 = r + \lambda_1\ (n-1) - n\lambda_2, \text{ with multiplicty } m - 1.$$

Now, for a semi-regular GD design, $\theta_2 = r + \lambda_1(n-1) - n\lambda_2 = 0$.
Therefore,

$$\begin{aligned}\text{rank}\ (NN') &= \text{no. of its non-zero characteristic roots}\\ &= 1 + v - m.\end{aligned}$$

Hence

$$v - m + 1 = \text{rank}(NN') = \text{rank}(N'N) \leq b$$

i.e., $b \geq v - m + 1$. Further, if the semi-regular GD design is resolvable, then

$$v - m + 1 = \text{rank}(NN') = \text{rank}(N'N) \leq b - (r - 1)$$

Hence, for a resolvable semi-regular GD design, we have

$$b \geq v - m + r.$$

96. Let the blocks be denoted by $B_1, B_2, \ldots, B_b$. Suppose B_i contains x_{ij} treatments from the jth group of the association scheme, $i = 1, 2, \ldots, b$ and $j = 1, 2, \ldots, m$. Then, we get for any j,

$$\sum_{i=1}^{b} x_{ij} = nr$$
$$\sum_{i=1}^{b} x_{ij}(x_{ij} - 1) = n(n-1)\lambda_1.$$

Hence

$$\sum_{i=1}^{b} x_{ij}^2 = n[(n-1)\lambda_1 + r].$$

Define $\bar{x}_j = \sum_i x_{ij}/b = nr/b = k/m$. Then

$$\begin{aligned}\sum_{i=1}^{b} (x_{ij} - \bar{x}_j)^2 &= n[(n-1)\lambda_1 + r] - \frac{b\,k^2}{m^2}\\ &= n[(n-1)\lambda_1 + r] - \frac{nrk}{m}\\ &= \frac{n}{m}[m(n-1)\lambda_1 + m\,r - rk].\end{aligned}$$

But $(n-1)\lambda_1 + n(m-1)\lambda_2 = r(k-1)$ and hence $(n-1)\lambda_1 = r(k-1) - n(m-1)\lambda_2$. Thus

$$\begin{aligned}\sum_i (x_{ij} - \bar{x}_j)^2 &= \frac{n}{m}[mr(k-1) - v(m-1)\lambda_2 + mr - rk]\\ &= \frac{n(m-1)(rk - v\lambda_2)}{m}\\ &= 0,\end{aligned}$$

since $rk - v\lambda_2 = 0$ for semi-regular GD design. Thus $x_{ij} = \bar{x}_j = \text{constant} = \frac{k}{m}$ for each i, and therefore k/m is an integer and each block contains k/m treatments from jth block. But since this result is independent of the jth block, it follows that each block contains k/m treatments from each group.

97. Let the blocks be denoted by $B_1, B_2, \ldots, B_b$. Suppose B_1 has $B_2, B_3, \ldots, B_{d+1}$ blocks disjoint with it and has x_i treatments common with the block B_i, $i = d+2, d+3, \ldots, b$. Then we get

$$\sum_{d+2}^{b} x_i = k(r-1).$$

Now, in Exercise 95, we have proved that each block contains k/m treatments from each group of the association scheme. Thus, B_1 contains k/m treatments from each group, which form pairs of first associates. Thus, we get

$$\begin{aligned}\Sigma\, x_i(x_i - 1) &= m(k/m)(k/m - 1)(\lambda_1 - 1)\\ &\quad + [k(k-1) - k(k/m - 1)](\lambda_2 - 1).\end{aligned}$$

Hence,

$$\begin{aligned}\sum_{i=d+2}^{b} x_i^2 &= \frac{k}{m}[(k-m)(\lambda_1 - 1) + k(m-1)(\lambda_2 - 1) + m(r-1)]\\ &= \frac{k}{m}[k(\lambda_1 - \lambda_2) + m(r - \lambda_1) + mk(\lambda_2 - 1)].\end{aligned}$$

Now, for a semi-regular GD design, $rk - v\lambda_2 = 0$. Hence

$$\lambda_2 = \frac{rk}{v},$$

and

$$(n-1)\lambda_1 + n(m-1)\lambda_2 = r(k-1),$$

which gives $\lambda_1 = r(k-m)/(v-m)$. Thus

$$\begin{aligned}r - \lambda_1 &= r(v-k)/(v-m) = k(b-r)/(v-m)\\ \lambda_1 - \lambda_2 &= m(k-v)/(v-m)v.\end{aligned}$$

Hence

$$\begin{aligned}\sum_{i=d+2}^{b} x_i^2 &= \frac{k}{m}\left[\frac{mk(k-v)}{(v-m)v}\frac{mk(b-r)}{(v-m)} + m\,k(\lambda_2 - 1)\right]\\ &= \frac{k^2}{v(v-m)}[(k-v) + v(b-r) + (v-m)(v\lambda_2 - v)]\\ &= \frac{k^2}{v(v-m)}[(b-r)(v-k) - (v-m)(v-rk)\Big].\end{aligned}$$

Hence, defining $\bar{x} = \Sigma\, x_i/(b-d-1)$, we get

$$\Sigma\,(x_i - \bar{x})^2 = \frac{k^2 T}{v(v-m)} - \frac{k^2(\gamma - 1)^2}{b-d-1},$$

where $T = (b - r)(v - k) - (v - m)(v - rk)$. Since $\Sigma\,(x_i - \bar{x})^2 \geq 0$, we get

$$d \leq b - 1 - \frac{v(v - m)(r - 1)^2}{T}.$$

Further if $d = b - 1 - \dfrac{v(v - m)(r - 1)^2}{T}$, then $\Sigma\,(x_i - \bar{x})^2 = 0$ and hence

$$x_i = \bar{x} = \frac{k(r - 1)}{b - d - 1} = \frac{k(r - 1)T}{v(v - m)(r - 1)^2} = \frac{kT}{v(v - m)(r - 1)},$$

which must be an integer.

98. Let the blocks be denoted by $B_1, B_2, \ldots, B_b$, and let B_1 have x_i treatments in common with B_i, $i = 2, 3, \ldots, b$. Then as in Exercise 97, we get

$$\sum_{i=2}^{b} x_i = k(r - 1)$$

and

$$\Sigma\, x_i^2 = \frac{k^2[(b - r)(v - k) - (v - m)(v - rk)]}{v(v - m)}.$$

Define $\bar{x} = \dfrac{1}{(b - 1)}\Sigma\, x_i = \dfrac{k(r - 1)}{(b - 1)}$. Then, we get

$$\Sigma\,(x_i - \bar{x})^2 = \frac{k^2[(b - r)(v - k) - (v - m)(v - rk)]}{v(v - m)} - \frac{k^2(r - 1)^2}{(b - 1)}$$
$$= \frac{k^2[(b - 1)(b - r)(v - k) - (v - m)(b - r)(v - k)]}{v(v - m)(b - 1)}.$$

Hence, the result follows. Since $\bar{x} = \dfrac{k(r - 1)}{b - 1} = \dfrac{k(r - 1)}{v - m}$ the second condition follows.

99. Consider a resolvable GD design, with $b = tr$, and $v = mn = tk$. Let B_{ij} denote the jth block in the ith replication, $i = 1, 2, \ldots, r$ and $j = 1, 2, \ldots, t$. Let the number of treatments common between B_{11} and B_{ij} be x_{ij}, $i = 2, 3, \ldots, r$; $j = 1, 2, \ldots, t$. Then, as in Exercise 97, we can prove that

$$\sum_{i=2}^{r} \sum_{j=1}^{t} x_{ij} = k(r - 1)$$

and

$$\sum_{i=2}^{r} \sum_{j=1}^{t} x_{ij}^2 = \frac{k^2[(b - r)(v - k) - (v - m)(v - rk)]}{v(v - m)}.$$

Define $\bar{x} = \frac{\Sigma\Sigma\, x_{ij}}{t(r-1)} = \frac{k}{t} = \frac{k^2}{v}$. Then

$$\begin{aligned}
\Sigma\Sigma\,(x_{ij} - \bar{x})^2 &= \frac{k^2[(b-r)(v-k)-(v-m)(v-rk)]}{v(v-m)} - \frac{k^2(-1)^2}{t(-1)} \\
&= \frac{k^2}{vt(v-m)}[t(b-r)(v-k)-(v-m)\{t(v-rk) \\
&\qquad +v(r-1)\}] \\
&= \frac{k^2(t-1)[tr(v-k)-(v-m)v]}{vt(v-m)} \\
&= \frac{k^2(t-1)[b(v-k)-v(v-m)]}{vt(v-m)} \\
&= \frac{k^2(t-1)(b-v+m-r)}{t(v-m)}.
\end{aligned}$$

Since $\Sigma\Sigma\,(x_{ij} - \bar{x})^2 \geq 0$, we get $b \geq v - m + r$. The other part of the exercise also follows from the consideration of

$$\Sigma\Sigma\,(x_{ij} - \bar{x})^2 = \frac{k^2(t-1)(b-v+m-r)}{t(v-m)}.$$

100. In Exercise 87, we have proved that the characteristic roots of NN' of a triangular design are given by

$\theta_0 = rk$, with multiplicity 1
$\theta_1 = r - 2\lambda_1 + \lambda_2$, with multiplicity $n\,(n-3)/2$
$\theta_2 = r + (n-4)\lambda_1 - (n-3)\lambda_2$, with multiplicity $(n-1)$.

(i) Let $r - 2\lambda_1 + \lambda_2 = 0$, then rank $(NN') = 1 + n - 1 = n$. Thus,

$n = \text{rank}\,(NN') = \text{rank}\,(N'N) \leq b.$

Hence $b \geq n$. Further if the design is resolvable, then

$n = \text{rank}\,(NN') = \text{rank}\,(N'N) \leq b - (r-1).$

Therefore, we get $b \geq n + r - 1$.

(ii) Let $r + (n-4)\lambda_1 - (n-3)\lambda_2 = 0$. Then

$\text{rank}\,(N\,N') = 1 + n(n-3)/2 = 1 + v - n.$

Hence,

$1 + v - n = \text{rank}\,(NN') = \text{rank}\,(N'N) \leq b.$

Therefore, we get $b \geq v - n + 1$. If in addition, the design is resolvable, then

$$1 + v - n = \text{rank}\,(NN') = \text{rank}\,(N'N) \le b - (r - 1).$$

Hence, we get

$$b \ge v - n + r.$$

101. Let the ith block contain x_{ij} treatments fromt he jth row of the association scheme, $i = 1, 2, \ldots, b$; $j = 1, 2, \ldots, n$. Then for any j, we get

$$\sum_{i=1}^{b} x_{ij} = (n-1)r$$

and

$$\sum_{i=1}^{b} x_{ij}(x_{ij} - 1) = (n-1)(n-2)\lambda_1.$$

Hence,

$$\sum_{i=1}^{b} x_{ij}^2 = (n-1)[(n-2)\lambda_1 + r].$$

Define $\bar{x}_j = \sum_1^b x_{ij}/b = \dfrac{(n-1)r}{b}$. Then

$$\sum_{i=1}^{b} (x_{ij} - \bar{x}_j)^2 = (n-1)[(n-2)\lambda_1 + r] - \frac{(n-1)^2 r^2}{b}.$$

Now, $b = vr/k = n(n-1)r/2\,k$, hence

$$\sum_{i=1}^{b} (x_{ij} - \bar{x}_j)^2 = \frac{(n-1)}{n}[n\,(n-2)\lambda_1 + nr - 2rk].$$

Now, $n_1\lambda_1 + n_2\lambda_2 = r\,(k-1)$, $n_1 = 2(n-2)$, and $n_2 = (n-2)(n-3)/2$, gives

$$2rk = 2r + 4(n-2)\lambda_1 + (n-2)(n-3)\lambda_2.$$

Hence,

$$\sum_{i=1}^{b} (x_{ij} - \bar{x}_j)^2 = \frac{(n-1)(n-2)}{n}[r + (n-4)\lambda_1 - (n-3)\lambda_2]$$
$$= 0,$$

since $r + (n-4)\lambda_1 - (n-3)\lambda_2 = 0$. Therefore, the ith block contains $x_{ij} = \bar{x}_j = (n-1)\,r/b = 2k/n$ treatments from the jth row of the association scheme. Since this does not depend on i and j, it follows that every block contains $2\,k/n$ treatments from every row of the association scheme.

102. Since $n_1 = 2(n-2)$, $n_2 = (n-2)(n-3)/2$, and $n_1\lambda_1 + n_2\lambda_2 = r(k-1)$ we get

$$4\,(n-2)\lambda_1 + (n-2)(n-3)\lambda_2 = 2r(k-1).$$

Further, we are given $r + (n-4)\lambda_1 - (n-3)\lambda_2 = 0$. Solving the above two equations for λ_1 and λ_2, we get

$$\lambda_1 = r\,(2k-n)/n(n-2)$$
$$\lambda_2 = 2\,r(nk+n-4k)/n(n-2)(n-3).$$

Now, let the blocks be denoted by $B_1, B_2, \ldots, B_b$ and let B_1 have d blocks $B_2, B_3, \ldots, B_{d+1}$ disjoint with it. Let B_1 have x_i treatments common with B_i, $i = d+2, d+3, \ldots b$. Then, we get

$$\sum_{i=d+2}^{b} x_i = k\,(r-1).$$

Since B_1 contains 2 k/n treatments from each row of the association scheme, we get

$$\sum_{i=d+2}^{b} x_i(x_i-1) = n(2\,k/n)(2\,k/n-1)(\lambda_1-1)$$
$$+\,[k(k-1) - n(2\,k/n)(2\,k/n-1)](\lambda_2-1).$$

Hence,

$$\Sigma\, x_i^2 = (k/n)[2(2k-n)(\lambda_1-1) + \{n(k-1) - 2(2k-n)\}(\lambda_2-1)$$
$$+\,k(r-1)]$$
$$= \frac{k}{n}[2\,(2\,k-n)(\lambda_1-\lambda_2) + n(k-1)(\lambda_2-1) + n\,(r-1)].$$

Substituting the values of λ_1 and λ_2, and simplifying, we get

$$\sum_{i=d+2}^{b} x_i^2 = \frac{k^2 \cdot T}{v\,(v-n)},$$

where $T = (v-k)(b-r) - (v-rk)(v-n)$. Therefore $\bar{x} = \dfrac{\Sigma x_i}{b-d-1} = \dfrac{k(r-1)}{b-d-1}$. Then,

$$\sum_{i=d+2}^{b} (x_i-\bar{x})^2 = \frac{k^2\,T}{v\,(v-n)} - \frac{k^2\,(r-1)^2}{b-d-1}.$$

Since $\Sigma\,(x_i-\bar{x})^2 \geq 0$, we get

$$d \leq b - 1 - \frac{v\,(v-n)(r-1)^2}{T}.$$

If $d = b - 1 - \dfrac{v(v-n)(r-1)^2}{T}$, then $\Sigma\,(x_i - \bar{x})^2 = 0$, which gives

$$x_i = \bar{x} = \frac{k(r-1)}{b-d-1} = \frac{kT}{v(v-n)(r-1)}.$$

Hence the result.

103. Let the blocks be denoted by $B_1, B_2, \ldots, B_b$. Suppose B_1 has x_i treatments in common with B_i, $i = 2, 3, \ldots, b$. Then as in Exercise 102, we can prove that

$$\sum_{i=2}^{b} x_i = k(r-1)$$

$$\sum_{i=2}^{b} x_i^2 = \frac{k^2[(b-r)(v-k) - (v-rk)(v-n)]}{v(v-n)}.$$

Define $\bar{x} = \Sigma\, x_i/(b-1) = k(r-1)/(b-1)$. Hence, we get

$$\begin{aligned}\Sigma\,(x_i-\bar{x})^2 &= \frac{k^2[(b-r)(v-k)-(v-rk)(v-n)]}{v(v-n)} - \frac{k^2(r-1)^2}{(b-1)}\\ &= \frac{k^2}{v(v-n)(b-1)}[(b-1)\{(b-r)(v-k)-(v-rk)(v-n)\}\\ &\qquad - v(v-n)(r-1)^2]\\ &= \frac{k^2}{v(v-n)(b-1)}[(b-1)(b-r)(v-k)\\ &\qquad - (v-n)(b-r)(v-k)]\\ &= \frac{k^2(b-r)(v-k)(b-v+n-1)}{v(v-n)(b-1)}.\end{aligned}$$

Hence, the result follows.

104. Consider a resolvable triangular design with $r + (n-4)\lambda_1 - (n-3)\lambda_2 = 0$ and $b = tr$, $v = n(n-1) = tk$, where t is a positive integer 1. Let B_{ij} denote the jth block in the ith replication, $i = 1, 2, \ldots, r$ and $j = 1, 2, \ldots, t$. Let the number of common treatments between B_{11} and B_{ij} be x_{ij}, $i = 2, 3, \ldots, r$; $j = 1, 2, \ldots, t$. Then, as in Exercise 102, we can prove that

$$\sum_{i=2}^{r}\sum_{j=1}^{t} x_{ij} = k(r-1)$$

$$\sum_{i=2}^{r}\sum_{j=1}^{t} x_{ij}^2 = \frac{k^2[(b-r)(v-k)-(v-rk)(v-n)]}{v(v-n)}.$$

Define $\bar{x} = \Sigma\Sigma\, x_{ij}/t(r-1) = k/t = k^2/v$. Then, we get

$$\Sigma\Sigma\,(x_{ij}-\bar{x})^2 = \frac{k^2[(b-r)(v-k)-(v-r\,k)(v-n)]}{v\,(v-n)} - \frac{k^2(r-1)}{t}$$

$$= \frac{k^2}{v\,t(v-n)}[t(b-r)(v-k)-v(v-n)(t-1)]$$

$$= \frac{k^2(t-1)[t\,r(v-k)-v(v-n)]}{v\,t(v-n)}$$

$$= \frac{k^2(t-1)[v(b-r)-v(v-n)]}{v\,t(v-n)}$$

$$= \frac{(k^2(t-1)(b-r-v+n)}{t\,(v-n)}.$$

Since $\Sigma\Sigma\,(x_{ij}-\bar{x})^2 \geq 0$, we get $b \geq v-n+r$. The other part of the exercise follows from the consideration of

$$\Sigma\Sigma\,(x_{ij}-\bar{x})^2 = \frac{k^2\,(t-1)(b-v-r+n)}{t\,(v-n)}.$$

105. In Exercise 86, we have proved that the distinct characteristic roots of NN' of a L_i design are given by

$\theta_0 = r\,k$, with multiplicity 1
$\theta_1 = r - i\lambda_1 + \lambda_2(i-1)$, with multiplicity $(s-1)\,(s-i-1)$
$\theta_2 = r + \lambda_1(s-i) - \lambda_2(s-i+1)$, with multiplicity $i\,(s-1)$.

(i) Let $r + i\lambda_1 + \lambda_2(i-1) = 0$. Then, since the rank of a matrix is equal to the number of its non-zero characteristic roots, we get

$\text{rank}\,(NN'') = 1 + i(s-1)$.

Hence,

$1 + i(s-1) = \text{rank}\,(NN') = \text{rank}\,(N'N) \leq b$.

Thus, $b \geq 1 + i(s-1)$. Further, if the design is resolvable, then

$1 + i(s-1) = \text{rank}\,(N'N) \leq b - (r-1)$.

Therefore, $b \geq r + i(s-1)$.

(ii) Let $r + \lambda_1(s - i - \lambda_2(s-i+1) = 0$. Then arguing as in (i), we get

$$1 + (s-1)(s-i+1) = \text{rank}\,(NN')$$
$$= \text{rank}\,(N'N) \leq b.$$

Hence, we get $b \geq 1 + (s-1)(s-i+1)$. If the design is resolvable, then

$1 + (s-1)(s-i+1) = \text{rank}\,(N'N) \leq b - (r-1)$.

Hence, $b \geq r + (s-1)(s-i+1)$.

106. Let the ith block contain x_{ij} treatments from the jth row (or column) of the association scheme, $i = 1, 2, \ldots, b$; $j = 1, 2, \ldots, s$. Then

$$\sum_{i=1}^{b} x_{ij} = s\,r, \text{ and } \sum_{i=1}^{b} x_{ij}(x_{ij} - 1) = s\,(s-1)\lambda_1.$$

Hence,

$$\sum_{i=1}^{b} x_{ij}^2 = s(s-1)\lambda_1 + s\,r.$$

Therefore, defining $\bar{x}_j = \sum_{i=1}^{b} x_{ij}/b = \frac{s\,r}{b} = \frac{k}{s}$, $j = 1, 2, \ldots, b$, we get

$$\begin{aligned}\sum_{i=1}^{b}(x_{ij} - \bar{x}_j)^2 &= s\,(s-1)\lambda_1 + s\,r - \frac{s^2\,r^2}{b}\\ &= s\,(s-1)\lambda_1 + s\,r - r\,k.\end{aligned}$$

Now, from $r(k-1) = n_1\lambda_1 + n_2\lambda_2$, we get

$$r\,(k-1) = 2(s-1)\lambda_1 + (s-1)^2\lambda_2.$$

Hence, $r\,k = r + 2(s-1)\lambda_1 + (s-1)^2\lambda_2$. Then, we get

$$\begin{aligned}\Sigma\,(x_{ij} - \bar{x}_j)^2 &= s\,(s-1)\lambda_1 + r(s-1) - 2(s-1)\lambda_1 - (s-1)^2\lambda_2\\ &= (s-1)[r + (s-2)\lambda_1 - (s-1)\lambda_2]\\ &= 0,\end{aligned}$$

since $r + (s-2)\lambda_1 - (s-1)\lambda_2 = 0$. Therefore, $x_{ij} = \bar{x} = k/s$ for all i and j. Hence the result.

107. Let the blocks be denoted by $B_1, B_2, \ldots, B_b$. Let the block B_1 have d blocks $B_2, B_3, \ldots, B_{d+1}$ disjoint with it and have x_i treatments in common with the block B_i, $i = d+2, d+3, \ldots, b$. Then, we get

$$\sum_{i=d+2}^{b} x_i = k\,(r-1).$$

Since B_1 contains k/s treatments from each row (column) of the association scheme, we get

$$\begin{aligned}\sum_{i=d+2}^{b} x_i(x_i - 1) &= 2\cdot s\cdot\left(\frac{k}{s}\right)\left(\frac{k}{s} - 1\right)(\lambda_1 - 1)\\ &\quad + \left[k\,(k-1) - 2\,s\left(\frac{k}{s}\right)\left(\frac{k}{s} - 1\right)\right](\lambda_2 - 1)\\ &= \frac{k}{s}[2(k-s)(\lambda_1 - \lambda_2) + s(k-1)(\lambda_2 - 1)].\end{aligned}$$

Hence,

$$\sum_{i=d+2}^{b} x_i^2 = \frac{k}{s}[2(k-s)(\lambda_1-\lambda_2)+s(k-1)\lambda_2+s(r-k)].$$

Now, we have

$$\begin{aligned} &r+(s-2)\lambda_1-(s-1)\lambda_2=0\\ &r\,(k-1)=2(s-1)\lambda_1+(s-1)^2\lambda_2. \end{aligned}$$

Solving these two equations for λ_1 and λ_2, we get

$$\begin{aligned} \lambda_1 &= \frac{r\,(k-s)}{s\,(s-1)}\\ \lambda_2 &= \frac{r\,(s+sk-2k)}{s\,(s-1)^2} \end{aligned}$$

and hence $\lambda_1-\lambda_2=-\dfrac{b\,r(v-k)}{s\,(s-1)^2}$. Substituting these values in $\Sigma\ x_i^2$, we obtain

$$\begin{aligned} \sum_{i=d+2}^{b} x_i^2 &= \frac{k}{s}\left[-\frac{2(k-s)r(v-k)}{s(s-1)^2}+\frac{rs(k-1)(s+sk-2k)}{s(s-1)^2}\right]\\ &= \frac{k}{s}\left[-\frac{(k-s)r(v-k)}{s\,(s-1)^2}-\frac{(k-s)r(v-k)}{s\,(s-1)^2}\right.\\ &\qquad\left.+\frac{r\,s(k-1)(s-k)}{s\,(s-1)^2}+\frac{r\,k(k-1)}{(s-1)}+s(r-k)\right]\\ &= \frac{k}{s}\left[-\frac{(k-s)k(b-r)}{s\,(s-1)^2}+\frac{r(s^2-k^2)}{s\,(s-1)}+\frac{rk\,(k-1)}{s-1}+s(r-k)\right]\\ &= \frac{k}{s}\left[-\frac{(k\,(b-r)(k-s)}{s\,(s-1)^2}+\frac{k\{-rk+rsk-rs+bs-vs+v\}}{s\,(s-1)}\right]\\ &= \frac{k^2}{v\,(s-1)^2}[-(b-r)(k-s)-(v-rk)(s-1)^2\\ &\qquad +(v-s)(b-r)]\\ &= \frac{k^2\,T}{v\,(s-1)^2}, \end{aligned}$$

where $T=(b-r)(v-k)-(v-rk)(s-1)^2$. Define $\bar{x}=\sum_{i=d+2}^{b} x_i/(b-d-1)=\dfrac{k\,(r-1)}{b-d-1}$. Then

$$\sum_{i=d+2}^{b}(x_i-\bar{x})^2=\frac{k^2\,T}{v\,(s-1)^2}-\frac{k^2(r-1)^2}{b-d-1}.$$

Since $\Sigma\,(x_i - \bar{x})^2 \geq 0$, we get

$$d \leq b - 1 - \frac{v\,(s-1)^2(r-1)^2}{T}.$$

If $d = b - 1 - \dfrac{v\,(s-1)^2(r-1)^2}{T}$, then $\Sigma\,(x_i - \bar{x})^2 = 0$. This gives

$$x_i = \bar{x} = \frac{k\,(r-1)}{b-d-1} = \frac{k\,T}{v\,(r-1)(s-1)^2},$$

$i = d+2, d+3, \ldots, b$. Hence the result.

108. Let the blocks be denoted by $B_1, B_2, \ldots, B_b$, and let B_1 have x_i treatments in common with B_i, $i = 2, 3, \ldots, b$. Then, as in Exercise 107, we can prove that

$$\sum_{i=2}^{b} x_i = k\,(r-1)$$

$$\sum_{i=2}^{b} x_i^2 = \frac{k^2\,T}{v\,(s-1)^2},$$

where $T = (b-r)(v-k) - (v-r\,k)(s-1)^2$.
Now, define $\bar{x} = \Sigma\,x_i/(b-1) = \dfrac{k\,(r-1)}{b-1}$. Then, we get

$$\begin{aligned}\Sigma\,(x_i - \bar{x})^2 &= \frac{k^2\,T}{v\,(s-1)^2} - \frac{k^2\,(r-1)^2}{(b-1)} \\ &= \frac{k^2}{v\,(s-1)^2(b-1)}[(b-1)T - v(s-1)^2(r-1)^2] \\ &= \frac{k^2}{v\,(s-1)^2(b-1)}[(b-1)(b-r)(v-k) \\ &\quad - (s-1)^2\{(b-1)(v-rk) + v(r-1)^2\}] \\ &= \frac{k^2[(b-1)(b-r)(v-k) - (s-1)^2(b-r)(v-k)]}{v\,(s-1)^2(b-1)} \\ &= \frac{k^2(b-r)(v-k)\{(b-1)-(s-1)^2\}}{v\,(s-1)^2(b-1)}.\end{aligned}$$

Consideration of $\Sigma\,(x_i - \bar{x})^2 = 0$ proves the result.

109. Consider a resolvable L_2 design with $r + (s-2)\lambda_1 + (s-1)\lambda_2 = 0$ and $b = t\,r$, $v = s^2 = t\,k$, where t is a positive integer greater than 1. Let B_{ij}

denote the jth block in the ith replication $j = 1, 2, \ldots, t,\ i = 1, 2, \ldots, r$. Let B_{11} have x_{ij} treatments in common with B_{ij}, $i = 2, 3, \ldots, r,\ j = 1, 2, \ldots, t$. Then, as in Exercise 107, we can prove that

$$\sum_{i=2}^{r}\sum_{j=1}^{t} x_{ij} = k\,(r-1)$$

$$\sum_{i=2}^{r}\sum_{j=1}^{t} x_{ij}^2 = \frac{k^2\,T}{v\,(s-1)^2}$$

Define

$$\bar{x} = \frac{\Sigma\Sigma\, x_{ij}}{t\,(r-1)} = \frac{k}{t} = \frac{k^2}{v}.$$

Then

$$\begin{aligned}
\sum_{i=2}^{r}\sum_{j=1}^{t}(x_{ij}-\bar{x})^2 &= \frac{k^2\,T}{v\,(s-1)^2} - \frac{k^2(r-1)^2}{b-t} \\
&= k^2\left[\frac{T}{v\,(s-1)^2} - \frac{(r-1)}{t}\right] \\
&= k^2\frac{[t\cdot T - v(s-1)^2(r-1)]}{v\,(s-1)^2 t} \\
&= \frac{k^2[t(b-r)(v-k) - v(t-1)(s-1)^2]}{v\,t(s-1)^2} \\
&= \frac{k^2(t-1)[r\,t(v-k) - v(s-1)^2]}{v\,t(s-1)^2} \\
&= \frac{k^2(t-1)[b\,(v-k) - v(s-1)^2]}{v\,t(s-1)^2} \\
&= \frac{k^2(t-1)v[b-r-(s-1)^2]}{v\,t(s-1)^2}.
\end{aligned}$$

Since $\Sigma\Sigma\,(x_{ij}-\bar{x})^2 \geq 0$, it follows that $b \geq r + (s-1)^2$. The necessary and sufficient condition for it to be affine resolvable follows from the consideration of $\Sigma\Sigma\,(x_{ij}-\bar{x})^2 = 0$.

110. We renumber the blocks so that the two given blocks occupy the first and second positions. Then, the distinct non-zero characteristic roots of $N'N$ are also $r\,k > \mu_0 > \mu_1 > \ldots, \mu_p$. Then it is easy to verify that the distinct non-zero characteristic roots of the matrix

$$A = N'N - \frac{r\,k - \mu_0}{b}E_{bb}$$

are $\mu_0 > \mu_1 > \ldots > \mu_p$. Then, if $\mathbf{y}$ is any real $b \times 1$ vector,

$$0 \leq \mathbf{y}'A\,\mathbf{y} \leq \mu_0$$

i.e.

$$0 \leq \mathbf{y}' \left[N'N - \frac{r\,k - \mu_0}{b} E_{bb} \right] \mathbf{y} \leq \mu_0.$$

Select $\mathbf{y}' = [1/\sqrt{2}, -1/\sqrt{2}, 0, 0, \ldots, 0]$, then we get

$$0 \leq k - x \leq \mu_0,$$

which gives

$$k - \mu_o \leq x.$$

Next, select $\mathbf{y}' = [1/\sqrt{2}, 1/\sqrt{2}, 0, 0, \ldots, 0]$. Then, we get

$$k + x - \frac{2(r\,k - \mu_0)}{b} \leq \mu_0,$$

which gives

$$x \leq \mu_0 - k + 2(r\,k - \mu_0)b^{-1}.$$

Thus, we get

$$k - \mu_0 \leq x \leq \mu_0 - k + 2(r\,k - \mu_0)b^{-1}.$$

Since $x \leq k$, and $0 \leq x$, one gets

$$\max[0, k - \mu_0] \leq x \leq \min[k, \mu_0 - k + 2(r\,k - \mu_0)b^{-1}].$$

Also, consider $\overset{*}{N} = E_{vb} - N$. Then $\overset{*}{N}{}'\overset{*}{N} = N'N + (v - 2k)E_{vb}$. Since, the elements of $\overset{*}{N}{}'\overset{*}{N}$ are non-negative, we obtain

$$x \geq 2\,k - v.$$

Hence, we get

$$\max[0, 2\,k - v, k - \mu_0] \leq x \leq \min[k, \mu_0 - k + 2(r\,k - \mu_0)b^{-1}].$$

111. In Exercise 65, we have proved that the characteristic roots of NN' of a BIBD are $r\,k$ and $r - \lambda$. Thus, in the notations of Exercise 110, we have $\mu_0 = r - \lambda$. Then, applying Exercise 110, one easily obtains the required result.

112. In a singular group divisible design, the characteristic roots of NN' are $r\,k, r - \lambda_1 = 0$ and $r\,k - v\,\lambda_2 > 0$. Thus, in the notation of Exercise 110, $\mu_0 = r\,k - v\lambda_2$. Now, from $n_1\lambda_1 + n_2\lambda_2 = r(k - 1)$, and $r = \lambda_1$, we get

$$\lambda_2 = \frac{r(k - n)}{v - n}.$$

Hence,

$$\mu_0 = r\,k - v\lambda_2 = k(b - r)/(m - 1),$$

since $v = m\,n$. Then

$$k - \mu_0 = -k(b - m - r + 1)/(m - 1).$$

$$\mu_0 - k + \frac{2(r\,k - \mu_0)}{b} = \frac{k(b - m - r + 1)}{(m - 1)} + \frac{2\,v\,\lambda_2}{b}$$
$$= \frac{k(b - m - r + 1)}{(m - 1)} + \frac{2\,v\,r(k - n)}{b(v - n)}$$
$$= \frac{k}{(v - n)}[n(b - m - r - 1) + 2k].$$

Applying Exercise 110, we get

$$\max[0, 2\,k - v, -k(b - m - r + 1)/(m - 1)]$$
$$\leq x \leq \min[k, k\{n(b - m - r - 1) + 2\,k\}/(v - n)].$$

113. In a semi-regular group divisible design, the characteristic roots of NN' are $r\,k, r - \lambda_1 > 0, r\,k - v\lambda_2 = 0$. Hence, in the notation of Exercise 110, we have $\mu_0 = r - \lambda_1$. Now, since $r\,k - v\lambda_2 = 0$, we have $\lambda_2 = rk/v$. Hence from $n_1\lambda_1 + n_2\lambda_2 = r(k - 1), n_1 = n - 1, n_2 = n(m - 1)$, and $\lambda_2 = rk/v$, we get

$$\lambda_1 = r(k - m)/(v - m)$$

and,

$$\mu_0 = r - \lambda_1 = k(b - r)/(v - m).$$

Therefore, we get

$$k = \mu_0 = -k(b - v + m - r)/(v - m).$$

Further,

$$\mu_0 - k + \frac{2(rk - \mu_0)}{b} = \frac{k(b - v + m - r)}{v - m} + \frac{2k}{b}\left[r - \frac{b - r}{v - m}\right]$$
$$= \frac{k}{b(v - m)}[(b - r)(b - 2) - (v - m)(b - 2r)].$$

Hence, using Exercise 110, we get

$$\max\,[0, 2k - v, -k(b - v + m - r)/(v - m)]$$
$$\leq x \leq \min[k, k\{(b - r)(b - 2) - (v - m)(b - 2r)\}/b(v - m)].$$

114. In a triangular design, the characteristic roots of NN' are $r\,k$, $r - 2\lambda_1 + \lambda_2$, and $r + (n - 4)\lambda_1 - (n - 3)\lambda_2$, (see Exercise 87).

(i) Consider $r - 2\lambda_1 + \lambda_2 = 0$. Then in the notation of Exercise 110, we have

$$\mu_0 = r + (n - 4)\lambda_1 - (n - 3)\lambda_2.$$
$$= r - 2\lambda_1 + \lambda_2 + (n - 2)(\lambda_1 - \lambda_2)$$
$$= (n - 2)(\lambda_1 - \lambda_2)$$
$$= (n - 2)(r - \lambda_1),$$

since $r - 2\lambda_1 + \lambda_2 = 0$. Now from $r - 2\lambda_1 + \lambda_2 = 0$, and $n_1\lambda_1 + n_2\lambda_2 = r(k-1)$, and $n_1 = 2(n-2)$, $n_2 = (n-2)(n-3)/2$, we obtain

$$\lambda_1 = \frac{r[(n-2)(n-3) + 2(k-1)]}{2(n-1)(n-2)}.$$

Hence,

$$\begin{aligned} r - \lambda_1 &= \frac{r[(n+1)(n-2) - 2(k-1)]}{2(n-1)(n-2)} \\ &= \frac{r[n(n-1) - 2k]}{2(n-1)(n-2)} \\ &= \frac{2r(v-k)}{2(n-1)(n-2)} = \frac{k(b-r)}{(n-1)(n-2)} \\ \mu_0 &= \frac{k(b-r)}{(n-1)} = \frac{k(b-r)}{(n-1)}, \end{aligned}$$

and

$$k - \mu_0 = k(n-1-b+r)/(n-1).$$

Also

$$\mu_0 - k + \frac{2(r\,k - \mu_0)}{b} = \frac{k(b+1-n-r)}{n-1} + \frac{2\,k}{b}\left\{r - \frac{b-r}{n-1}\right\}$$

$$= \frac{k}{b(n-1)}[b(b+1-n-r) + 2\{r(n-1) - (b-r)\}]$$

$$= \frac{k}{b(n-1)}[(b-r)(b-2) - (n-1)(b-2r)].$$

Hence, applying Exercise 110, we get

$$\max[0, 2k - v, -k(b-n-r+1)/(n-1)]$$
$$\leq x \leq \min[k, k\{(b-r)(b-2) - (n-1)(b-2\,r)\}/b(n-1)].$$

(ii) Now consider $r + (n-4)\lambda_1 - (n-3)\lambda_2 = 0$. Hence, in the notation of Exercise 110, we have

$$\mu_0 = r - 2\lambda_1 + \lambda_2.$$

But since $r + (n-4)\lambda_1 - (n-3)\lambda_2 = 0$, we have

$$r - 2\,\lambda_1 + \lambda_2 + (n-2)(\lambda_1 - \lambda_2) = 0.$$

Hence

$$r - 2\,\lambda_1 + \lambda_2 = -(n-2)(\lambda_1 - \lambda_2),$$

and

$$\mu_0 = -(n-2)(\lambda_1 - \lambda_2).$$

Now, from $n_1\lambda_1 + n_2\lambda_2 = r(k-1)$, and $r + (n-4)\lambda_1 - (n-3)\lambda_2 = 0$, we get

$$\lambda_1 = r(2k-n)/n(n-2)$$
$$\lambda_2 = 2\,r(n\,k + n - 4k)/n(n-2)(n-3).$$

Therefore

$$\begin{aligned}\lambda_1 - \lambda_2 &= \frac{r[(n-3)(2\,k-n) - 2(n\,k+n-4\,k)]}{n(n-2)(n-3)}\\ &= \frac{r[2\,k - n(n-1)]}{n(n-2)(n-3)} = \frac{2\,r(k-v)}{n(n-2)(n-3)}\\ &= -\frac{2\,k(b-r)}{n(n-2)(n-3)},\end{aligned}$$

and

$$\begin{aligned}\mu_0 &= \frac{2\,k(b-r)}{n(n-3)} = \frac{2\,k(b-r)}{2\,v-2\,n}\\ &= \frac{k(b-r)}{v-n}.\end{aligned}$$

Then

$$k - \mu_0 = -\frac{k(b-v-r+n)}{(v-n)}$$

and

$$\begin{aligned}&\mu_0 - k + \frac{2(r\,k-\mu_0)}{b}\\ &= \frac{k(b-v-r+n)}{v-n} + \frac{2\,k}{b}\left[r - \frac{(b-r)}{(v-n)}\right]\\ &= \frac{k}{b(v-n)}[b(b-v-r+n) + 2\,r(v-n) - 2(b-r)]\\ &= \frac{k}{b(v-n)}[(b-r)(b-2) - (v-n)(b-2\,r)].\end{aligned}$$

Hence, using Exercise 110, we get

$$\begin{aligned}&\max[0, 2\,k-v, -k(b-v-r+n)/(v-n)]\\ &\quad\le x \le \min[k, k\{(b-2\,r)(b-2) - (v-n)(b-2\,r)\}/b(v-n)]\end{aligned}$$

115. In a triangular design, the characteristic roots of NN' are $\theta_0 = r\,k$, $\theta_1 = r + (n-4)\lambda_1 - (n-3)\lambda_2$, $\theta_2 = r - 2\lambda_1 + \lambda_2$. Now, we can easily verify that

$$\theta_1 - \theta_2 = (n-2)(\lambda_1 - \lambda_2).$$

Now if $\lambda_1 > \lambda_2$, then $\theta_1 > \theta_2$. Hence in the notation of Exercise 110,

$\mu_0 = \theta_1$. Applying Exercise 110, we get

$$\max[0, 2\,k - v, k - \theta_1] \leq x \leq \min[k, \theta_1 - k + 2(r\,k - \theta_1)b^{-1}].$$

Further, if $\lambda_1 < \lambda_2$, then $\theta_1 < \theta_2$. Hence in the notation of Exercise 110, $\mu_0 = \theta_2$. Then, we get

$$\max[0, 2\,k - v, k - \theta_2] \leq x \leq \min[k, \theta_2 - k + 2(r\,k - \theta_2)b^{-1}].$$

116. In an L_i design, the characteristic roots of NN' are $\theta_0 = r\,k$, $\theta_1 = r + (s - i)\lambda_1 - (s - i + 1)\lambda_2$ and $\theta_2 = r - i\lambda_1 + (i - 1)\lambda_2$. Therefore, we find

$$\theta_1 - \theta_2 = s(\lambda_1 - \lambda_2)$$

Thus, if $\lambda_1 > \lambda_2$, then $\theta_1 > \theta_2$ and hence in the notation of Exercise 110, $\mu_0 = \theta_1$. If $\lambda_1 < \lambda_2$, then $\theta_1 < \theta_2$ and hene in the notation of Exercise 110, $\mu_0 = \theta_2$. Therefore using Exercise 110, we get the required result.

117. In an L_2 design, the characteristic roots of NN' are $r\,k$, $r + (s - 2)\lambda_1 - (s - 1)\lambda_2$ and $r - 2\lambda_1 + \lambda_2$. Since $r + (s - 2)\lambda_1 - (s - 1)\lambda_2 = 0$, we get

$$r - 2\,\lambda_1 + \lambda_2 + s(\lambda_1 - \lambda_2) = 0$$

i.e.

$$r - 2\,\lambda_1 + \lambda_2 = -s(\lambda_1 - \lambda_2).$$

Now, in the notation of Exercise 110, we have

$$\mu_0 = r - 2\,\lambda_1 + \lambda_2 = -s(\lambda_1 - \lambda_2)$$

and in Exercise 107, we have proved that

$$\lambda_1 - \lambda_2 = -\frac{r(v - k)}{s(s - 1)^2} = -\frac{k(b - r)}{s(s - 1)^2}.$$

Hence,

$$\mu_0 = \frac{k(b - r)}{(s - 1)^2},$$

and

$$k - \mu_0 = \frac{k\{(s - 1)^2 - (b - r)\}}{(s - 1)^2}.$$

Also

$$\mu_0 - k + \frac{2}{b}(r\,k - \mu_0) = \frac{k\{(b - r) - (s - 1)^2\}}{(s - 1)^2} + \frac{2k}{b}\left\{r - \frac{(b - r)_2}{s - 1}\right\}$$

$$= \frac{k}{b(s - 1)^2}[(b - r)(b - 2) - (s - 1)^2(b - 2r)].$$

Hence using Exercise 110, we get the requird result.

118. Let us renumber the blocks so that the first two blocks B_1 and B_2 have x treatments in common. The other blocks are denoted by $B_3, B_4, \ldots, B_b$. Let B_1 have x_i treatments common with the block B_i, $i = 3, 4, \ldots, b$. Then, as in Exercise 96, we can prove that

$$x + \sum_{i=3}^{b} x_i = k(r-1)$$

$$x^2 + \sum_{i=3}^{b} x_i^2 = \frac{k^2 T}{v(v-m)},$$

where $T = (b-r)(v-k) - (v-m)(v-r\,k)$. Then, we get

$$\sum_{i=3}^{b} (x_i - \bar{x})^2 = \left[\frac{k^2 T}{v(v-m)} - x^2\right] - \frac{[k(r-1)-x]^2}{b-2}.$$

Since $\Sigma\,(x_i - \bar{x})^2 \geq 0$, we get

$$x^2(b-1) - 2\,x\,k(r-1) + k^2(r-1)^2 - \frac{k^2 T(b-2)}{v(v-m)} \leq 0$$

i.e.

$$\left[x - \frac{k(r-1)}{b-1}\right] \leq \frac{k^2(b-2)}{v(v-m)(b-1)^2}[T(b-1) - v(v-m)(r-1)^2].$$

Now

$$\begin{aligned} &T(b-1) - v(v-m)(r-1)^2 \\ &\quad = (b-1)(b-r)(v-k) - (v-m)[(b-1)(v-r\,k) + v(r-1)^2] \\ &\quad = (b-r)(v-k)(b-1) - (v-m)(b-r)(v-k) \\ &\quad = (b-r)(v-k)(b-v+m-1). \end{aligned}$$

Hence

$$\begin{aligned} \left[x - \frac{k(r-1)}{b-1}\right] &\leq \frac{k^2(b-2)(b-r)(v-k)(b-v+m-1)}{v(v-m)(b-1)^2} \\ &\leq \frac{A^2}{(b-1)^2}, \end{aligned}$$

where $A^2 = k^2(b-2)(b-r)(v-k)(b-v+m-1)/v(v-m)$. Therefore, we get

$$\frac{k(r-1)-A}{b-1} \leq x \leq \frac{k(r-1)+A}{b-1}.$$

119. Let us renumber the blocks as $B_1, B_2, \ldots, B_b$, so that the first two B_1 and B_2 have x treatments in common. Further let B_1 and $B_i (i = 3, 4, \ldots, b)$ have x_i

treatments in common. Then, as in Exercise 102, we can prove that

$$x + \sum_{i=3}^{b} x_i = k(r-1)$$

$$x^2 + \sum_{i=3}^{b} x_i^2 = \frac{k^2 T}{v(v-n)},$$

where $T = (b-r)(v-k) - (v-n)(v-r\,k)$. Therefore, we get

$$\sum_{i=3}^{b} (x_i - \bar{x})^2 = \left[\frac{k^2 T}{v(v-n)} - x^2\right] - \frac{[k(r-1)-x]^2}{b-2}.$$

Since $\Sigma\,(x_i - \bar{x})^2 \geq 0$, we get

$$x^2(b-1) - 2k(r-1)x + k^2(r-1)^2 - \frac{k^2 T(b-2)}{v(v-n)} \leq 0$$

i.e.

$$\left[x - \frac{k(r-1)}{b-1}\right]^2 \leq \frac{k^2(b-2)}{v(v-n)(b-1)^2}[T(b-1) - v(v-n)(r-1)^2].$$

Now

$$\begin{aligned} &T(b-1) - v(v-n)(r-1)^2 \\ &\quad = (b-1)(b-r)(v-k) - (v-n)[(b-1)(v-r\,k) - v(r-1)^2] \\ &\quad = (b-1)(b-r)(v-k) - (v-n)(b-r)(v-k) \\ &\quad = (b-r)(v-k)(b-v+n-1) \end{aligned}$$

and therefore

$$\left[x - \frac{k(r-1)}{b-1}\right]^2 \leq \frac{A^2}{(b-1)^2},$$

where $A^2 = k^2(b-2)(b-r)(v-k)(b-v+n-1)/v(v-n)$.
Hence, the result follows.

120. Let us renumber the blocks as $B_1, B_2, \ldots, B_b$ so that the first two blocks B_1 and B_2 have x treatments in common. Let B_1 and $B_i (i = 3, 4, \ldots, b)$ have x_i treatments in common. Then, as in Exercise 107, we can prove that

$$x + \sum_{i=3}^{b} x_i = k(r-1)$$

$$x^2 + \sum_{i=3}^{b} x_i^2 = \frac{k^2 T}{v(s-1)^2},$$

where $T = (b-r)(v-k) - (v-r\,k)(s-1)^2$. Then from the above, we get

$$\Sigma\,(x_i - \bar{x})^2 = \frac{k^2 T}{v(s-1)^2} - x^2 - \frac{[k(r-1)-x]^2}{b-2}.$$

Since $\Sigma (x_i - \bar{x})^2 \geq 0$, we get

$$x^2(b-1) - 2\,x\,k(r-1) + k^2(r-1)^2 - \frac{k^2T(b-2)}{v(s-1)^2} \leq 0$$

i.e.

$$\left[x - \frac{k(r-1)}{b-1}\right]^2 \leq \frac{k^2(b-2)}{v(s-1)^2(b-1)^2}[T(b-1) - v(s-1)^2(r-1)^2].$$

Now

$$\begin{aligned} &T(b-1) - v(s-1)^2(r-1)^2 \\ &\quad = (b-1)(b-r)(v-k) - (s-1)^2[(b-1)(v-r\,k) + v(r-1)^2] \\ &\quad = (b-1)(b-r)(v-k) - (s-1)^2(b-r)(v-k) \\ &\quad = (b-r)(v-k)[b-1-(s-1)^2]. \end{aligned}$$

Therefore, we get

$$\left[x - \frac{k(r-1)}{(b-1)}\right]^2 \leq \frac{A^2}{(b-1)^2},$$

where $A^2 = \dfrac{k^2(b-2)(b-r)(v-k)\{b-1-(s-1)^2\}}{v(s-1)^2}$. Hence, the result follows.

121. Let the blocks of a group divisible design be denoted by $B_1, B_2, \ldots, B_b$. Let B_i contain x_{ij} treatments from jth group of treatments, $i = 1, 2, \ldots, b, j = 1, 2, \ldots, m$. Then we have shown in Exercise 96 that

$$\sum_{i=1}^{b}(x_{ij} - \bar{x})^2 = \frac{n(m-1)(rk - v\lambda_2)}{m}.$$

Since $\sum_{i=1}^{b}(x_{ij} - \bar{x}_j)^2 \geq 0$, we get $r\,k - v\lambda_2 \geq 0$.

122. Let the ith block contain x_{ij} treatments from the jth row of the association scheme, $i = 1, 2, \ldots, b$, and $j = 1, 2, \ldots, n$. In Exercise 101, we have proved that

$$\sum_{i=1}^{b}(x_{ij} - \bar{x}_j)^2 = \frac{(n-1)(n-2)}{n}[r + (n-4)\lambda_1 - (n-3)\lambda_2]$$

Now $n_1\lambda_1 + n_2\lambda_2 = r(k-1)$, $n_1 = 2(n-2)$, $n_2 = (n-2)(n-3)/2$, give

$$(n-3)\lambda_2 = \frac{2r(k-1)}{n-2} - 4\lambda_1.$$

Hence

$$\begin{aligned} r+(n-4)\lambda_1-(n-3)\lambda_2 &= r+(n-4)\lambda_1-\frac{2r(k-1)}{n-2}+4\lambda_1 \\ &= \frac{1}{(n-2)}[r(n-2)+n(n-2)\lambda_1-2r(k-1)] \\ &= \frac{1}{(n-2)}[rn+n(n-1)\lambda_1-n\lambda_1-2rk] \\ &= \frac{1}{(n-2)}[n(r-\lambda_1)-2(rk-v\lambda_1)] \end{aligned}$$

and

$$\sum_{i=1}^{b}(x_{ij}-\bar{x}_j)^2=\frac{(n-1)}{n}[n(r-\lambda_1)-2(rk-v\lambda_1)].$$

Since $\Sigma\,(x_{ij}-\bar{x}_j)^2\geq 0$, it follows that a necessary condition for the existence of a triangular design is

$$n(r-\lambda_1)\geq 2(rk-v\lambda_1)$$

i.e. $rk-v\lambda_1\leq n(r-\lambda_1)/2$.

Equivalently, the condition can be stated as

$$r+(n-4)\lambda_1-(n-3)\lambda_2\geq 0.$$

123. Let the ith block contain x_{ij} treatments from the jth row of the association scheme, $i=1,2,\ldots,b$, $j=1,2,\ldots,s$. Then, in Exercise 106, we have proved that

$$\Sigma\,(x_{ij}-\bar{x}_j)^2=(s-1)[r+(s-2)\lambda_1-(s-1)\lambda_2].$$

Now, $n_1\lambda_1+n_2\lambda_2=r(k-1)$, $n_1=2(s-1)$, and $n_2=(s-1)^2$, give

$$\lambda_2=\frac{r(k-1)}{(s-1)^2}-\frac{2\lambda_1}{(s-1)}.$$

Hence,

$$\begin{aligned} r+(s-2)\lambda_1-(s-1)\lambda_2 &= r+(s-2)\lambda_1-\frac{r(k-1)}{s-1}+2\lambda_1 \\ &= \frac{1}{s-1}[r(s-1)+s(s-1)\lambda_1-r(k-1)] \\ &= \frac{1}{s-1}[s(r-\lambda_1)-(rk-v\lambda_1)] \end{aligned}$$

and

$$\Sigma\,(x_{ij}-\bar{x}_j)^2=s(r-\lambda_1)-(rk-v\lambda_1).$$

Since $\Sigma\,(x_{ij}-\bar{x}_j)^2\geq 0$, it follows that a necessary condition for the existence of a L_2 design is that $rk-v\lambda_1\leq s(r-\lambda_1)$. Equivalently, the condition can be stated as

$$r + (s-2)\lambda_1 - (s-1)\lambda_2 \geq 0.$$

124. Due to the interchange treatment 2 occurs twice in the first column and treatment 1 does not occur in the first column, and also treatment 2 does not occur int he second column while treatment 1 occurs twice in the second column. Hence

$$L = E_{vv},$$

$$M = \begin{bmatrix} 0 & 2 & 1 & \dots & 1 \\ 2 & 0 & 1 & \dots & 1 \\ 1 & 1 & 1 & \dots & 1 \\ \cdot & & & & \\ \cdot & & & & \\ \cdot & & & & \\ 1 & 1 & 1 & \dots & 1 \end{bmatrix} = \begin{bmatrix} 2E_{22} - 2I_2 & E_{2(v-2)} \\ E_{(v-2)2} & E_{(v-2)(v-2)} \end{bmatrix}.$$

Let the vector $\mathbf{c} = \{c_1, c_2, \dots, c_v\}$ be partitioned as $\mathbf{c} = \{\mathbf{A}, \mathbf{B}\}$, where $\mathbf{A} = \{c_1, c_2\}$ and $\mathbf{B} = \{c_3, c_4, \dots, c_v\}$. Also the vector of treatment effects $\mathbf{t} = \{t_1, \dots, t_v\}$ is partitioned as $\mathbf{t} = \{\mathbf{a}, \mathbf{b}\}$, where $\mathbf{a} = \{t_1, t_2\}$ and $\mathbf{b} = \{t_3, t_4, \dots, t_v\}$. Then, we have

$$\begin{aligned}
\mathbf{Q} &= \mathbf{T} - \frac{1}{v}\mathbf{L}\mathbf{R} - \frac{1}{v}\mathbf{M}\mathbf{C} + \frac{1}{v^2}\mathbf{L}E_{v1}G \\
&= \mathbf{T} - \frac{G}{v}E_{v1} - \frac{1}{v}\begin{bmatrix} 2E_{22} - 2I_2 & E_{2(v-2)} \\ E_{(v-2)2} & E_{(v-2)(v-2)} \end{bmatrix}\begin{bmatrix} \mathbf{A} \\ \mathrm{B} \end{bmatrix} + \frac{G}{v}E_{v1} \\
&= \mathbf{T} - \frac{1}{v}\begin{bmatrix} 2GE_{21} - 2\mathbf{A} \\ GE_{(v-2)1} \end{bmatrix}.
\end{aligned}$$

Hence, we have

$$\begin{aligned}
Q_1 &= T_1 - \frac{2G}{v} + \frac{2}{v}c_1 \\
Q_2 &= T_2 - \frac{2G}{v} + \frac{2}{v}c_2 \\
Q_j &= T_j - \frac{G}{v}, j = 3, 4, \dots, v.
\end{aligned}$$

Further,

$$\begin{aligned}
F &= \mathrm{diag}(r, r, \dots, r) - \frac{1}{v}LL' - \frac{1}{v}MM' + \frac{1}{v^2}LE_{vv}L' \\
&= vI_v - E_{vv} - \frac{1}{v}\begin{bmatrix} 4I_2 + (v-2)E_{22} & vE_{2(v-2)} \\ vE_{(v-2)2} & vE_{(v-2)(v-2)} \end{bmatrix} + E_{vv} \\
&= vI_v - \begin{bmatrix} \dfrac{4I_2 + (v-2)E_{22}}{v} & E_{2(v-2)} \\ E_{(v-2)2} & E_{(v-2)(v-2)} \end{bmatrix}.
\end{aligned}$$

Hence, the equations for estimating $\mathbf{t}$ are given by

$$\begin{aligned}\mathbf{Q} &= \mathrm{F}\mathbf{t}\\ &= \mathrm{v}\mathbf{t} - \begin{bmatrix} \frac{1}{\mathrm{v}}(4\mathrm{I}_2 + (\mathrm{v}-2)\mathrm{E}_{22}) & \mathrm{E}_{2(\mathrm{v}-2)} \\ \mathrm{E}_{(\mathrm{v}-2)2} & \mathrm{E}_{(\mathrm{v}-2)(\mathrm{v}-2)} \end{bmatrix} \begin{bmatrix} \mathbf{a} \\ \mathbf{b} \end{bmatrix}.\end{aligned}$$

We assume $\mathrm{E}_{1\mathrm{v}}\mathbf{t} = 0$. Hence, the above equations can be written as

$$\mathbf{Q} = \begin{bmatrix} \frac{\mathrm{v}^2-4}{\mathrm{v}}\mathbf{a} + \frac{2}{\mathrm{v}}(\mathrm{E}_{12}\mathbf{a})\mathrm{E}_{21} \\ \mathrm{v}\mathbf{b} \end{bmatrix}.$$

Thus, we get

$$\begin{aligned}\mathrm{Q}_1 &= \frac{\mathrm{v}^2-4}{\mathrm{v}}\hat{\mathrm{t}}_1 + \frac{2}{\mathrm{v}}(\hat{\mathrm{t}}_1 + \hat{\mathrm{t}}_2)\\ \mathrm{Q}_2 &= \frac{\mathrm{v}^2-4}{\mathrm{v}}\hat{\mathrm{t}}_2 + \frac{2}{\mathrm{v}}(\hat{\mathrm{t}}_1 + \hat{\mathrm{t}}_2)\\ \mathrm{Q}_j &= \mathrm{v}\hat{\mathrm{t}}_j, j = 3, 4, \ldots, \mathrm{v}.\end{aligned}$$

Solving these equations, we get

$$\begin{aligned}&\hat{\mathrm{t}} = [(\mathrm{v}^2-2)\mathrm{Q}_1 - 2\mathrm{Q}_2]/\mathrm{v}(\mathrm{v}^2-4)\\ &\hat{\mathrm{t}}_2 = [(\mathrm{v}^2-2)\mathrm{Q}_2 - 2\mathrm{Q}_1]/\mathrm{v}(\mathrm{v}^2-4)\\ &\hat{\mathrm{t}}_j = \mathrm{Q}_j/\mathrm{v}, j = 3, 4, \ldots, \mathrm{b}.\end{aligned}$$

Hence, we get

$$\begin{aligned}&\mathrm{var}\,(\hat{\mathrm{t}}_i) = \sigma^2(\mathrm{v}^2-2)/\mathrm{v}(\mathrm{v}^2-4), i = 1, 2\\ &\mathrm{var}\,(\hat{\mathrm{t}}_i) = \sigma^2/\mathrm{v}, i = 3, 4, \ldots, \mathrm{v}\\ &\mathrm{cov}\,(\hat{\mathrm{t}}_1, \hat{\mathrm{t}}_2) = -2\sigma^2/\mathrm{v}(\mathrm{v}^2-4)\\ &\mathrm{cov}\,(\hat{\mathrm{t}}_i, \hat{\mathrm{t}}_j) = 0, i = 1, 2, j = 3, 4, \ldots, \mathrm{v}\\ &\mathrm{cov}\,(\hat{\mathrm{t}}_i, \hat{\mathrm{t}}_j) = 0, i \neq j = 3, 4, \ldots, \mathrm{v}.\end{aligned}$$

Then, we obtain

$$\begin{aligned}&\mathrm{var}\,(\hat{\mathrm{t}}_1 - \hat{\mathrm{t}}_2) = 2\sigma^2/\mathrm{v}\\ &\mathrm{var}\,(\hat{\mathrm{t}}_1 - \hat{\mathrm{t}}_i) = 2\sigma^2(\mathrm{v}^2-3)/\mathrm{v}(\mathrm{v}^2-4), i = 3, 4, \ldots, \mathrm{v}\\ &\mathrm{var}\,(\hat{\mathrm{t}}_2 - \hat{\mathrm{t}}_i) = 2\sigma^2(\mathrm{v}^2-3)/\mathrm{v}(\mathrm{v}^2-4), i = 3, 4, \ldots, \mathrm{v}\\ &\mathrm{var}\,(\hat{\mathrm{t}}_i - \hat{\mathrm{t}}_j) = 2\sigma^2/\mathrm{v}, i \neq j = 3, 4, \ldots, \mathrm{v}.\end{aligned}$$

When the treatments 1 and 2 are not interchanged,

$$\mathrm{var}\,(\hat{\mathrm{t}}_i - \hat{\mathrm{t}}_j) = 2\sigma^2/\mathrm{v}, i \neq j = 1, 2, \ldots, \mathrm{v}.$$

When the treatment 1 and 2 are interchanged, the average variance of BLUEs of elementary treatment contrasts is given by

$$\bar{V} = \frac{2\sigma^2}{v}\left[2 + \frac{4(v-2)(v^2-3)}{(v^2-4)} + (v-2)(v-3)\right]/v(v-1)$$
$$= 2\sigma^2(v^3+v^2-2v+4)/v^2(v-1)(v+2).$$

Hence, the efficiency of the modified Latin square design relative to the original Latin square is given by

$$E = v(v-1)(v+2)/(v^3+v^2-2v+4).$$

The loss in efficiency is

$$1 - E = 4/(v^3+v^2-2v+4).$$

125. In an m-ple lattice design, the solutions for t_s are given by

$$\hat{t}_s = \left[Q_s + \sum_{i=1}^{m} S_i(Q_s)/k(m-1)\right]/r, \; s = 1, 2, \ldots, v$$

where $v = k^2$ and

$S_1(Q_s)$ = sum Q's over treatments which occur in the same row as t_s.

$S_2(Q_s)$ = sum of Q's over treatments which occur in the same column as t_s.

$S_i(Q_s)$ = sum of Q's over treatments which correspond to the same letter as t_s in the $(i-2)$-th Latin Square, $i = 3, 4, \ldots, m$.

Then, we have

$$\text{var}(\hat{t}_s) = \frac{\sigma^2}{r}\left[1 + \frac{m}{k(m-1)}\right], \; s = 1, 2, \ldots, v$$

$\text{cov}(\hat{t}'_s, \hat{t}_{s'}) = \sigma^2/rk\,(m-1)$, if the treatments t_s and $t_{s'}$ occur together in a block,

$= 0$, if the treatments t_s and $t_{s'}$ do not occur together in a block.

Hence,

$\text{var}(\hat{t}_s - \hat{t}_{s'}) = \dfrac{2\sigma^2}{kr}\left[k + \dfrac{m}{m-1}\right]$, if the treatments t_s and $t_{s'}$ do not occur together in a block

$= \dfrac{2\sigma^2(k+1)}{kr}$, if the treatments t_s and $t_{s'}$ occur together in a block.

Now let us find the number of pairs of treatments which occur together as a block. From each row of treatments, we can form $k(k-1)$ pairs, from each column of treatments, we can form $k(k-1)$ pairs, from the treatments which correspond to the same letter in the $(i-2)$th Latin square, we can form $k(k-1)$ pairs. Thus, the total number of pairs of treatments which occur together in a block is

$$k(k-1)[k+k+(m-2)k] = k^2(k-1)m.$$

Then, the number of pairs of treatments which do not occur together in a block is

$$k^2(k^2-1) - mk^2(k-1) = k^2(k-1)(k+1-m).$$

Hence, the average variance of the BLUE's of elementary treatment contrasts is given by

$$\begin{aligned}\bar{V} &= \frac{2\sigma^2}{k\,r\,k^2(k^2-1)}[(k+1)mk^2(k-1) + \left(k + \frac{m}{m-1}\right) \\ &\qquad k^2(k-1)(k+1-m)] \\ &= \frac{2\sigma^2[(m-1)(k+1)+m]}{r(m-1)(k+1)}.\end{aligned}$$

Now the efficiency of this design relative to RBD is given by

$$\begin{aligned}E &= (m-1)(k+1)/[(m-1)(k+1)+m] \\ &= 1 - m/[(m-1)(k+1)+m].\end{aligned}$$

When the lattice design is balanced, then $m = k+1$, and in this case

$$E = 1 - \frac{1}{k+1} = k/(k+1).$$

126. The normal equations for estimating t_s in the intrablock analysis are given by

$$Q_s = r\left(1 - \frac{1}{k}\right)\hat{t}_s - \frac{1}{k}\sum_1^m S_i'(\hat{t}_s), \text{ where}$$

$S_s'(t_s)$ = sum of treatments in the same row as t_s but excluding t_s,

$S_2'(t_s)$ = sum of treatments in the same column as t_s but excluding t_s and

$S_i'(t_s)$ = sum of treatments which correspond to the same letter as t_s in the $(i-2)$-th Latin Square, $i = 3, 4, \ldots, m$.

Applying the result of Exercise 53, we find that the normal equations for estimating t_s in the analysis with recovery of interblock information are given by

$$P_s = r\left[w_1 + \frac{w_2}{k-1}\right]\left(1 - \frac{1}{k}\right)t_s - \frac{(w_1-w_2)}{k}\sum_{i=1}^m S_i'(\hat{t}_s),$$
$$s = 1, 2, \ldots, v.$$

Thus

$$P_s = rw_1\hat{t}_s - \frac{w_1 - w_2}{k}\sum_{i}^{m} S_i(\hat{t}_s)$$

since r = m. Hence, summing over the treatments in the group $S_i(\hat{t}_s)$, $\sum^*$, we get

$$S_i(P_s) = rw_1 S_i(\hat{t}_s) - \frac{w_1 - w_2}{k}\sum^*\sum_{i=1}^{m} S_i(\hat{t}_s)$$

$$= rw_1 S_i(\hat{t}_s) - \frac{w_1 - w_2}{k}[kS_i(\hat{t}_s) + (m-1)\sum^* \hat{t}_s]$$

$$= [(r-1)w_1 + w_2]S_i(\hat{t}_s),$$

since $\sum^* \hat{t}_s = 0$. Thus, we get

$$S_i(\hat{t}_s) = S_i(P_s)/[(m-1)w_1 + w_2],$$

for r = m. Hence

$$P_s = rw_1\hat{t}_s - \frac{(w_1 - w_2)}{k\{(m-1)w_1 + w_2\}}\sum_{i=1}^{m} S_i(P_s),$$

$$\hat{t}_s = \frac{1}{rw_1}\left[P_s + \frac{(w_1 - w_2)\sum_{1}^{m} S_i(P_s)}{k\{(m-1)w_1 + w_2\}}\right],$$

and

$$\text{var}\,(\hat{t}_s) = \frac{1}{r\,w_1}\left[1 + \frac{m(w_1 - w_2)}{k\{(m-1)w_1 + w_2\}}\right].$$

If t_s and $t_{s'}$ occur together in a block, then

$$\text{cov}\,(\hat{t}_s, \hat{t}_{s'}) = \frac{(w_1 - w_2)}{r\,w_1\,k\{(m-1)w_1 + w_2\}}.$$

If t_s and $t_{s'}$ do not occur together in a block, then

$$\text{cov}\,(\hat{t}_s, \hat{t}_{s'}) = 0.$$

Hence, if t_s and $t_{s'}$ occur together in a block, then

$$\text{var}\,(\hat{t}_s - \hat{t}_{s'}) = \frac{2}{r\,w_1}\left[1 + \frac{(m-1)(w_1 - w_2)}{k\{(m-1)w_1 + w_2\}}\right].$$

If t_s and $t_{s'}$ do not occur together in a block, then

$$\text{var}\,(\hat{t}_s - \hat{t}_{s'}) = \frac{2}{r\,w_1}\left[1 + \frac{m(w_1 - w_2)}{k\{(m-1)w_1 + w_2\}}\right].$$

The number of pairs of treatments which occur together in a block (see Exercise 125) is $m\,k^2(k-1)$. Hence, the average variance of BLUEs of elementary treatment contrasts is given by

$$\bar{V} = \frac{1}{k^2(k^2-1)}\left[mk^2(k-1)\frac{2}{r\,w_1}\left\{1+\frac{(m-1)(w_1-w_2)}{k(m-1)w_1+w_2)}\right\}\right.$$
$$\left.+k^2(k-1)(k+1-m)\frac{2}{r\,w_1}\left\{1+\frac{m(w_1-w_2)}{k(m-1)w_1+w_2}\right\}\right]$$
$$=\frac{2}{r\,w_1(k+1)}\left[k+1+\frac{m(w_1-w_2)}{(m-1)w_1+w_2}\right].$$

Hence the efficiency is given by

$$E=\frac{2/r\,w_1}{\dfrac{2}{r\,w_1(k+1)}\left[k+1+\dfrac{m(w_1-w_2)}{(m-1)w_1+w_2}\right]}$$
$$=\frac{\{(m-1)w_1+w_2\}(k+1)}{\{(m-1)w_1+w_2\}(k+1)+m(w_1-w_2)}$$
$$=1-\frac{m(w_1-w_2)}{(k+1)\{(m-1)w_1+w_2\}+m(w_1-w_2)}.$$

When the lattice design is balanced, then $m = k + 1$ and then

$$E=1-\frac{w_1-w_2}{m\,w_1-w_2}=\frac{k\,w_1}{(k+1)w_1-w_2}.$$

127. Clearly, the sum of squares due to F_{1t} is given by

$$[\hat{F}_{1t}]^2/\frac{1}{r}\left[\sum_{x=0}^{S_1-1}X_t^2(x)\right]\cdot S_2S_3\ldots S_m.$$

We now find $\sum_{x=0}^{S_1-1} X_t^2(x)$. Note that $X_t(x)$ are orthogonal polynomials. Hence $\sum_x X_t(x)X_{t'}(x) = 0$ for $t \neq t'$. Therefore, we obtain, by multiplying the relation about X_t by X_{t-1} and summing

$$\Sigma\,X_t^2(x)=\Sigma\,X_1(x)X_t(x)X_{t-1}(x). \tag{1}$$

Also, multiplying the relation about X_t by X_{t-2} and summing gives

$$0=\Sigma\,X_1(x)X_{t-1}(x)X_{t-2}(x)-\frac{(t-1)^2\{S_1^2-(t-1)^2\}}{4(2t-1)(2t-3)}\Sigma\,X_{t-2}^2(x). \tag{2}$$

Replacing t by $(t - 1)$ in (1), we get

$$\Sigma\,X_{t-1}^2(x)=\Sigma\,X_1(x)X_{t-1}(x)X_{t-2}(x). \tag{3}$$

From (2) and (3), we get

$$\Sigma\,X_{t-1}^2(x)=\frac{(t-1)^2\{S_1^2-(t-1)^2\}}{4(2t-1)(2t-3)}\Sigma\,X_{t-2}^2(x). \tag{4}$$

Replacing (t − 1) by t in (4), we get

$$\Sigma X_t^2(x) = \frac{t^2\{S_1^2 - t^2\}}{4(2t+1)(2t-1)} \Sigma X_{t-1}^2(x).$$

Repeated application of the above relation gives

$$\Sigma X_t^2(x) = \frac{t^2(t-1)^2 \ldots 1^2(S_1^2 - t^2)\ldots(S_1^2 - 1^2)}{4^t\{(2t+1)(2t-1)\ldots 3(2t-1)(2t-3)\ldots 1\}} \sum_{x=0}^{S_1-1} X_0(x)$$

$$= \frac{(t!)^2(S_1^2 - 1)(S_1^2 - 2^2)\ldots(S_1^2 - t^2)\cdot S_1}{4^t\left\{\frac{(2t+1)!}{2^t(t!)}\right\}\left\{\frac{(2t)!}{2^t(t!)}\right\}}$$

$$= \frac{(t!)^4(S_1^2 - 1)(S_1^2 - 2^2)\ldots(S_1^2 - t^2)\cdot S_1}{(2t+1)!(2t)!}.$$

Hence, the sum of squares due to F_{1t} is given by

$$[\hat{F}_{1t}]^2 / \frac{[S_1 S_2 \ldots S_m (t!)^4 (S_1^2 - 1)\ldots(S_1^2 - t^2)]}{r(2t)!(2t+1)!}.$$

128. We have

$$X_i = \begin{bmatrix} X_{i-1} \\ a_i X_{i-1} \end{bmatrix} = \begin{bmatrix} 1 \\ a_i \end{bmatrix} \otimes X_{i-1}$$

$$= \begin{bmatrix} 1 \\ a_i \end{bmatrix} \otimes \begin{bmatrix} 1 \\ a_{i-1} \end{bmatrix} \otimes \ldots \otimes \begin{bmatrix} 1 \\ a_1 \end{bmatrix}, i = 1, 2, \ldots, m$$

= column vector of treatments in a 2^i design.

Similarly

$$Y_i = \begin{bmatrix} (a_i + 1)Y_{i-1} \\ (a_i - 1)Y_{i-1} \end{bmatrix} = \begin{bmatrix} a_i + 1 \\ a_i - 1 \end{bmatrix} \otimes Y_{i-1}$$

$$= \begin{bmatrix} a_i + 1 \\ a_i - 1 \end{bmatrix} \otimes \begin{bmatrix} a_{i-1} + 1 \\ a_{i-1} - 1 \end{bmatrix} \otimes \ldots \otimes \begin{bmatrix} a_1 + 1 \\ a_1 - 1 \end{bmatrix}$$

= column vector of S_i and all factorial effects in a 2^i design.

Also, note that

$$H_1 = \begin{bmatrix} 1 & 1 \\ -1 & 1 \end{bmatrix}$$

and

$$H_i = H_1 \otimes H_{i-1} = H_1 \otimes H_1 \otimes H_{i-2} = H_1 \otimes H_1 \otimes \ldots \otimes H_1.$$

Yates' method aims at obtaining Y_i from X_i and inversely X_i from Y_i.

Consider H_iX_i, $i = 1, 2, \ldots, m$.

$$\begin{aligned}
H_iX_i &= [H_1 \otimes H_1 \otimes \ldots \otimes H_1]\left\{\begin{bmatrix} 1 \\ a_i \end{bmatrix} \otimes \begin{bmatrix} 1 \\ a_{i-i} \end{bmatrix} \otimes \ldots \otimes \begin{bmatrix} 1 \\ a_1 \end{bmatrix}\right\} \\
&= H_1\begin{bmatrix} 1 \\ a_i \end{bmatrix} \otimes H_1\begin{bmatrix} 1 \\ a_{i-i} \end{bmatrix} \otimes \ldots \otimes H_1\begin{bmatrix} 1 \\ a_1 \end{bmatrix} \\
&= \begin{bmatrix} a_i + 1 \\ a_i - 1 \end{bmatrix} \otimes \begin{bmatrix} a_{i-1} + 1 \\ a_{i-1} - 1 \end{bmatrix} \otimes \ldots \otimes \begin{bmatrix} a_1 + 1 \\ a_1 - 1 \end{bmatrix} \\
&= Y_i
\end{aligned}$$

Note the operator H_1. When H_1 is applied on a pair of values, it is equivalent to taking the sum and difference (lower minus upper) of the two values of the pair. We thus note that $H_i\,X_i$ is thus equivalent to repeating i times in succession the operation of taking the sum and difference of pairs of the treatments in X_i.

We now prove the inverse Yates' method which obtains X_i from Y_i. Consider

$$\begin{aligned}
H_iH_i' &= [H_1 \otimes H_1 \otimes \ldots \otimes H_1][H_1' \otimes H_1' \otimes \ldots \otimes H_1'] \\
&= H_1H_1' \otimes H_1H_1' \otimes \ldots \otimes H_1H_1' \\
&= 2I_2 \otimes 2I_2 \otimes \ldots \otimes 2I_2 \\
&= 2^iI_{2i} = H_i'H_i
\end{aligned}$$

Hence,

$$H_i^{-1} = H_i'/2^i, \ i = 1, 2, \ldots, m.$$

From $H_iX_i = Y_i$, we then get

$$\begin{aligned}
X_i &= H_i^{-1}Y_i \\
&= \frac{1}{2^i}H_i'Y_i \\
&= \frac{1}{2^i}[H_1' \otimes H_1' \otimes \ldots \otimes H_1']\left[\begin{pmatrix} a_i + 1 \\ a_i - 1 \end{pmatrix} \otimes \ldots \otimes \begin{pmatrix} a_1 + 1 \\ a_1 - 1 \end{pmatrix}\right] \\
&= \frac{1}{2^i}\left[H_1'\begin{pmatrix} a_i + 1 \\ a_i - 1 \end{pmatrix} \otimes H_1'\begin{pmatrix} a_{i-1} + 1 \\ a_{i-1} - 1 \end{pmatrix} \otimes \ldots \otimes H_1'\begin{pmatrix} a_1 + 1 \\ a_1 - 1 \end{pmatrix}\right].
\end{aligned}$$

Note the operator H_1'. When H_1' is applied to a pair of values, it is equivalent to taking the difference (upper minus lower) and the sum of the pair of values. Thus, X_i is obtained by repeating i times in succession the operation of taking the difference (upper minues lower) and the sum of pair of values in the column vector of Y_i and dividing by 2^i. This establishes the inverse method of Yates.

129. We have

$$X_i = \begin{bmatrix} X_{i-1} \\ a_i X_{i-1} \\ a_i^2 X_{i-1} \end{bmatrix} = \begin{bmatrix} 1 \\ a_i \\ a_i^2 \end{bmatrix} \otimes X_{i-1}$$

$$= \begin{bmatrix} 1 \\ a_i \\ a_i^2 \end{bmatrix} \otimes \begin{bmatrix} 1 \\ a_{i-1} \\ a_{i-1}^2 \end{bmatrix} \otimes \ldots \otimes \begin{bmatrix} 1 \\ a_1 \\ a_1^2 \end{bmatrix}$$

= column vector of treatments in a 3^i design

Also

$$Y_i = \begin{bmatrix} a_i^2 + a_i + 1 \\ a_i^2 - 1 \\ a_i^2 - 2a_i + 1 \end{bmatrix} \otimes Y_{i-1}$$

$$= \begin{bmatrix} a_i^2 + a_i + 1 \\ a_i^2 - 1 \\ a_i^2 - 2a_i + 1 \end{bmatrix} \otimes \begin{bmatrix} a_{i-1}^2 + a_{i-1} + 1 \\ a_{i-1}^2 - 1 \\ a_{i-1}^2 - 2a_{i-1} + 1 \end{bmatrix} \otimes \ldots \otimes \begin{bmatrix} a_1^2 + a_1 + 1 \\ a_1^2 - 1 \\ a_1^2 - 2a_1 + 1 \end{bmatrix}$$

= column vector consisting of S_i and contrasts belonging to factorial effects in a 3^i design.

Also

$$H_i = \begin{bmatrix} H_{i-1} & H_{i-1} & H_{i-1} \\ -H_{i-1} & 0 & H_{i-1} \\ H_{i-1} & -2H_{i-1} & H_{i-1} \end{bmatrix} = H_i \otimes H_{i-1}$$

$$= H_1 \otimes H_1 \otimes \ldots \otimes H_1$$

Hence

$$H_i X_i = \{H_1 \otimes H_1 \otimes \ldots \otimes H_1\} \left\{ \begin{bmatrix} 1 \\ a_i \\ a_i^2 \end{bmatrix} \otimes \begin{bmatrix} 1 \\ a_{i-1} \\ a_{i-1}^2 \end{bmatrix} \otimes \ldots \otimes \begin{bmatrix} 1 \\ a_1 \\ a_1^2 \end{bmatrix} \right\}$$

$$= \left\{ H_1 \begin{bmatrix} 1 \\ a_i \\ a_i^2 \end{bmatrix} \otimes \ldots \otimes H_1 \begin{bmatrix} 1 \\ a_1 \\ a_1^2 \end{bmatrix} \right\}$$

$$= \begin{bmatrix} a_i^2 + a_i + 1 \\ a_i^2 - 1 \\ a_i^2 - 2a_i + 1 \end{bmatrix} \otimes \ldots \otimes \begin{bmatrix} a_1^2 + a_1 + 1 \\ a_1^2 - 1 \\ a_1^2 - 2a_1 + 1 \end{bmatrix}$$

$$= Y_i.$$

This establishes extended Yates' rule. In order to obtain Y_m in a 3^m design we proceed as follows. We write the column vector X_m of treatments as under

$$X_m = \{1, a_1, a_1^2, a_2, a_1a_2, a_1^2a_2, a_2^2, a_1a_2^2, a_1^2a_2^2, \ldots, a_m, a_1a_m, a_1^2a_m, \ldots, a_1^2a_2^2a_m, \ldots, a_m^2, a_1a_m^2, a_1^2a_m^2, \ldots, a_1^2a_2^2a_{m-1}^2a_m^2\}$$

We repeat the following operation m times in succession on triplets of X_m:

(i) add three values
(ii) subtract the first from third
(iii) subtract two times the second value from the sum of the first and third values.

130. Consider a RBD with v treatments and b replications. Let the missing yield be denoted by x. Now

$$\mathcal{E}(x) = \mu + \alpha_i + t_j$$

Hence x is estimated by

$$x = \hat{\mu} + \hat{\alpha}_j + \hat{t}_i$$

Now for a RBD,

$$\hat{\mu} = (G + x)/bv$$

$$\hat{\alpha}_j = \frac{(B_j + x)}{v} - \frac{(G + x)}{bv}$$

$$\hat{t}_i = \frac{(T_i + x)}{b} - \frac{G + x}{bv}.$$

We have assumed that $B_1, B_2, \ldots, B_b, T_1, T_2, \ldots, T_v$ and G denote respectively the actual block totals, treatments totals and the total yield. Hence, we get

$$x = \frac{G + x}{bv} + \frac{(B_j + x)}{v} - \frac{(G + x)}{bv} + \frac{T_i + x}{b} - \frac{G + x}{bv}$$

Solving for x, we get the estimate of x as

$$\hat{x} = \frac{(bB_j + vT_i - G)}{(b - 1)(v - 1)}.$$

Now bias in estimated treatment SS is given by

$$\begin{aligned} &\text{bias (est. treatment SS)} \\ &\quad = [\text{Est. conditional error SS with estimated missing yield}] \\ &\qquad - [\text{minimum value of conditional error SS}] \end{aligned}$$

The conditional ($t_1 = t_2 = \ldots = t_v = 0$) error SS is

$$\begin{aligned} &(\text{Total SS}) - (\text{Block SS}) \\ &\quad = \left[\Sigma\, y^2 + x^2 - \frac{(G + x)^2}{bv}\right] \\ &\qquad - \left[\frac{B_1^2 + \ldots + (B_j + x)^2 + \ldots + B_b^2}{v} - \frac{(G + x)^2}{bv}\right] \end{aligned}$$

Hence estimated conditional error SS is

$$\left[\Sigma\, y^2 + \hat{x}^2 - \frac{(G+\hat{x})^2}{bv}\right]$$
$$-\left[\frac{B_1^2 + \ldots + (B_j+\hat{x})^2 + \ldots + B_b^2}{v} - \frac{(G+\hat{x})^2}{bv}\right]$$

where $\hat{x} = (bB_j + vT_i - G)/(b-1)(v-1)$.

Now, we shall find the minimum value of conditional error SS. We equate to zero the derivative of conditional error SS with repeat to x. We then obtain the estimate of x as

$$x_0 = B_j/(v-1).$$

Hence, minimum value of conditional error SS is

$$\Sigma\, y^2 + x_0^2 - \frac{B_1^2 + \ldots + (B_j + x)^2 + \ldots + B_b^2}{v}$$

and the bias in the estimated treatment SS is

$$\hat{x}^2 - x_0^2 - \frac{(B_j+\hat{x})^2 - (B_j + x_0)^2}{v}$$
$$= (\hat{x} - x_0)\left[(\hat{x} + x_0) - \frac{(2\,B_j + \hat{x} + x_0)}{v}\right]$$
$$= (\hat{x} - x_0)\left[\frac{(\hat{x} + x_0)(v-1) - 2B_j}{v}\right]$$
$$= (\hat{x} - x_0)\left[\frac{(\hat{x} + x_0)(v-1) - 2x_0(v-1)}{v}\right]$$
$$= \frac{(\hat{x} - x_0)^2(v-1)}{v} = (B_j + vT_i - G)^2/v(v-1)(b-1)^2$$

Since B_j contains $(v-1)$ yields, T_i contains $(b-1)$ yields and G contains $(bv-1)$ yields, we get

$$\begin{aligned}
&\text{var}\,(B_j) = \sigma^2(v-1)\\
&\text{var}\,(T_i) = \sigma^2(b-1)\\
&\text{var}\,(G) = \sigma^2(bv-1)\\
&\text{cov}\,(B_j, T_i) = 0\\
&\text{cov}\,(B_j, G) = \sigma^2(v-1)\\
&\text{cov}\,(T_i, G) = \sigma^2(b-1).
\end{aligned}$$

Hence

$$\text{var}\,(\hat{x}) = \frac{(\sigma^2(v+b-1)}{(v-1)(b-1)}.$$

Consider an elementary treatment contrast $t_i - t_s, i \neq s$. This contrast is estimated by

$$\hat{t}_i - \hat{t}_s = (T_i + \hat{x} - T_s)/b.$$

Hence,

$$\operatorname{var}(\hat{t}_j - \hat{t}_s) = \frac{\sigma^2}{b^2}\left[(b-1) + \frac{v+b-1}{(v-1)(b-1)} + b + \frac{2\operatorname{cov}(T_i, \hat{x})}{\sigma^2} - \frac{2\operatorname{cov}(T_i, T_s)}{\sigma^2} - \frac{2\operatorname{cov}(\hat{x}, T_s)}{\sigma^2}\right].$$

Now, one can easily verify that

$$\operatorname{cov}(\hat{x}, T_i) = \sigma^2$$
$$\operatorname{cov}(\hat{x}, T_s) = 0$$
$$\operatorname{cov}(T_i, T_s) = 0.$$

Hence, we obtain

$$\operatorname{var}(\hat{t}_i - \hat{t}_s) = \frac{2\sigma^2}{b}\left[1 + \frac{v}{2(v-1)(b-1)}\right].$$

For any other elementary treatment contrast not involving missing yield, such as $t_s - t_{s'}$, we get

$$\operatorname{var}(\hat{t}_s - \hat{t}_{s'}) = 2\sigma^2/b.$$

Hence, the average variance $\bar{V}$ of the estimates of elementary treatment contrasts is given by

$$\bar{V} = \frac{\frac{2\sigma^2}{b}\left[1 + \frac{v}{2(v-1)(b-1)}\right]2(v-1) + \frac{(v-1)(v-2)}{2}\cdot\frac{2\sigma^2}{b}}{v(v-1)}$$
$$= \frac{2\sigma^2}{b}\left[\frac{1}{(v-1)(b-1)}\right].$$

If there was no missing yield, the variance of the BLUE of an elementary treatment contrastis $2\sigma^2/b$. Hence, the efficiency of RBD with one missing yield relative to that of RBD with no missing yield is given by

$$E = \frac{2\sigma^2/b}{\bar{V}} = \left[1 + \frac{1}{(v-1)(b-1)}\right]^{-1}.$$

Hence, the loss in efficiency is

$$1 - E = [1 + (v-1)(b-1)]^{-1}.$$

131. Consider a BIBD $(v, b, \gamma, k, \lambda)$ and suppose the yield $x = x_{ij}$ corresponding to the ith treatment in the jth block is missing. Then x is estimated by

$$x = \hat{\mu} + \hat{\alpha}_j + \hat{t}_i.$$

Solving the above equation for x and denoting the solution by $\hat{x}$, we get

$$\hat{x} = (\lambda\, vB_j - kQ_j^{'} + k^2 Q_i)/(k-1)(\lambda\, v - k).$$

We shall now find the bias in the estimated adjusted treatment SS.

The conditional error SS is

$$\Sigma\, y^2 + x^2 - \frac{(G+x)^2}{b\,k} - \frac{B_1^2 + \ldots + (B_j + x)^2 + \ldots + B_b^2}{k} + \frac{(G+x)^2}{b\,k}$$

Hence estimated conditional error SS is

$$\Sigma\, y^2 + \hat{x}^2 - \frac{B_1^2 + \ldots + (B_j + \hat{x})^2 + \ldots + B_b^2}{k},$$

where $\hat{x} = (\lambda\, vB_j - kQ_j^{'} + k^2Q_i)^2/(k-1)(\lambda v - k)$.

We shall now find the minimum value of the conditional error SS. Equating the derivative of the conditional error SS to zero, we get

$$x = B_j/(k-1) = x_0, \text{ say,}$$

and, the minimum value of the conditional error SS is

$$\Sigma\, y^2 + x_0^2 - \frac{B_1^2 + \ldots + (B_j + x_0)^2 + \ldots + B_b^2}{k}$$

Hence, bias in the estimated adjusted treatment SS is given by

$$\begin{aligned}
\text{Bias} &= [\text{estimated conditional error SS}] - [\text{minimum conditional error SS}] \\
&= \hat{x}^2 - x_0^2 - \frac{(B_j + \hat{x})^2 - (B_j + x_0)^2}{k} \\
&= (\hat{x} - x_0)\left[\hat{x} + x_0 - \frac{(2B_j + \hat{x} + x_0)}{k}\right] \\
&= \frac{(\hat{x} - x_0)}{k}[(k-1)(\hat{x} + x_0) - 2B_j] \\
&= \frac{(\hat{x} - x_0)}{k}[(k-1)(\hat{x} + x_0) - 2(k-1)x_0] \\
&= \frac{(\hat{x} - x_0)^2(k-1)}{k} \\
&= \frac{k(B_j - Q_j^{'} + kQ_i)^2}{(k-1)(\lambda v - k)^2}.
\end{aligned}$$

Now for a BIBD,

$$\begin{aligned}
\hat{\mu} &= (G + x)/bk \\
\hat{\alpha}_j &= \frac{1}{k}(B_j + x) - \hat{\mu} - \frac{1}{k}\sum_{p=1}^{v} n_{pj}\hat{t}_p \\
\hat{t}_p &= \frac{k}{\lambda v}Q_p^*,\ p = 1, 2, \ldots, v,
\end{aligned}$$

when for $p \neq i$,

$$Q_p^* = T_p - \frac{1}{k}\sum_{s\neq j} n_{ps}B_s - \frac{1}{k}n_{pj}(B_j + x)$$
$$= T_p - \frac{1}{k}\sum_{s=1}^{b} n_{ps}B_s - \frac{x}{k}n_{pj}$$
$$= Q_p - \frac{x}{k}n_{pj}$$

and

$$Q_i^* = T_i + x - \frac{1}{k}\sum_{s\neq j} n_{ps}B_s - \frac{1}{k}n_{ij}(B_j + x)$$
$$= T_i - \frac{1}{k}\sum_{s=1}^{b} n_{ps}B_s + x\left(1 - \frac{1}{k}\right)$$
$$= Q_i + x\left(1 - \frac{1}{k}\right).$$

Hence

$$\hat{t}_p = \frac{k}{\lambda v}\left[Q_p - \frac{x \cdot n_{pj}}{k}\right], \text{ if } p \neq i$$
$$= \frac{k}{\lambda v}\left[Q_i + x\left(1 - \frac{1}{k}\right)\right], \text{ if } p = i.$$

Thus, x is estimated by

$$\hat{x} = \hat{\mu} + \frac{1}{k}(B_j + x) - \hat{\mu} - \frac{1}{k}\sum_{p=1}^{v} n_{pj}\hat{t}_p + \frac{k}{\lambda\, v}\left[Q_i + x\left(1 - \frac{1}{k}\right)\right]$$
$$= \frac{1}{k}(B_j + x) - \frac{1}{k}\sum_{p\neq i} n_{pj}\frac{k}{\lambda\, v}\left[Q_p - \frac{1}{k}xn_{pj}\right]$$
$$-\frac{1}{k}\cdot\frac{k}{\lambda\, v}\left[Q_i + x\left(1 - \frac{1}{k}\right)\right] + \frac{k}{\lambda\, v}\left[Q_i + x\left(1 - \frac{1}{k}\right)\right]$$
$$= \frac{1}{k}(B_j + x) - \frac{1}{\lambda\, v}\sum_{p=1}^{v} n_{pj}Q_p + \frac{x}{k\,\lambda\, v}\sum_{p\neq i} n_{pj}^2$$
$$-\frac{x}{k\,\lambda\, v}(k-1) + \frac{k}{\lambda\, v}Q_i + \frac{x}{\lambda\, v}(k-1)$$
$$= \frac{1}{k}(B_j + x) - \frac{1}{\lambda\, v}Q_j' + \frac{x}{k\,\lambda\, v}(k-1) - \frac{x}{k\,\lambda\, v}(k-1)$$
$$+\frac{k}{\lambda\, v}Q + \frac{x}{\lambda\, v}(k-1)$$
$$= \frac{1}{k}(B_j + x) - \frac{1}{\lambda\, v}Q_j' + \frac{k}{\lambda\, v}Q_i + \frac{x(k-1)}{\lambda\, v}$$

132. The missing yield corresponding to the ith treatment in the jth block is denoted by $x_{ij} = x$. This will be estimated by

$$x = \hat{\mu} + \hat{\alpha}_j + \hat{t}_i.$$

Now for a binary block design,

$$\hat{\mu} = (G + x)/n$$

$$\hat{\alpha}_j = (B_j + x)/k_j - \hat{\mu} - \frac{1}{k_j}\sum_{p=1}^{v} n_{pj}\hat{t}_p$$

$$\hat{t}_i = h_{i1}Q_1^* + h_{i2}Q_2^* + \ldots + h_{iv}Q_v^*$$

where $[h_{ij}]$ = a g-inverse of the C-matrix and

$$\begin{aligned}
Q_p^* &= T_p - \sum_{q\neq j} n_{pq}B_q/k_q - \frac{n_{pj}(B_j + x)}{k_j} \\
&= T_p - \sum_{q=1}^{b} n_{pq}B_q/k_q - \frac{n_{pj}x}{k_j} \\
&= Q_p - n_{pj}x/k_j,\ p \neq i \\
Q_i^* &= T_i + x - \sum_{q\neq j} n_{iq}B_q/k_q - \frac{n_{ij}(B_j + x)}{k_j} \\
&= T_i + x - \sum_{q=1}^{b} n_{iq}B_q/k_q - \frac{x}{k_j} \\
&= Q_i + x\left(1 - \frac{1}{k_j}\right).
\end{aligned}$$

Hence

$$\begin{aligned}
\hat{t}_p &= \sum_{s=1}^{v} h_{ps}Q_s^* = \sum_{s\neq i} h_{ps}\left(Q_s - \frac{1}{k_j}n_{sj}x\right) + h_{pi}\left(Q_i + x\left(1 - \frac{1}{k_j}\right)\right) \\
&= \sum_{s=1}^{v} h_{ps}Q_s - \frac{x}{k_j}\sum_{s=1}^{v} n_{sj}h_{ps} + xh_{pi}, \quad p = 1, 2, \ldots, v
\end{aligned}$$

and x is estimated by

$$\begin{aligned}
x = \hat{\mu} &+ \frac{1}{k_j}(B_j + x) - \hat{\mu} \\
&- \frac{1}{k_j}\sum_{p=1}^{v} n_{pj}\left[\sum_{s=1}^{v} h_{ps}Q_s - \frac{1}{k_j}x\sum_{s=1}^{v} n_{sj}h_{ps} + xh_{pi}\right] \\
&+ \sum_{s=1}^{v} h_{is}Q_s - \frac{1}{k_j}x\sum_{s=1}^{v} n_{sj}h_{is} + xh_{ii}.
\end{aligned}$$

Denoting the solution of x by $\hat{x}$, we get

$$\hat{x} = \frac{B_j - \sum_{p=1}^{v}\sum_{s=1}^{v} n_{pj}h_{ps}Q_s + k_j \sum_{s=1}^{v} h_{is}Q_s}{k_j - 1 - k_j h_{ii} + \sum_{s=1}^{v} n_{sj}h_{is} + \sum_{p=1}^{v} n_{pj}\left(h_{pi} - \frac{1}{k_j}\sum_{s=1}^{v} n_{sj}h_{ps}\right)}.$$

Under $\mathbf{t} = \mathbf{0}$, the conditional error SS is given by

$$S_{e0}^2(\mathbf{y}, x) = \Sigma\, y^2 + x^2 - B_1^2/k_1 - B_2^2/k_2 - \ldots - (B_j + x)^2/k_j \\ - \ldots - B_b^2/k_b.$$

Hence, the estimated conditional error SS is given by

$$S_{e0}^2(\mathbf{y},\hat{x}) = \Sigma\, y^2 + \hat{x}^2 - B_1^2/k_1 - \ldots - (B_j + \hat{x})^2/k_j - \ldots - B_b^2/k_b$$

where $\hat{x}$ is given as above.

Now the minimum value of the conditional error SS is the actual conditional error SS. Hence,

$$\min_x S_{e0}^2(\mathbf{y}, x) = \Sigma\, y^2 - B_1^2/k_1 - \ldots - B_j^2/(k_j - 1) - \ldots - B_b^2/k_b.$$

Hence bias in the estimated treatment SS is given by

$$\begin{aligned} \text{Bias} &= [\text{Estimated Conditional Error SS}] - [\text{Min. value of} \\ &\qquad \text{Conditional Error SS}] \\ &= \hat{x}^2 - (B_j + \hat{x})^2/k_j + B_j^2/(k_j - 1) \\ &= \frac{(k_j - 1)}{k_j}\left[\hat{x} - B_j/(k_j - 1)\right]^2. \end{aligned}$$

For a RBD, $k_j = v$ for all j, and $h_{pp} = \frac{1}{b}$ for all p and $h_{ps} = 0$ for $p \neq s$, $n_{pq} = 1$, for all p and q. Hence

$$\hat{x} = \frac{B_j - \frac{1}{b}\sum_p Q_p + v \cdot \frac{1}{b} Q_i}{v - 1 - \frac{k}{b} + \frac{1}{b} + 0}.$$

Since $\sum_p Q_p = 0$, we get $\hat{x} = \dfrac{bB_j + vQ_i}{(k-1)(b-1)} = \dfrac{bB_j + vT_i - G}{(b-1)(v-1)}$.

For a BIBD, $k_j = k$ for all j and $h_{pp} = k/\lambda v$ for all p and $h_{ps} = 0$ for $p \neq s$. Hence

$$\begin{aligned} \hat{x} &= \frac{B_j - \frac{k}{\lambda\, v} Q_j' + k^2 Q_i/\lambda\, v}{k - 1 - k^2/\lambda\, v + k/\lambda\, v + 0} \\ &= \frac{\lambda\, vB_j - kQ_j' + k^2 Q_i}{(k-1)(\lambda\, v - k)}, \end{aligned}$$

where $Q'_j = \Sigma\, n_{pj}Q_p$ = sum of Q's over treatments of the jth block. For a RBD, the bias in the estimated treatment SS is given by

$$\text{Bias} = \frac{v-1}{v}\left[\frac{b\,B_j + v\,T_i - G}{(b-1)(v-1)} - \frac{B_j}{v-1}\right]^2$$
$$= [B_j + vT_i - G]^2/v(v-1)(b-1)^2.$$

For a BIBD, the bias in the estimated adjusted treatment SS is given by

$$\text{Bias} = \frac{(k-1)}{k}\left[\frac{\lambda\,vB_j - kQ'_j + k^2Q_i}{(k-1)(\lambda\,v-k)} - \frac{B_j}{k-1}\right]^2$$
$$= k(B_j - kQ'_j + k^2Q_i)^2/(k-1)(\lambda\,v-k)^2.$$

133. Suppose in a $v \times v$ Latin square the yield $x_{ijk} = x$ corresponding to the ith treatment in the jth row and k-th collumn is missing. Then x is estimated by

$$x = \hat{\mu} + \hat{\alpha}_j + \hat{\beta}_k + \hat{t}_i$$

In a Latin square, the estimates of μ, α_j, β_k and t_i are given by

$$\hat{\mu} = (G+x)/v^2$$
$$\hat{\alpha}_j = \frac{1}{v}(R_j + x) - \hat{\mu}$$
$$\hat{\beta}_k = \frac{1}{v}(C_k + x) - \hat{\mu}$$
$$\hat{t}_i = \frac{1}{v}(T_i + x) - \hat{\mu}.$$

Hence x is estimated by

$$x = \hat{\mu} + \frac{1}{v}(R_j + x) - \hat{\mu} + \frac{1}{v}(C_k + x) - \hat{\mu} + \frac{1}{v}(T_i + x) - \hat{\mu}$$
$$= \frac{1}{v}(R_j + x) + \frac{1}{v}(C_k + x) + \frac{1}{v}(T_i + x) - 2\frac{(G+x)}{v^2}.$$

Denoting the solution of the above equation by $\hat{x}$, we obtain

$$\hat{x} = \frac{vR_j + vC_k + vT_i - 2G}{(v-1)(v-2)}.$$

We shall now find the bias in the estimated treatment SS. Under $\mathbf{t} = \mathbf{0}$, the conditional error SS in a Latin square design is given by

$$S^2_{e0}(\mathbf{y}, x) = \left[\Sigma\, y^2 + x^2 - \frac{(G+x)^2}{v^2}\right]$$
$$- \left[\frac{R_1^2 + \ldots + (R_j + x)^2 + \ldots + R_v^2}{v} - \frac{(G+x)^2}{v^2}\right]$$
$$- \left[\frac{\sum_{\ell \neq k} C_\ell^2 + (X_k + x)^2}{v} - \frac{(G+x)^2}{v^2}\right].$$

Hence, the estimated conditional error is given by

$$S_{e0}^2(\mathbf{y}, \hat{x}) = \Sigma\, y^2 + \hat{x}^2 - \frac{\sum_{b\neq j} R_b^2 + (R_j + \hat{x})^2}{v} - \frac{\sum_{\ell\neq k} C_\ell^2 + (C_k + \hat{x})^2}{v} + \frac{(G + \hat{x})^2}{v^2}.$$

Now, equating the derivative of conditional error SS to zero, we get

$$2x - \frac{2}{v}(R_j + \hat{x}) - \frac{2}{v}(C_k + x) + \frac{2(G + x)}{v^2} = 0.$$

Denoting the solution of the above equation by x_0, we get

$$x_0 = \frac{vR_j + vC_k - G}{(v-1)^2}.$$

Hence, the minimum balue of the conditional error SS is

$$\Sigma\, y^2 + x_0^2 - \frac{1}{v}\left[\sum_{p\neq j} R_p^2 + (R_j + x_0)^2\right] - \frac{1}{v}\left[\sum_{\ell\neq k} C_\ell^2 + (C_k + x_0)^2\right] + \frac{(G + x_0)^2}{v^2}$$

and bias in treatment SS is given by

$$\begin{aligned}
\text{Bias} &= S_{e0}^2(\mathbf{y}, \hat{x}) - \min_x S_{e0}^2(\mathbf{y}, x) \\
&= \hat{x}^2 - x_0^2 - \frac{1}{v}[(R_j + \hat{x})^2 - (R_j + x_0)^2] - \frac{1}{v}[(C_k + \hat{x})^2 \\
&\quad - (C_k + x_0)^2] + \frac{1}{v^2}[(G + \hat{x})^2 - (G + x_0)^2] \\
&= (\hat{x} - x_0)\left[(\hat{x} + x_0) - \frac{1}{v}(2R_j + \hat{x} + x_0) - \frac{1}{v}(2C_k + \hat{x} + x_0) \right. \\
&\quad \left. + \frac{1}{v^2}(2G + \hat{x} + x_0)\right] \\
&= (\hat{x} - x_0)\left[(\hat{x} + x_0)\left(1 - \frac{1}{v} + \frac{1}{v^2}\right) - \frac{2}{v^2}(vR_j + vC_k - G)\right] \\
&= (\hat{x} - x_0)\left[(\hat{x} + x_0) - \frac{(v-1)^2}{v^2} - \frac{2}{v^2}x_0(v-1)^2\right] \\
&= \frac{(v-1)^2}{v^2}(\hat{x} - x_0)^2 \\
&= \frac{[R_j + C_k + (v-1)T_i - G]^2}{(v-1)^2(v-2)^2}.
\end{aligned}$$

We shall now find the loss in efficiency. Since R_j, C_k, T_i and G contain (v − 1), (v − 1), (v − 1) and (v^2 − 1) yields respectively, we can easily verify that

$$\begin{aligned}
&\text{var}\,(R_j) = \sigma^2(v-1),\ \ \text{var}\,(C_k) = \sigma^2(v-1),\\
&\text{var}\,(T_i) = \sigma^2(v-1),\ \ \text{var}\,(G) = \sigma^2(v^2-1)\\
&\text{cov}\,(R_j, G) = \sigma^2(v-1),\ \text{cov}\,(R_j, C_k) = 0,\\
&\text{cov}\,(R_j, G) = \sigma^2(v-1),\ \text{cov}\,(R_j, C_k) = 0,\\
&\text{cov}\,(C_k, T_i) = 0,\ \ \text{cov}\,(T_i, G) = \sigma^2(v-1).
\end{aligned}$$

Hence, we can find

$$\text{var}\,(\hat{x}) = \frac{\sigma^2(3v-2)}{(v-1)(v-2)}.$$

Consider an elementary contrast between two treatments, one containing the missing yield, $t_i - t_s$; say. This is estimated by $\hat{t}_i - \hat{t}_s = (T_i + x - T_s)/v$. Hence

$$\begin{aligned}
\text{var}\,(\hat{t}_i - \hat{t}_s) = \frac{\sigma^2}{v^2}\Bigg[&(v-1) + \frac{(3\,v\ -2)}{(v-1)(v-2)} + v\\
&+ \frac{2\text{cov}\,(T_i, \hat{x})}{\sigma^2} - \frac{2\text{cov}\,(T_i, T_s)}{\sigma^2} - \frac{2\ \text{cov}\,(T_s, \hat{x})}{\sigma^2}\Bigg]
\end{aligned}$$

Now, we can verify that

$$\begin{aligned}
&\text{cov}\,(T_i, \hat{x}) = \sigma^2,\ \ \text{cov}\,(T_s, \hat{x}) = 0\\
&\text{cov}\,(T_i, T_s) = 0.
\end{aligned}$$

Hence,

$$\text{var}\,(\hat{t}_i - \hat{t}_s) = \frac{2\sigma^2}{v}\left[1 + \frac{v}{2(v-1)(v-2)}\right].$$

Further if $t_s - t_{s'}$ is an elementary contrast between treatments not containing the missing yield, then

$$\text{var}\,(\hat{t}_s - \hat{t}_{s'}) = 2\sigma^2/v.$$

Therefore, the average variance $\bar{V}$ of the estimates of all elementary treatment contrasts is given by

$$\begin{aligned}
\bar{V} &= \frac{\dfrac{2\sigma^2}{v}\left[1 + \dfrac{v}{2(v-1)(v-2)}\right]2(v-1) + (v-1)(v-2)\dfrac{2\sigma^2}{v}}{v(v-1)}\\
&= \frac{2\sigma^2}{v}\left[1 + \frac{1}{(v-1)(v-2)}\right].
\end{aligned}$$

Hence, the efficiency of the Latin square design with one missing yield relative to that of the Latin square design with no missing yield is

$$\frac{2\sigma^2/v}{\bar{V}} = [1 + \{(v-1)(v-2)\}^{-1}]^{-1}$$

And the loss of efficiency is

$$1 - [1 + \{(v-1)(v-2)\}^{-1}]^{-1} = [1 + (v-1)(v-2)]^{-1}.$$

134. Consider a Youden square in which the columns form the blocks of a SBIBD ($v = b, r = k, \lambda$). Each row is a complete replication of all treatments. Suppose that the yield $x_{ijk} = x$ corresponding to the ith treatment in the jth row and the kth column is missing. Then x is estimated by

$$x = \hat{\mu} + \hat{\alpha}_j + \hat{\beta}_k + \hat{t}_i,$$

where

$$\begin{aligned}
\hat{\mu} &= (G + x)/vk \\
\hat{\alpha}_j &= \text{Estimate of the jth row effect} \\
&= \frac{1}{v}(R_j + x) - \hat{\mu} \\
\hat{\beta}_k &= \text{Esstimate of the kth column effect} \\
&= \text{Estimate of kth block of a SBIBD} \\
&= \frac{1}{k}(C_k + x) - \hat{\mu} - \frac{1}{k}\sum_{p=1}^{v} n_{pk}\hat{t}_p \\
\hat{t}_p &= \text{Estimate of pth treatment effect} \\
&= \frac{k}{\lambda v}Q_p^*,
\end{aligned}$$

where for $p \neq i$,

$$\begin{aligned}
Q_p^* &= T_p - \frac{1}{k}\sum_{q\neq k} n_{pq}C_q - \frac{1}{k}n_{pk}(C_k + x) \\
&= T_p - \frac{1}{k}\sum_{q} n_{pq}C_q - \frac{1}{k}n_{pk}x \\
&= Q_p - \frac{1}{k}n_{pk}x
\end{aligned}$$

and

$$\begin{aligned}
Q_i^* &= T_i - x - \frac{1}{k}\sum_{q\neq k} n_{iq}C_q - \frac{1}{k}n_{ik}(C_k + x) \\
&= T_i - \frac{1}{k}\sum_{q} n_{iq}C_q + x\left(1 - \frac{1}{k}\right) \\
&= Q_i + x\left(1 - \frac{1}{k}\right).
\end{aligned}$$

Hence,

$$
\begin{aligned}
\hat{\beta}_k &= \frac{1}{k}(C_k + x) - \hat{\mu} - \frac{1}{k}\sum_{p\neq i} n_{pk}\hat{t}_p - \frac{1}{k}n_{ik}\hat{t}_i \\
&= \frac{1}{k}(C_k + x) - \hat{\mu} - \frac{1}{k}\sum_{p\neq i} n_{pk}\cdot\frac{k}{\lambda\,v}\left(Q_p - \frac{1}{k}n_{pk}x\right) \\
&\quad - \frac{1}{k}\cdot\frac{k}{\lambda\,v}\left\{Q_i + \frac{x(k-1)}{k}\right\} \\
&= \frac{1}{k}(C_k + x) - \hat{\mu} - \frac{1}{\lambda\,v}\sum_{p} n_{pk}Q_p + \frac{x}{\lambda\,v\,k}\sum_{p\neq i} n_{pk}^2 - \frac{x(k-1)}{\lambda\,v\,k} \\
&= \frac{1}{k}(C_k + x) - \hat{\mu} - \frac{1}{\lambda\,v}Q'_k + \frac{x}{\lambda\,v\,k}(k-1) - \frac{x}{\lambda\,v\,k}(k-1) \\
&= \frac{1}{k}(C_k + x) - \hat{\mu} - \frac{1}{\lambda\,v}Q'_k.
\end{aligned}
$$

Therefore, x is estimated by

$$
\begin{aligned}
x = \hat{\mu} + \frac{1}{v}(R_j + x) - \hat{\mu} + \frac{1}{k}(C_k + x) - \frac{G + x}{v\,k} - \frac{1}{\lambda\,v}Q'_k \\
+ \frac{k}{\lambda\,v}\left\{Q_i + \frac{x(k-1)}{k}\right\}.
\end{aligned}
$$

Denoting the solution of the above equation by $\hat{x}$, we obtain

$$
\hat{x} = \frac{\lambda(k\,R_j + vC_k - G) - kQ'_k + k^2Q_i}{k(k-1)(k-2)}.
$$

Since $r = k$, the above result can be also written in terms of r by replacing k by r as follows.

$$
\hat{x} = \frac{\lambda(r\,R_j + vC_k - G) - r\,Q'_k + r^2Q_i}{r(r-1)(r-2)}
$$

We shall now find the bias in the estimated treatment SS. Under $\mathbf{t} = \mathbf{0}$, the conditional error SS adjusted is given by

$$
\begin{aligned}
S_{e0}^2(\mathbf{y}, x) = \left[\Sigma\, y^2 + x^2 - \frac{(G+x)^2}{v\,r}\right] - \left[\frac{\sum_{p\neq j} R_p^2 + (R_j + x)^2}{v} - \frac{(G+x)^2}{v\,r}\right] \\
- \left[\frac{\sum_{\ell\neq k} C_\ell^2 + (C_k + x)^2}{r} - \frac{(G+x)^2}{v\,r}\right].
\end{aligned}
$$

Hence, the estimated conditional error is given by

$$S_{e0}^2(\mathbf{y}, \hat{x}) = \Sigma y^2 + \hat{x}^2 - \frac{\sum_{p\neq j} R_p^2 + (R_j + \hat{x})^2}{v} - \frac{\sum_{\ell\neq j} C_\ell^2 + (C_k + \hat{x})^2}{r} + \frac{(G+\hat{x})^2}{v\,r}.$$

Equating the derivative of the conditional error SS to zero, we get

$$2\hat{x} - \frac{2}{v}(R_j + \hat{x}) - \frac{2}{v\,r}(C_k + \hat{x}) + \frac{2}{v\,r}(G + x) = 0.$$

Denoting the solution of the above equation by x_0, we obtain

$$x_0 = \frac{r\,R_j + v\,C_k - G}{(v-1)(r-1)} = \frac{\lambda(r\,R_j + vC_k - G)}{r(r-1)^2}$$

since $\lambda(v-1) = \lambda(r-1)$. Hence, the minimum balue of the conditional error SS is given by

$$\min_x S_{e0}^2(\mathbf{y}, x) = \Sigma y^2 + x_0^2 - \frac{\sum_{p\neq j} R_p^2 + (R_j + x_0)^2}{v} - \frac{\sum_{\ell\neq k} C_\ell^2 + (C_k + x_0)^2}{r} + \frac{(G+x_0)^2}{v\,r}.$$

Hence, the bias in the estimated adjusted treatment SS is given by

$$\begin{aligned}
\text{Bias} &= [\text{estimated conditional error SS}] \\
&\quad - [\text{minimum value of conditional error SS}] \\
&= \hat{x}^2 - x_0^2 - \frac{(R_j+\hat{x})^2 - (R_j+x_0)^2}{v} - \frac{(C_k+\hat{x})^2 - (C_k+x_0)^2}{r} \\
&\quad + \frac{(G+\hat{x})^2 - (G+x_0)^2}{v\,r} \\
&= (\hat{x} - x_0)\left[(\hat{x}+x_0) - \frac{(2\,R_j + \hat{x} + x_0)}{v} - \frac{(2\,C_k + \hat{x} + x_0)}{r}\right. \\
&\quad \left. + \frac{(2\,G + \hat{x} + x_0)}{v\,r}\right] \\
&= \frac{(\hat{x} - x_0)}{v\,r}[(\hat{x} - x_0)(v-1)(r-1) - 2(r\,R_j + v\,C_k - G)] \\
&= \frac{(\hat{x} - x_0)}{v\,r}[(\hat{x} + x_0)(v-1)(r-1) - 2(v-1)(r-1)x_0] \\
&= \frac{(v-1)(r-1)}{v\,r}(\hat{x} - x_0)^2 \\
&= \frac{(v-1)[\lambda(r\,R_j + v\,C_k - G) + r(r-1)(r\,Q_i - Q_k)]^2}{v\,r^3(r-1)^3(r-2)^2}.
\end{aligned}$$

135. The missing yields in the first row are denoted by $x_1, x_2, \ldots, x_v$. Note that x_i corresponds to the ith treatment and the ith column, $i = 1, 2, \ldots, v$. Here x_i's are estimated by

$$x_i = \hat{\mu} + \hat{\alpha} + \hat{\beta}_i + \hat{t}_i, \ i = 1, 2, \ldots, v$$

where $\hat{\mu} = (G + X)/v^2, \ X = \sum_1^v x_i,$

$$\begin{aligned} \hat{\alpha} &= \text{estimate of the effect of the first row} \\ &= \frac{1}{v}X - \hat{\mu}, \\ \hat{\beta}_i &= \text{estimate of the effect of the ith column} \\ &= \frac{1}{v}(C_i + x_i) - \hat{\mu}, \\ \hat{t}_i &= \text{estimate of the ith treatment effect} \\ &= \frac{1}{v}(T_i + x_i) - \hat{\mu}. \end{aligned}$$

Hence, we get

$$vx_i - X = \frac{v(C_i + T_i) - 2G}{(v-2)}, i = 1, 2, \ldots v.$$

The above equations can be written as

$$A\mathbf{x} = \mathbf{h}$$

where A is a $v \times v$ matrix with elements

$$\begin{aligned} a_{is} &= v - 1, \ \text{if } i = s \\ &= -1 \qquad \text{if } i \neq s \end{aligned}$$

and $\mathbf{x}' = [x_1, x_2, \ldots, x_v], \mathbf{h}' = [h_1, h_2, \ldots, h_v]$, when $h_i = (v\,C_i + v\,T_i - 2\,G)/(v-2)$. The matrix A can be written as

$$A = v\,I_v - E_{vv},$$

which is singular. The Moore-Penrose inverse of A is seen to be

$$A^+ = (1/v)I_v - (1/v^2)E_{vv}.$$

Hence, a solution of $A\mathbf{x} = \mathbf{h}$ is obtained as

$$\begin{aligned} \hat{\mathbf{x}} &= A^+\mathbf{h} = [(1/v)I_v - (1/v^2)E_{vv}]\mathbf{h} \\ &= \frac{1}{v}\mathbf{h}, \end{aligned}$$

since $E_{1v}\mathbf{h} = 0$. Hence, we get

$$\hat{x}_i = h_i/v = (v\,C_i + v\,T_i - 2\,G)/v(v-2), i = 1, 2, \ldots, v.$$

We shall now find the bias in the estimated treatment SS. Under $\mathbf{t} = \mathbf{0}$, the conditional error SS is given by

$$S_{e0}^2(\mathbf{y}, x) = \left[\Sigma\, y^2 + \sum_1^v x_i^2 - (G+X)^2/v^2\right] - \left[\frac{X^2 + \sum_{j\neq 1} R_j^2}{v} \frac{(G+X)^2}{v^2}\right] - \left[\frac{\sum_1^v (C_i + x_i)^2}{v} - \frac{(G+X)^2}{v^2}\right].$$

Hence, the estimated conditional error SS is given by

$$S_{e0}^2(\mathbf{y}, \hat{\mathbf{x}}) = \Sigma\, y^2 + \sum_1^v \hat{x}_i{}^2 - \frac{\hat{X}^2 + \sum_{j\neq 1} R_j^2}{v} - \frac{\Sigma(C_i + \hat{x}_i)^2}{v} + \frac{(G+\hat{x})^2}{v^2}$$

$$= \Sigma\, y^2 + \sum_1^v \hat{x}_i{}^2 - \frac{1}{v}\sum_{j\neq 1} R_j^2 - \frac{1}{v}\sum_1^v (C_i + \hat{x}_i)^2 + G/v^2,$$

since $\hat{X} = 0$.

Equating the derivative of the conditional error SS to zero, we get

$$v\, x_i - X = \frac{vC_i - G}{v-1}, \quad i = 1, 2, \ldots, v.$$

These equations can be written as $A\mathbf{x} = \mathbf{b}$, where A has the same meanings as in $A\mathbf{x} = \mathbf{h}$ and $\mathbf{b}' = [b_1, b_2, \ldots, b_v]$, where $b_i = (vC_i - G)/(v-1)$. Then a solution of $A\mathbf{x} = \mathbf{b}$ is given by

$$\overset{*}{x} = A^{+}\mathbf{b}$$

and hence we obtain

$$\overset{*}{x}_i = (v\, C_i - G)/v(v-1), \; i = 1, 2, \ldots, v.$$

Hence, the minimum value of conditional error SS is given by

$$S_{e0}{}^2(\mathbf{y}, \overset{*}{\mathbf{x}}) = \Sigma\, y^2 + \sum_1^v \overset{*}{x}_i{}^2 - \frac{1}{v}\sum_{j\neq 1} R_j^2 - \frac{1}{v}\sum_1^v (C_i + \overset{*}{x}_i)^2 + G^2/v^2$$

Thus, the bias in the estimated treatment SS is given by

$$\begin{aligned}
\text{Bias} &= [\text{estimated conditional error SS}] \\
&\quad - [\text{min. value of conditional error SS}] \\
&= \sum_{i=1}^v (\hat{x}_i{}^2 - \overset{*}{x}_i{}^2) - \frac{1}{v}\Sigma[(C_i + \hat{x}_i)^2 - (C_i + \overset{*}{x}_i)^2] \\
&= \frac{1}{v}\sum_{i=1}^v (\hat{x}_i - \overset{*}{x}_i)[(v-1)(\hat{x}_i + \overset{*}{x}_i) - 2C_i]
\end{aligned}$$

But $[v(v-1)\overset{*}{x}_i + G]/v = C_i$. Hence,

$$\begin{aligned}
\text{Bias} &= \frac{1}{v}\sum_{i=1}^{v}(\hat{x}_i - \overset{*}{x}_i)\left[(v-1)(\hat{x}_i + \overset{*}{x}_i) - 2(v-1)\overset{*}{x}_i - \frac{2\,G}{v}\right] \\
&= \frac{1}{v}\Sigma(\hat{x}_i - \overset{*}{x}_i)\left[(v-1)(\hat{x}_i - \overset{*}{x}_i) - \frac{2\,G}{v}\right] \\
&= \frac{(v-1)}{v}\sum_{i=1}^{v}(\hat{x}_i - \overset{*}{x}_i)^2, \text{ since } \Sigma\, x_i = \Sigma^*_{x_i} = 0 \\
&= \sum_{i=1}^{v}[C_i + (v-1)T_i - G]^2/v(v-1)(v-2)^2.
\end{aligned}$$

136. Let the yields corresponding to treatments 1 and 2 in the first block which are mixed up be denoted by x_1 and x_2. Their sum is $x_1 + x_2 = u$. We estimate x_1 and x_2 by minimizing the error SS subject to the condition $x_1 + x_2 = u$. Thus we consider the minimization of

$$\begin{aligned}
\phi = \Sigma\, y^2 + x_1^2 + x_2^2 &- \frac{1}{v}\left[(B_1 + u)^2 + \sum_{j\neq 1} B_j^2\right] \\
&- \frac{1}{r}\left[(T_1 + x_1)^2 + (T_2 + x_2)^2 + \sum_{i\neq 1,2} T_i^2\right] + \frac{(G+u)^2}{v\,r} \\
&+ 2\lambda(u - x_1 - x_2),
\end{aligned}$$

where λ is the Lagrangian Multiplier. Then

$$\begin{aligned}
\frac{\partial\,\phi}{\partial\,x_1} &= 2\,x_1 - \frac{2}{r}(T_1 + x_1) - 2\lambda = 0 \\
\frac{\partial\,\phi}{\partial\,x_2} &= 2\,x_2 - \frac{2}{r}(T_2 + x_2) - 2\lambda = 0.
\end{aligned}$$

We then obtain the equations

$$\begin{aligned}
x_1(r-1) &= T_1 + r\lambda \\
x_2(r-1) &= T_2 + r\lambda.
\end{aligned}$$

Adding these two equations, we obtain

$$\lambda = \frac{u(r-1)}{2\,r} - \frac{T_1 + T_2}{2\,r}.$$

Denoting the solutions of x_1 and x_2 by $\hat{x}_1$ and $\hat{x}_2$, we obtain

$$\hat{x}_1 = \frac{(T_1 - T_2)}{2(r-1)} + \frac{u}{2}, \quad \hat{x}_2 = \frac{(T_2 - T_1)}{2(r-1)} + \frac{u}{2}.$$

These provide the estimates of the mixed up yields. To find the bias in the treatment sum of squares, we consider the conditional error SS. Under $\mathbf{t} = \mathbf{0}$.

$$\text{Conditional Error SS} = \Sigma\, y^2 + x_1^2 + x_2^2 - \frac{1}{v}\left[(B_1 + u)^2 + \sum_{j=2}^{r} B_j^2\right].$$

The estimated conditional error SS is

$$\Sigma\, y^2 + \hat{x}_1^2 + \hat{x}_2^2 - \frac{1}{v}\left[(B_1 + u)^2 + \sum_{j=2}^{r} B_j^2\right].$$

Now we shall find the minimum value of the conditional error SS subject to $x_1 + x_2 = u$. We minimize

$$f = \Sigma\, y^2 + x_1^2 + x_2^2 - \frac{1}{v}\left[(B_1 + u)^2 + \sum_{2}^{r} B_j^2\right] + 2\lambda(u - x_1 - x_2)$$

Then,

$$\frac{\partial f}{\partial x_1} = 2x_1 - 2\lambda = 0$$
$$\frac{\partial f}{\partial x_2} = 2x_2 - 2\lambda = 0.$$

Hence, we get $x_1 = x_2 = \lambda$. Since $x_1 + x_2 = u$, we obtain the solutions of x_1 and x_2 as

$$\overset{*}{x}_1 = \overset{*}{x}_2 = u/2$$

and the minimum value of the conditional error SS is

$$\Sigma\, y^2 + \overset{*}{x}{}_1^2 + \overset{*}{x}{}_2^2 - \frac{1}{v}\left[(B_1 + u)^2 + \sum_{2}^{r} B_j^2\right]$$

Thus, the bias in the treatment sum of squares is given by

$$\begin{aligned}
\text{Bias} &= (\hat{x}_1^2 - \overset{*}{x}{}_1^2) + (\hat{x}_2^2 - \overset{*}{x}{}_2^2) \\
&= \frac{(T_1 - T_2)}{2(r-1)}\left[\frac{(T_1 - T_2)}{2(r-1)} + u\right] + \frac{(T_2 - T_1)}{2(r-1)}\left[\frac{(T_2 - T_1)}{2(r-1)} + u\right] \\
&= \frac{(T_1 - T_2)}{2(r-1)}\left[\frac{(T_1 - T_2)}{(r-1)}\right] \\
&= \frac{(T_1 - T_2)^2}{2(r-1)^2}.
\end{aligned}$$

Now we shall find the loss in efficiency. Clearly,

$$\text{var}\,(\hat{x}_1) = \text{var}\left[\frac{T_1 - T_2}{2(r-1)} + \frac{u}{2}\right] = \frac{r\,\sigma^2}{2(r-1)}$$

$$\text{var}\,(\hat{x}_2) = \text{var}\left[\frac{T_2 - T_1}{2(r-1)} + \frac{u}{2}\right] = \frac{r\,\sigma^2}{2(r-1)}$$

$$\text{cov}\,(\hat{x}_1, \hat{x}_2) = \frac{(r-2)\sigma^2}{2(r-1)}.$$

The different types of elementary treatment contrasts are

(i) $t_1 - t_2, t_2 - t_1$
(ii) $t_1 - t_\ell, t_\ell - t_1, \ell = 3, 4, \ldots, v$
(iii) $t_2 - t_\ell, t_\ell - t_2, \ell = 3, 4, \ldots, v$
(iv) $t_i - t_j, t_j - t_i, i \neq j = 3, 4, \ldots, v$

One can easily verify that

(i) $\text{var}\,(\hat{t}_1 - \hat{t}_2) = \text{var}(\hat{t}_2 - \hat{t}_1) = 2\sigma^2/(r-1)$

(ii) $\text{var}\,(\hat{t}_1 - \hat{t}_\ell) = \text{var}(\hat{t}_\ell - \hat{t}_1) = \dfrac{2\sigma^2}{r}\left[1 + \dfrac{1}{4(r-1)}\right]$

(iii) $\text{var}\,(\hat{t}_2 - \hat{t}_\ell) = \text{var}(\hat{t}_\ell - \hat{t}_2) = \dfrac{2\sigma^2}{r}\left[1 + \dfrac{1}{4(r-1)}\right]$

(iv) $\text{var}\,(\hat{t}_i - \hat{t}_j) = \text{var}(\hat{t}_j - \hat{t}_i) = 2\sigma^2/r.$

Hence, the average variance $\bar{V}$ of the estimates of all elementary treatment contrasts is given by

$$\bar{V} = \frac{2\,\sigma^2}{r}\left[\frac{2\,r}{r-1} + 4(v-2)\left\{1 + \frac{1}{4(r-1)}\right\} + (v-2)(v-3)/v(v-1)\right]$$
$$= \frac{2\,\sigma^2}{r}\left[1 + \frac{1}{(v-1)(r-1)}\right].$$

Therefore the efficiency of the RBD with mixed up yields relative to that of the RBD without mixed up yields is

$$\left[1 + \frac{1}{(v-1)(r-1)}\right]^{-1}$$

And the loss in efficiency is $[1 + (v-1)(r-1)]^{-1}$.

137. (a) Here, $p = 2, n = 5$ and

$$X'X = \begin{bmatrix} 5 & 1 \\ 1 & 5 \end{bmatrix},$$
$$(X'X)^{-1} = \frac{1}{24}\begin{bmatrix} 5 & -1 \\ 1 & 5 \end{bmatrix},$$

and

$$\text{tr}\,(X'X)^{-1} = 10/24.$$

Hence, efficiency of this weighing design is

$$p/n\ \text{tr}(X'X)^{-1} = 24/25.$$

(b) Here $p = 4, n = 4$ and

$$X'X = \begin{bmatrix} 3 & 2 & 2 & 2 \\ 2 & 3 & 2 & 2 \\ 2 & 2 & 3 & 2 \\ 2 & 2 & 2 & 3 \end{bmatrix} = I_4 + 2E_{44}$$
$$(X'X)^{-1} = I_4 - (2/9)E_{44}$$

and

$$\text{tr}\,(X'X)^{-1} = 4 - 8/9 = 28/9.$$

Hence, efficiency of this design is

$$p/n\ \text{tr}(X'X)^{-1} = 9/28$$

138. Here

$$X'X = (N-1)I_p + E_{pp}.$$

Hence,

$$(X'X)^{-1} = \frac{1}{N-1}I - \frac{1}{(N-1)(N-1+p)}E_{pp}.$$

Therefore, $\text{tr}\,(X'X)^{-1} = p(N-2+p)/(N-1)(N-1+p)$.
Hence, the efficiency of this weighing design is

$$p/n\ \text{tr}(X'X)^{-1} = \frac{(N-1)(N-1+p)}{N(N-2+p)}.$$

139. (i) Here no. of objects to be weighed is equal to p and the no. of weighings is equal to $N - 1 + r$. Hence, the efficiency of this design is

$$p/(N-1+r)\ \text{tr}\,(X'X)^{-1}.$$

Now,

$$X'X = (N-1)I_p + rE_{pp}.$$

Hence,

$$(X'X)^{-1} = \frac{1}{N-1}I_p - \frac{r}{(N-1)(N-1+pr)}E_{pp}.$$

Therefore, $\text{tr}\,(X'X)^{-1} = p(N-1-r+pr)/(N-1)(N-1+pr)$, and

$$\text{efficiency} = \frac{(N-1)(N-1+pr)}{(N-1+r)(N-1-r+pr)}.$$

(ii) For this design, the number of objects to be weighed is p and the number of weighings is equal to $(N-1+r)$. Hence, the efficiency of this design is

$$p/(N-1+r)\ \text{tr}\,(X'X)^{-1}.$$

We now find $\text{tr}\,(X'X)^{-1}$. Clearly,

$$X'X = \begin{bmatrix} rE_{11} & 0_{1(p-1)} \\ 0_{(p-1)1} & 0_{(p-1)(p-1)} \end{bmatrix} + (N-1)I_p$$
$$= \text{diag}[(N-1+r), (N-1), \ldots, (N-1)].$$

Hence,

$$(X'X)^{-1} = \text{diag}\,[(N-1+r)^{-1}, (N-1)^{-1}, \ldots, (N-1)^{-1}]$$

and

$$\text{tr}\,(X'X)^{-1} = \frac{p(N-1)+r(p-1)}{(N-1)(N-1+r)}.$$

Therefore, the efficiency of the design is

$$p(N-1)/[p(N-1)+r(p-1)].$$

140. We know that

$$\hat{\mathbf{w}} = (X'X)^{-1}X'\mathbf{y}.$$

Denote the ith column of X by $\mathbf{x}_i, i = 1, 2, \ldots, b$ and the jth collumn of $X(X'X)^{-1}$ by $\mathbf{c}_j, j = 1, 2, \ldots, p$. Hence, we see that

$$\hat{w}_i = \mathbf{c}_i'\mathbf{y}, i = 1, 2, \ldots, p$$

and

$$\text{var}(\hat{w}_i) = \sigma^2\mathbf{c}_i'\mathbf{c}_i.$$

Now, since $(X'X)^{-1}(X'X) = I_p$, we have $\mathbf{c}_i'\mathbf{x}_j = \delta_{ij}$, where $\delta_{ij} = 1$ if $i = j$ and $\delta_{ij} = 0$ if $i \neq j$.
By Schwarz inequality we have

$$(\mathbf{x}_i'\mathbf{x}_i)(\mathbf{c}_i'\mathbf{c}_i) \geq (\mathbf{x}_i'\mathbf{c}_i)^2 = 1$$

Hence, $\mathbf{c}_i'\mathbf{c}_i \geq \dfrac{1}{\mathbf{x}_i'\mathbf{x}_0} \geq \dfrac{1}{n}$.
Since $\mathbf{x}_i$ is a column vector consisting of $+1, -1$ or 0, we have

$$\text{var}(\hat{w}_i) \geq \sigma^2/n, \text{ for } i = 1, 2, \ldots, p.$$

141. Suppose the variances of all the estimated weights are minimum, i.e.

$$\text{var}(\hat{w}_i) = \sigma^2/n, i = 1, 2, \ldots, p.$$

Defining $\mathbf{x}_i, i = 1, 2, \ldots, p$ and $\mathbf{c}_j, j = 1, 2, \ldots p$ as in Exercise 140, we see that

$$\text{var}(\hat{w}_i) = \sigma^2\mathbf{c}_i\mathbf{c}_i.$$

By Schwarz inequality, we get

$$(\mathbf{x}_i'\mathbf{x}_i)(\mathbf{c}_i'\mathbf{c}_i) \geq (\mathbf{x}_i'\mathbf{c}_i)^2 = 1,$$

since $\mathbf{c}_i\mathbf{x}_j = \delta_{ij}$, the Kronecker delta, in view of $(X'X)^{-1}(X'X) = I_p$. Hence

$$\mathbf{c}_i'\mathbf{c}_i \geq 1/\mathbf{x}_i'\mathbf{x}_i \geq 1/n$$

since $\mathbf{x}_i$ is a vector of $+1, -1$ or 0. Thus,

$$\text{var}(\hat{w}_i) \geq \sigma^2/n.$$

Further equality occurs if and only if $\mathbf{c}_i = k\mathbf{x}_i$ for some constant k, and $\mathbf{x}_i'\mathbf{x}_i = n$ for $i = 1, 2, \ldots, n$. Since $\mathbf{c}_i'\mathbf{x}_j = 0$ for $i \neq j$ we immediately get $\mathbf{x}_i'\mathbf{x}_j = 0$. Hence

$$X'X = \begin{bmatrix} \mathbf{x}_1' \\ \mathbf{x}_2' \\ \cdot \\ \cdot \\ \cdot \\ \mathbf{x}_p' \end{bmatrix} [\mathbf{x}_1, \mathbf{x}_2, \ldots, \mathbf{x}_p]$$

$$= \begin{bmatrix} \mathbf{x}_1'\mathbf{x}_1 & 0 & \ldots & 0 \\ 0 & \mathbf{x}_2'\mathbf{x}_2 & \ldots & 0 \\ \cdot & & & \\ \cdot & & & \\ \cdot & & & \\ 0 & 0 & \ldots & \mathbf{x}_p'\mathbf{x}_p \end{bmatrix}$$

$$= nI_p$$

Conversely, if $X'X = n\,I_p$, then $(X'X)^{-1} = \frac{1}{n}I_p$, and hence

$$\text{var}(\hat{w}_i) = \sigma^2/n, \; i = 1, 2, \ldots, p.$$

This proves the sufficiency.

142. In a spring-balance design, the weighing matrix $X = [x_{ij}]$ is defined as:

$$\begin{aligned} x_{ij} &= 1, \text{ if the jth object is weighed in the ith weighing} \\ &= 0, \text{ if the jth object is not weighed in the ith weighing} \end{aligned}$$

Let there be v objects and suppose k objects are weighed in each weighing. The total number of weighings is b and each object is weighed r times. Hence, by identifying the object with treatments and weighings by objects, we see that $X = N'$, where N is the incidence matrix of a BIBD (v, b, r, k, λ). Hence, $X'X = NN' = (r - \lambda)I_v + \lambda E_{vv}$ and

$$(X'X)^{-1} = (1/(r - \lambda))I_v - (\lambda/(r - \lambda)rk)E_{vv}.$$

The variance-covariance matrix of the estimated weights is given by

$$\text{var}(\hat{\mathbf{w}}) = \sigma^2 \left[\frac{1}{(r - \lambda)} I_v - \frac{\lambda}{(r - \lambda)rk} E_{vv} \right].$$

Defining the efficiency of a weighing design as $p/n\,\text{tr}(X'X)^{-1}$, where p = no. of objects to be weighed, n = no. of weighings, we see that the efficiency of the above design is $k^2(r - \lambda)/(rk - \lambda)$. Hence, the efficiency is maximum if $(rk - \lambda)/k^2(r - \lambda)$ is minimum.

143. Consider a Hadamard matrix H_{n+1} of order (n + 1) which is assumed to exist. A Hadamard matrix remains as Hadamard matrix if any of its rows (or columns) are multiplied by −1. Hence without loss of generality we

can assume H_{n+1} as the matrix whose first row and first column contain the element + 1. Now subtract the first row from each of the other rows and multiply 2nd, 3rd, ..., (n + 1)-st rows by −1/2. Then we shall get the following matrix. $\begin{bmatrix} 1 & E_{1(n)} \\ 0 & L_{n\times n} \end{bmatrix}$.

The matrix L is easily seen to be the incidence matrix of a SBIBD with parameters

$$v = b = n, \ r = k = (n-1)/2, \ \lambda = (n-3)/4.$$

Now, clearly

$$|H_{n+1}| = (-1)^n \cdot |L|$$

Since H_{n+1} has the maximum value of $|H_{n+1}|$, it follows that L has maximum value of $|L|$. Thus, L is the weighing design of the spring-balance type involving n weighings of n objects and having maximum efficiency. Now

$$H_4 = H_2 \otimes H_2 = \begin{bmatrix} 1 & 1 \\ 1 & -1 \end{bmatrix} \otimes \begin{bmatrix} 1 & 1 \\ 1 & -1 \end{bmatrix} = \begin{bmatrix} 1 & 1 & 1 & 1 \\ 1 & -1 & 1 & -1 \\ 1 & 1 & -1 & -1 \\ 1 & -1 & -1 & 1 \end{bmatrix}.$$

Subtracting the first row from the other rows, we get the matrix

$$\begin{bmatrix} 1 & 1 & 1 & 1 \\ 0 & -2 & 0 & -2 \\ 0 & 0 & -2 & -2 \\ 0 & -2 & -2 & 0 \end{bmatrix}.$$

Multiplying the 2nd, 3rd and 4th rows by −1/2, we get the matrix

$$\begin{bmatrix} 1 & 1 & 1 & 1 \\ 0 & 1 & 0 & 1 \\ 0 & 0 & 1 & 1 \\ 0 & 1 & 1 & 0 \end{bmatrix}.$$

Hence the design

$$\begin{bmatrix} 1 & 0 & 1 \\ 0 & 1 & 1 \\ 1 & 1 & 0 \end{bmatrix}$$

is the required weighing design.

144 We know that the variance-covariance matrix of the BLUEs of the weights is given by (see Exercise 142)

$$\text{var}(\hat{\mathbf{w}}) = \sigma^2 \left[\frac{1}{(r-\lambda)} I_v - \frac{\lambda}{(r-\lambda)rk} E_{vv} \right].$$

The BLUE of total weight of all objects is given by

$$\hat{W} = E_{1v}\hat{\mathbf{w}}.$$

Hence,

$$\begin{aligned}
\mathrm{var}(\hat{W}) &= E_{1v}\mathrm{var}(\hat{\mathbf{w}})E_{v1} \\
&= \sigma^2 E_{1v}[I_v - \frac{\lambda}{rk}E_{vv}]E_{v1}/(r-\lambda) \\
&= \sigma^2\left[v - \frac{\lambda v^2}{rk}\right]/(r-\lambda) \\
&= \sigma^2 v(rk - \lambda v)/(r-\lambda)rk \\
&= \sigma^2 v/rk.
\end{aligned}$$

145. Clearly L_i consists of s subscripts 0, 1, 2,..., s − 1. We shall now prove that each subscript occurs exactly once in each row and exactly once in each column.

Suppose, if possible, the same subscript occurs more than once, say twice in the αth row, that is, it occurs in the cells (α, β) and (α, γ), say. Hence

$$u_i u_\alpha + u_\beta = u_i u_\alpha + u_\gamma,$$

which gives $u_\beta = u_\gamma$. Thus β = γ. Hence each subscript occurs exactly once in each row. Further, suppose the same subscript occurs twice in the same column, say in the cells (α, β) and (γ, β). Then we get

$$u_i u_\alpha + u_\beta = u_i u_\gamma + u_\beta$$

which gives

$$u_i(u_\alpha - u_\gamma) = 0.$$

Since $u_i \neq 0$, it follows that $u_\alpha = u_\gamma$. Thus α = γ. Hence each element occurs exactly once in each column. Thus, L_i i = 1, 2, ..., s − 1 is a Latin Square. We now prove that L_i and L_j, $i \neq j = 1, 2, \ldots, s-1$ are orthogonal Latin squares.

Let the (x, y)th cell of L_i and L_j contain respectively the subscripts m and n. Hence, when L_j is superimposed upon L_i, we get an ordered pair (m, n) in the cell (x, y). We have now to prove that the ordered pair (m, n) occurs exactly once. If possible, suppose the pair (m, n) occurs in another cell (α, β). Then, we get

$$u_i u_x + u_y = u_i u_\alpha + u_\beta$$
$$u_j u_x + u_y = u_j u_\alpha + u_\beta$$

Subtracting the second equation from the first, we get

$$(u_i - u_j)u_x = (u_i - u_j)u_\alpha.$$

Since $i \neq j$; $u_i \neq u_j$, hence $u_x = u_\alpha$, that is x = α. Then from the first equation we get $u_y = u_\beta$, that is, y = β. Hence the cell (α, β) is the same as the cell (x, y). Thus, the pair (m, n) occurs exactly once. Hence L_i and L_j are orthogonal.

146. (i) Consider the cell (α, β) of $L_{i+1}, \alpha = 1, 2, \ldots, s-2$. It contains the subscript of element

$$u_{i+1}u_\alpha + u_\beta = x^{i+\alpha-1} + u_p$$

Now consider the cell $(\alpha+1, \beta)$ of L_i. It contains the subscript of the element

$$u_i u_{\alpha+1} + u_\beta = x^{i+\alpha-1} + u_\beta$$

Thus, we see that the subscript in the cell (α, β) of L_{i+1} is the same as the subscript in the cell $(\alpha+1, \beta)$ of L_i. Thus, we see that the αth row of L_{i+1} is the same as the $(\alpha+1)$-st row of L_i. Consider the cell $(s-1, \beta)$ of L_{i+1}. It contains the subscript of the element

$$\begin{aligned} u_{i+1}u_{s-1} + u_\beta &= x^{i+s-2} + u_\beta \\ &= x^{i-1} + u_\beta, \end{aligned}$$

since $x^{s-1} = 1$.
Now consider the cell $(1, \beta)$ of L_i. It contains the subscript of the element

$$u_i u_1 + u_\beta = x^{i-1} + u_\beta.$$

Thus, we see that the subscript in the cell $(s-1, \beta)$ of L_{i+1} is equal to the subscript of the cell $(1, \beta)$ of L_i, $= 0, 1, 2, \ldots, s-1$. Hence, the last row of L_{i+1} is the same as the first row of L_i.

(ii) Since, the subscript in the cell (α, β) of L_i is j, we get

$$u_j = u_i u_\alpha + u_\beta = x^{i+\alpha-2} + x^{\beta-1}.$$

Now, consider the cell $(\alpha+1, \beta+1)$. It will contain the subscript of the element

$$u_i u_{\alpha+1} + u_{\beta+1}.$$

Then,

$$\begin{aligned} u_i u_{\alpha+1} + u_{\beta+1} &= x^{i+\alpha-1} + x^\beta \\ &= (x^{i+\alpha-2} + x^{\beta-1}) \\ &= u_2 u_j, \end{aligned}$$

since $u_j = x^{i+\alpha-2} + x^{\beta-1}$. Clearly

$$\begin{aligned} u_2 u_j &= u_0, && \text{if } j = 0 \\ &= u_{j+1} && \text{if } j = 1, 2, \ldots, s-2 \\ &= u_1 && \text{if } j = s-1. \end{aligned}$$

Hence, the results follow.

147. (i) Side 5. The 0th row is constructed by writing the subscripts 0, 1, 2, 3, 4. The first row is constructed by filling its $(1, \beta)$ cell by the subscript of the element $u_1 u_1 + u_\beta = 1 + u_\beta, \beta = 0, 1,$

2, 3, 4. Now the elements of GF (5) are $u_0 = 0, u_1 = 1, u_2 = x$, $u_3 = x^2$, where $x = 2$. Therefore, the elements of GF (5) are given by $u_0 = 0, u_1 = 1, u_2 = 2, u_3 = 4, u_4 = 3$. Thus, the subscripts in the first row of L_1 are the subscripts of elements $1 + u_0 = u_1$, $1 + u_1 = u_2$, $1 + u_2 = 1 + 2 = 3 = u_4$, $1 + u_3 = 1 + 4 = 5 = u_0$, $1 + u_4 = 1 + 3 = 4 = u_3$, thus the first row of L_1 is

1, 2, 4, 0, 3

Hence L_1 is

0	1	2	3	4
1	2	4	0	3
2	4	3	1	0
3	0	1	4	1
4	3	0	2	0

(ii) Side 7. The elements of GF (7) are $u_0 = 0, u_1 = 1, u_2 = x, u_3 = x^2, u_4 = x^3, u_5 = x^4, u_6 = x^5$, where $x = 3$. Thus the elements are $u_0 = 0, u_1 = 1, u_2 = 3, u_3 = 2, u_4 = 6, u_5 = 4, u_6 = 5$. The 0th row of L_1 is 0, 1, 2, 3, 4, 5, 6. The first row of L_1 is constructed by filling its $(1, \beta)$-th cell by the subscript of $u_1 u_1 + u = u_1 + u$. For $\beta = 0, 1, 2, 3, 4, 5, 6$, the elements $u_1 + u_\beta$ are $u_1 + u_0 = 1 = u_1, u_1 + u_1 = 2 = u_3, u_1 + u_2 = 1 + 3 = u_5, u_1 + u_3 = 1 + 2 = 3 = u_2, u_1 + u_4 = 1 + 6 = 7 = u_0, u_1 + u_5 = 1 + 4 = u_6, u_1 + u_6 = 1 + 5 = 6 = u_4$. Thus, the first row of L_1 is

1, 3, 5, 2, 0, 6, 4,

Hence, we obtain L_1 as

0	1	2	3	4	5	6
1	3	5	2	0	6	4
2	5	4	6	3	0	1
3	2	6	5	1	4	0
4	0	3	1	6	2	5
5	6	0	4	2	1	3
6	4	1	0	5	3	2

(iii) Side 8. The elements of GF (8) are $u_0 = 0, u_1 = 1, u_2 = x, u_3 = x^2, u_4 = x^3 = x + 1, u_5 = x^4 = x^2 + x, u_6 = x^5 = x^2 + x + 1, u_7 = x^6 = x^2 + 1$.
The 0th row of L_1 is

0, 1, 2, 3, 4, 5, 6, 7.

The first row of L_1 is constructed by filling its $(1, \beta)$-th cell by the subscript of the element $u_1 + u_\beta$. Now for $\beta = 0, 1, 2, 3, 4, 5, 6, 7$, the elements $u_1 + u_\beta$ are

$$u_1 + u_0 = u_1, u_1 + u_1 = u_0, u_1 + u_2 = 1 + x = u_4,$$
$$u_1 + u_3 = 1 + x^2 = u_7, u_1 + u_4 = 1 + x + 1 = x = u_2,$$
$$u_1 + u_5 = 1 + x^2 + x = u_6,$$
$$u_1 + u_6 = 1 + x^2 + x + 1 = x^2 + x = u_5,$$
$$u_1 + u_7 = 1 + x^2 + 1 = x^2 = u_3.$$

Hence the 1st row of L_1 is

1, 0, 4, 7, 2, 6, 5, 3.

Thus, L_1 is

0	1	2	3	4	5	6	7
1	0	4	7	2	6	5	3
2	4	0	5	1	3	7	6
3	7	5	0	6	2	4	1
4	2	1	6	0	7	3	5
5	6	3	2	7	0	1	4
6	5	7	4	3	1	0	2
7	3	6	1	5	4	2	0

(iv) Side 9. The element of GF (3^2) are

$$u_0 = 0, u_1 = 1, u_2 = x, u_3 = x^2 = 2x + 1,$$
$$u_4 = x^3 = 2x + 2, u_5 = x^4 = 2, u_6 = x^5 = 2x,$$
$$u_7 = x^6 = x + 2, u_8 = x^7 = x + 1.$$

The 0th row of L_1 is

0 1 2 3 4 5 6 7 8

The first row of L_1 is constructed by filling its $(1, \beta)$th cell by the subscript of the element $u_1 + u_\beta$. Now for $\beta = 0, 1, 2, 3, 4, 5, 6, 7, 8$, the elements $u_1 + u_\beta$ are

$$u_1 + u_0 = u_1, u_1 + u_1 = 2 = u_5, u_1 + u_2 = 1 + x = u_8,$$
$$u_1 + u_3 = 1 + 2x + 1 = u_4, u_1 + u_4 = 1 + 2x + 2 = u_6,$$
$$u_1 + u_5 = 1 + 2 = u_0, u_1 + u_6 = 1 + 2x = u_3$$
$$u_1 + u_7 = 1 + x + 2 = u_2, u_1 + u_8 = 1 + 2x + 1 = u_7.$$

Hence, the first row of L_1 is

1 5 8 4 6 0 3 2 7

Thus, L_1 is

0	1	2	3	4	5	6	7	8
1	5	8	4	6	0	3	2	7
2	8	6	1	5	7	0	4	3
3	4	1	7	2	6	8	0	5
4	6	5	2	8	3	7	1	0
5	0	7	6	3	1	4	8	2
6	3	0	8	7	4	2	5	1
7	2	4	0	1	8	5	3	6
8	7	3	5	0	2	1	6	4

(v) Side 16. The elements of GF (2^4) are

$u_0 = 0, u_1 = 1, u_2 = x, u_3 = x^2, u_4 = x^3,$
$u_5 = x^4 = x + 1, u_6 = x^5 = x^2 + x, u_7 = x^3 + x^2,$
$u_8 = x^3 + x + 1, u_9 = x^2 + 1, u_{10} = x^3 + x,$
$u_{11} = x^2 + x + 1, u_{12} = x^3 + x^2 + x,$
$u_{13} = x^3 + x^2 + x + 1, u_{14} = x^3 + x^2 + 1,$
$u_{15} = x^3 + 1.$

The 0th row of L_1 is

0 1 2 3 4 5 6 7 8 9 10 11 12 13 14 15 16

The first row of L_1 is constructed by filling its $(1, \beta)$-th cell by the subscript of the element $u_1 + u$. Now for $\beta = 0, 1, 2, 3, 4, 5, 6, 7, 8, 9, 10, 11, 12, 13, 14, 15$, the elements $u_1 + u_\beta$ are

$u_1 + u_0 = u_1, u_1 + u_1 = u_0, u_1 + u_2 = 1 + x = u_5,$
$u_1 + u_3 = 1 + x^2 = u_9, u_1 + u_4 = 1 + x^3 = u_{15},$
$u_1 + u_5 = 1 + x + 1 = u_2, u_1 + u_6 = 1 + x^2 + x = u_{11},$
$u_1 + u_7 = 1 + x^3 + x^2 = u_{14}, u_1 + u_8 = 1 + x^3 + x + 1 = u_{10}$
$u_1 + u_9 = 1 + x^2 + 1 = u^3, u_1 + u_{10} = 1 + x^3 + x = u_8,$
$u_1 + u_{11} = 1 + x^2 + x + 1 = u_6,$
$u_1 + u_{12} = 1 + x^3 + x^2 + x = u_{13},$
$u_1 + u_{13} = 1 + x^3 + x^2 + x + 1 = u_{12},$
$u_1 + u_{14} = 1 + x^3 + x^2 + 1 = u_7,$
$u_1 + u_{15} = 1 + x^3 + 1 = u_4.$

Hence, the first row of L_1 is

1 0 5 9 15 2 11 14 10 3 8 6 13 12 7 4

and L_1 is

0	1	2	3	4	5	6	7	8	9	10	11	12	13	14	15
1	0	5	9	15	2	11	14	10	3	8	6	13	12	7	4
2	5	0	6	10	1	3	12	15	11	4	9	7	14	13	8
3	9	6	0	7	11	2	4	13	1	12	5	10	8	15	14
4	15	10	7	0	8	12	3	5	14	2	13	6	11	9	1
5	2	1	11	8	0	9	13	4	6	15	3	14	7	12	10
6	11	3	2	12	9	0	10	14	5	7	1	4	15	8	13
7	14	12	4	3	13	10	0	11	15	6	8	2	5	1	9
8	10	15	13	5	4	14	11	0	12	1	7	9	3	6	2
9	3	11	1	14	6	5	15	12	0	13	2	8	10	4	7
10	8	4	12	2	15	7	6	1	13	0	14	3	9	11	5
11	6	9	5	13	3	1	8	7	2	14	0	15	4	10	12
12	13	7	10	6	14	4	2	9	8	3	15	0	1	5	11
13	12	14	8	11	7	15	5	3	10	9	4	1	0	2	6
14	7	13	15	9	12	8	1	6	4	11	10	5	2	0	3
15	4	8	14	1	10	13	9	2	7	5	12	11	6	3	0

(vi) Side 25. The elements of GF (25) are

$u_0 = 0, u_1 = 1, u_2 = x, u_3 = x^2 = 2x + 2,$

$u_4 = x + 4, u_5 = x + 2, u_6 = 4x + 2, u_7 = 3,$

$u_8 = 3x, u_9 = x + 1, u_{10} = 3x + 2, u_{11} = 3x + 1,$

$u_{12} = 2x + 1, u_{13} = 4, u_{14} = 4x, u_{15} = 3x + 3,$

$u_{16} = 4x + 1, u_{17} = 4x + 3, u_{18} = x + 3,$

$u_{19} = 2, u_{20} = 2x, u_{21} = 4x + 4, u_{22} = 2x + 3,$

$u_{23} = 2x + 4, u_{24} = 3x + 4$

The 0th row of L_1 is

0 1 2 3 4 5 6 7 8 9 10 11 12 13 14 15
16 17 18 19 20 21 22 23 24.

The first row of L_1 is constructed by filling its $(1, \beta)$-th element by the subscript of the element $u_1 + u$. Now $\beta = 0, 1, 2, 3, 4, 5, 6, 7, 8, 9, 10, 11, 12, 13, 14, 15, 16, 17, 18, 19, 20, 21, 22, 23, 24$, and the elements $u_1 + u$ are

$u_1 + u_0 = u_1, u_1 + u_1 = 2 = u_{19}, u_1 + u_2 = 1 + x = u_9,$

$u_1 + u_3 = 1 + 2x + 2 = u_{22}, u_1 + u_4 = 1 + x + 4 = u_2,$

$u_1 + u_5 = 1 + x + 2 = u_{18}, u_1 + u_6 = 1 + 4x + 2 = u_{17},$

$u_1 + u_7 = 1 + 3 = u_{13}, u_1 + u_8 = 1 + 3x = u_{11},$

$u_1 + u_9 = 1 + x + 1 = u_5, u_1 + u_{10} = 1 + 3x + 2 = u_{15},$

$u_1 + u_{11} = 1 + 3x + 1 = u_{10}, u_1 + u_{12} = 1 + 2x + 1 = u_3,$

$u_1 + u_{13} = 1 + 4 = u_0, u_1 + u_{14} = 1 + 4x = u_{16},$

$$u_1 + u_{15} = 1 + 3x + 3 = u_{24}, u_1 + u_{16} = 1 + 4x + 1 = u_6,$$
$$u_1 + u_{17} = 1 + 4x + 3 = u_{21}, u_1 + u_{18} = 1 + x + 3 = u_4,$$
$$u_1 + u_{19} = 1 + 2 = u_7, u_1 + u_{20} = 1 + 2x = u_{12},$$
$$u_1 + u_{21} = 1 + 4x + 4 = u_{14}, u_1 + u_{22} = 1 + 2x + 3 = u_{23}$$
$$u_1 + u_{23} = 1 + 2x + 4 = u_{20}, u_1 + u_{24} = 1 + 3x + 4 = u_8.$$

Hence the first row of L_1 is

1 91 9 22 2 18 17 13 11 5 15 10 3 0 16 24
6 21 4 7 12 14 23 20 8

and L_1 is

0	1	2	3	4	5	6	7	8	9	10	11	12	13	14	15	16	17	18	19	20	21	22	23	24
1	19	9	22	2	18	17	13	11	5	15	10	3	0	16	24	6	21	4	7	12	14	23	20	8
2	9	20	10	23	3	19	18	14	12	6	16	11	4	0	17	1	7	22	5	8	13	15	24	21
3	22	10	21	11	24	4	20	19	15	13	7	17	12	5	0	18	2	8	23	6	9	14	16	1
4	2	23	11	22	12	1	5	21	20	16	14	8	18	13	6	0	19	3	9	24	7	10	15	17
5	18	3	24	12	23	13	2	6	22	21	17	15	9	19	14	7	0	20	4	10	1	8	11	16
6	17	19	4	1	13	24	14	3	7	23	22	18	16	10	20	15	8	0	21	5	11	2	9	12
7	13	18	20	5	2	14	1	15	4	8	24	23	19	17	11	21	16	9	0	22	6	12	3	10
8	11	14	19	21	6	3	15	2	16	5	9	1	24	20	18	12	22	17	10	0	23	7	13	4
9	5	12	15	20	22	7	4	16	3	17	6	10	2	1	21	19	13	23	18	11	0	24	8	14
10	15	6	13	16	21	23	8	5	17	4	18	7	11	3	2	22	20	14	24	19	12	0	1	9
11	10	16	7	14	17	22	24	9	6	18	5	19	8	12	4	3	23	21	15	1	20	13	0	2
12	3	11	17	8	15	18	23	1	10	7	19	6	20	9	13	5	4	24	22	16	2	21	14	0
13	0	4	12	18	9	16	19	24	2	11	8	20	7	21	10	14	6	5	1	23	17	3	22	15
14	16	0	5	13	19	10	17	20	1	3	12	9	21	8	22	11	15	7	6	2	24	18	4	23
15	24	17	0	6	14	20	11	18	21	2	4	13	10	22	9	23	12	16	8	7	3	1	19	5
16	6	1	18	0	7	15	21	12	19	22	3	5	14	11	23	10	24	13	17	9	8	4	2	20
17	21	7	2	19	0	8	16	22	13	20	23	4	6	15	12	24	11	1	14	18	10	9	5	3
18	4	22	8	3	20	0	9	17	23	14	21	24	5	7	16	13	1	12	2	15	19	11	10	6
19	7	5	23	9	4	21	0	10	18	24	15	22	1	6	8	17	14	2	13	3	16	20	12	11
20	12	8	6	24	10	5	22	0	11	19	1	16	23	2	7	9	18	15	3	14	4	17	21	13
21	14	13	9	7	1	11	6	23	0	12	20	2	17	24	3	8	10	19	16	4	15	5	18	22
22	23	15	14	10	8	2	12	7	24	0	13	21	3	18	1	4	9	11	20	17	5	16	6	19
23	20	24	16	15	11	9	3	13	8	1	0	14	22	4	19	2	5	10	12	21	18	6	17	7
24	8	21	1	17	16	12	10	4	14	9	2	0	15	23	5	20	3	6	11	13	22	19	7	18

148. Denote the elements of $GF(p_i^{e_i})$, $i = 1, 2, \ldots, m$ by

$$g_{i0} = 0,\ g_{i1} = 1, g_{i2} = \alpha_i,\ g_{i3} = \alpha_i^2,\ \ldots, g_i p_i^{e_i} = \alpha_i p_i^{e_i} - 2,$$

where α_i is a primitive root of $GF(p_i^{e_i})$. Consider the set of $s = p_i^{e_i} p_2^{e_2} \cdots p^{e_m}$ elements

$$w = g_{11_1}, g_{21_2}, \ldots, g_{m1_m},$$

where $g_{i1_i} \in GF(p_i^{e_i})$, $i = 1, 2, \ldots, m$. We donote the set of elements by $\{w\}$. Define the addition and multiplication among the elements of this set $\{w\}$

as follows. If $w_1 = (g_{1i_1}, g_{2i_2}, \ldots, g_{mi_m})$ and $w_2 = (g_{1j_1}, g_{2j_2}, \ldots, g_{mj_m})$ are two elements of the set {w}, then

$$w_1 + w_2 = (g_{1i_1} + g_{1j_1}, g_{2i_2} + g_{2j_2}, \ldots, g_{mi_m} + g_{mj_m})$$
$$w_1 w_2 = (g_{1i_1} \cdot g_{1j_1}, g_{2i_2} \cdot g_{2j_2}, \ldots, g_{mi_m} \cdot g_{mj_m})$$

The set {w} is not a field, since for instance, the element (0, 1, 1, ..., 1) has no multiplicative inverse. All elements of {w} that have no 0 among their coordinates possess inverses.

Now, we number the elements of the set {w} such that the first $n(s) + 1$ elements of the set {w} are

$$w = (g_{1j}, g_{2j}, \ldots, g_{mj}), \; j = 0, 1, 2, \ldots, n(s)$$

while the rest of the elements of {w} are numbered arbitrarily. Note that the elements $w_1, w_2, \ldots w_{n(s)}$ posses inverses and so do their differences $w_i - w_j, i \neq j = 1, 2, \ldots, n(s)$.

We construct the jth Latin square $L_j, j = 1, 2, \ldots, n(s)$ by filling its (α, β)th cell by the element

$$w_j w_\alpha + w_\beta, \alpha, \beta = 0, 1, \ldots, s-1, \qquad j = 1, 2, \ldots, n(s).$$

We now prove that (i) $L_j, j = 1, 2, \ldots, n(s)$ is a Latin square and (ii) L_i and L_j are orthogonal, $i \neq j = 1, 2, \ldots, n(s)$.

To prove that L_j is a Latin square, suppose that the same element occurs twice in the jth row, say in the cells (α, β) and (α, β'). Then we have

$$w_j w_\alpha + w_\beta = w_j w_\alpha + w_{\beta'},$$

which shows that $w_\beta = w_{\beta'}$. Hence each element occurs exactly once in each row. Suppose now that the same element occurs twice in the jth column, say in the cells (α, β) and (α', β). Then we get

$$w_j w_\alpha + w_\beta = w_j w_{\alpha'} + w_\beta.$$

Then we get $w_j w_\alpha = w_j w_{\alpha'}$. Since w_j possesses an inverse, we get $w_\alpha = w_{\alpha'}$. Thus each element occurs exactly once in each column. Thus $L_j, j = 1, 2, \ldots, n(s)$ is a Latin Square.

Consider now L_i and L_j, $i \neq j = 1, 2, \ldots, n(s)$. Suppose the same ordered pair of elements obtained by superimposing L_j on L_i occurs in two cells (α, β) and (γ, δ). Then we get

$$w_i w_\alpha + w_\beta = w_i w_\gamma + w_\delta$$
$$w_j w_\alpha + w_\beta = w_j w_\gamma + w_\delta,$$

which on subtaction gives

$$(w_i - w_j) w_\alpha = (w_i - w_j) w_\gamma$$

Since $w_i - w_j, i \neq j = 1, 2, \ldots, n(s)$, possesses an inverse, we get $w_\alpha = w_\gamma$. And hence from the first equation we get $w_\beta = w_\delta$. Thus, when L_j is superimposed on L_i, then each ordered pair of elements occurs only once. Hence L_i and L_j are orthogonal.

149. Here $s = 2^2 \cdot 3^1$, and $n(s) = \min(2^2, 3^1) - 1 = 2$. The elements of GF (2^2) and GF (3) are as follows:

$$\mathrm{GF}(2^2) : g_{10} = 0, g_{11} = 1, g_{12} = \alpha_1, g_{13} = \alpha_1^2 = \alpha_1 + 1$$
$$\mathrm{G}(3) : g_{20} = 0, g_{21} = 1, g_{22} = 2.$$

Consider the set of elements

$$w = (g_{11_1}, g_{21_2})$$

where $g_{11_1} \in \mathrm{GF}(2^2)$ and $g_{21_2} \in \mathrm{GF}(3)$. The set $\{w\}$ contains 12 elements. We number the first $n(s) + 1 = 3$ elements of the set $\{w\}$ as

$$w_0 = (0, 0), w_1 = (1, 1), w_2 = (\alpha_1, 2),$$

while the other elements of the set $\{w\}$ are numbered arbitrarily as

$$w_3 = (0, 1), \quad w_4 = (0, 2), \quad w_5 = (1, 0),$$
$$w_6 = (1, 2), \quad w_7 = (\alpha_1, 0), \quad w_8 = (\alpha_1, 1),$$
$$w_9 = (\alpha_1 + 1, 0), \quad w_{10} = (\alpha_1 + 1, 1),$$
$$w_{11} = (\alpha_1 + 1, 2).$$

The Latin square L_1 is constructed by filling its (α, β)th cell by the element $w_1 w_\alpha + w_\beta = w_\alpha + w_\beta, \alpha, \beta = 0, 1, 2, \ldots, 11$. Thus L_1 is obtained as

0	1	2	3	4	5	6	7	8	9	10	11
1	4	9	6	5	3	0	10	11	8	2	7
2	9	3	7	8	11	10	4	0	6	5	1
3	6	7	4	0	1	5	8	2	10	11	9
4	5	8	0	3	6	1	2	7	11	9	10
5	3	11	1	6	0	4	9	10	7	8	2
6	0	10	5	1	4	3	11	9	2	7	8
7	10	4	8	2	9	11	0	3	5	1	6
8	11	0	2	7	10	9	3	4	1	6	5
9	8	6	10	11	7	2	5	1	0	3	4
10	2	5	11	9	8	7	1	6	3	4	0
11	7	1	9	10	2	8	6	5	4	0	3

The second Latin square L_2 is constructed by filling its (α, β)th cell by the element $w_2 w_\alpha + w_\beta$. Thus L_2 is obtained as

0	1	2	3	4	5	6	7	8	9	10	11
2	9	3	7	8	11	10	4	0	6	5	1
10	2	5	11	9	8	7	1	6	3	4	0
4	5	8	0	3	6	1	2	7	11	9	10
3	6	7	4	0	1	5	8	2	10	11	9
7	10	4	8	2	9	11	0	3	5	1	6
8	11	0	2	7	10	9	3	4	1	6	5
9	8	6	10	11	7	2	5	1	0	3	4
11	7	1	9	10	2	8	6	5	4	0	3
5	3	11	1	6	0	4	9	10	7	8	2
6	0	10	5	1	4	3	11	9	2	7	8
1	4	9	6	5	3	0	10	11	8	2	7

150. Consider the finite projective geometry PG (m,s). Identify the points of PG(m,s) with the treatments and the g-flats ($1 \leq g \leq m-1$) of PG(m,s) with the blocks.

Since there are $(s^{m+1}-1)/(s-1)$ points in PG (m, s), we see that the number of treatments is given by

$$v = (s^{m+1}-1)/(s-1).$$

Now, the number of g-flats of PG (m, s) is ϕ (m, g, s). Hence, the number of blocks is given by

$$b = \phi\,(m, g, s) = \frac{(s^{m+1}-1)(s^m-1)\ldots(s^{m-g+1}-1)}{(s^{g+1}-1)(s^g-1)\ldots(s-1)}.$$

The number of points in a g-flat of PG (m, s) is $(s^{g+1}-1)/(s-1)$. Hence, the number of treatments in a block will be given by

$$k = (s^{g+1}-1)/(s-1).$$

In the PG (m, s), a given point (i.e. a o-flat) is contained in $\phi\,(m-1, g-1, s)$ g-flats. Hence, a given treatment will occur in

$$r = \phi\,(m-1, g-1, s) = \frac{(s^m-1)(s^{m-1}-1)\ldots(s^{m-g+1}-1)}{(s^g-1)(s^{g-1}-1)\ldots(s-1)}$$

blocks. Also in PG (m, s), a given pair of points (i.e., a 1-flat) is contained in $\phi\,(m-2, g-2, s)$ g-flats. Hence, a given pair of treatments will occur in

$$\lambda = \phi\,(m-2, g-2, s) = \frac{(s^{m-1}-1)(s^{m-2}-1)\ldots(s^{m-g+1}-1)}{(s^{g-1}-1)(s^{g-2}-1)\ldots(s-1)}$$

blocks.

151. We consider a PG (2, 3). The points of PG (2, 3) are

$$\begin{aligned} &p_1 = (0\,0\,1), && p_2 = (0\,1\,0), && p_3 = (0\,1\,1),\\ &p_4 = (0\,1\,2), && p_5 = (1\,0\,0), && p_6 = (1\,0\,1),\\ &p_7 = (1\,0\,2), && p_8 = (1\,1\,0), && p_9 = (1\,1\,1),\\ &p_{10} = (1\,1\,2), && p_{11} = (1\,2\,0), && p_{12} = (1\,2\,1),\\ &p_{13} = (1\,2\,2). \end{aligned}$$

Note that we have written the points so that the first non-zero coordinate is always 1. We take the points as treatments and 1-flats as blocks. The 1-flat is determined by a pair of points. The 1-flat passing through the points p_1 and p_2 consists of points $p_1 + \lambda p_2$, where λ is 0, 1 or 2, in addition to the points p_1 and p_2. Thus, the 1-flat through p_1 and p_2 consists of points $p_1, p_2, p_1 + p_2 = (0\,1\,1) = p_3$ and $p_1 + 2p_2 = (0\,2\,1) = (0\,1\,2) = p_4$. The 1-flats through p_1 and p_3, p_1 and p_4, p_2 and p_3, p_2 and p_4, and p_3 and p_4 are all the same. We now take the pair of points p_1 and p_5. The 1-flat through p_1 and p_5 consists of points $p_1, p_5, p_1 + p_5 = (1\,0\,1) = p_6$ and $p_1 + 2p_5 = (2\,0\,1) = (1\,0\,2) = p_7$. The 1-flats through p_1 and p_5, p_1 and p_6, p_1 and p_7, p_5 and p_6, p_5 and p_7, and p_6 and p_7 are all the same. In this way we go on determining the 1-flats. The 13 1-flats so determined are obtained as

1	p_1	p_2	p_3	p_4	8	p_3	p_5	p_9	p_{13}
2	p_1	p_5	p_6	p_7	9	p_3	p_6	p_{10}	p_{11}
3	p_1	p_8	p_9	p_{10}	10	p_3	p_9	p_8	p_{12}
4	p_1	p_{11}	p_{12}	p_{13}	11	p_4	p_5	p_{10}	p_{12}
5	p_2	p_5	p_8	p_{11}	12	p_4	p_6	p_8	p_{13}
6	p_2	p_6	p_9	p_{12}	13	p_4	p_7	p_9	p_{11}
7	p_2	p_7	p_{10}	p_{13}					

152. Consider the finite Euclidean Geometry EG (m,s). Identify the points of this geometry with treatments and the g-flats $(1 \le g \le m - 1)$ with blocks. Now, there are s^m points in EG (m,s). Hence, the number of treatments is given by

$$v = s^m.$$

The number of g-flats in EG (m,s) is equal to $\phi(m, g, s) - \phi(m - 1, g, s)$. Hence the number of blocks is given by

$$\begin{aligned} b &= \phi(m, g, s) - \phi(m - 1, g, s)\\ &= \frac{s^{m-g}(s^m - 1)(s^{m-1} - 1)\ldots(s^{m-g+1} - 1)}{(s^g - 1)(s^{g-1} - 1)\ldots(s - 1)}. \end{aligned}$$

The number of points on a g-flat of EG (m, s) is s^g. Hence, the number of treatments in a block is given by $k = s^g$. Further in EG (m, s), the number of g-flats containing a given o-flat (i.e. a point) is $\phi(m - 1, g - 1, s)$. Hence, the number of blocks containing a given treatment is given by

$$r = \phi(m-1, g-1, s) = \frac{(s^m - 1)(s^{m-1} - 1)\dots(s^{m-g+1} - 1)}{(s^g - 1)(s^{g-1} - 1)\dots(s - 1)}.$$

Finally in EG (m, s), the number of g-flats containing a given 1-flat (i.e. a pair of points) is $\phi(m-2, g-2, s)$. Hence, the number of blocks containing a given pair of treatments is given by

$$\lambda = \phi(m-2, g-2, s) = \frac{(s^{m-1} - 1)(s^{m-2} - 1)\dots(s^{m-g+1} - 1)}{(s^{g-1} - 1)(s^{g-2} - 1)\dots(s - 1)}.$$

153. Clearly the incidence matrix of new design is $\overset{*}{N} = E_{vb} - N$, where N is the incidence matrix of the given BIBD. Further, it is easy to see that the new design has v treatments, and b blocks and each block of the new design contains v − k treatments. Thus for the new design, its parameters are given by

$$\overset{*}{v} = v, \ \overset{*}{b} = b, \ \overset{*}{k} = v - k.$$

Now

$$\begin{aligned}
\overset{*}{N}\,\overset{*}{N}' &= (E_{vb} - N)(E_{vb} - N)' \\
&= b\,E_{vv} - 2rE_{vv} + N\,N' \\
&= (b - 2r)E_{vv} + [(r - \lambda)I_v + \lambda E_{vv}] \\
&= (r - \lambda)I_v + (b - 2r + \lambda)E_{vv} \\
&= (\overset{*}{r} - \overset{*}{\lambda})I_v + \overset{*}{\lambda}\,E_{vv},
\end{aligned}$$

where $\overset{*}{\lambda} = b - 2r + \lambda$ and $\overset{*}{r} = b - r$. Thus, we see that $\overset{*}{N}$ is the incidence matrix of a BIBD with parameters

$$\overset{*}{v} = v, \ \overset{*}{b} = b, \ \overset{*}{r} = b - r, \ \overset{*}{k} = v - k \quad \text{and} \quad \overset{*}{\lambda} = b - 2r + \lambda.$$

154. Let the blocks of D be denoted by $B_1, B_2, \dots, B_b$. Without any loss of generality, assume that D_1 is obtained by omitting the block B_1 and all its treatments from the $B_2, \dots, B_b$. Clearly, for the design, D_1, we have $v_1 = v - k$, $b_1 = b - 1$. Since D is a SBIBD, the number of common treatments between B_1 and B_i, $i = 2, 3, \dots, b$ is λ. Hence, each block of D_1 will contain $k - \lambda$ treatments. Thus, $k_1 = k - \lambda$. Consider a treatment which does not belong to the block B_1. It occurs exactly in r blocks of the set $B_2, \dots, B_b$ of D. Hence, it will also occur in exactly r blocks of D_1, and $r_1 = r$. Further, consider a pair of treatments which do not belong to the block B_1. This pair of treatments will occur together in exactly λ blocks of the set $B_2, B_3, \dots, B_b$. Hence this pair of treatments will occur together in exactly λ blocks of D_1. Thus, $\lambda_1 = \lambda$. Hence D_1 is BIBD with parameters $v_1 = v - k$, $b_1 = b - 1$, $r_1 = r$, $k_1 = k - \lambda$, $\lambda_1 = \lambda$.

155. Let the blocks of D be denoted by $B_1, B_2, \ldots, B_b$. Without loss of generality, assume that B_1 is omitted and that its treatments are only retained in the blocks $B_2, B_3, \ldots, B_b$. These new blocks, are denoted by $\overset{*}{B}_2, \overset{*}{B}_3, \ldots, \overset{*}{B}_b$ and they form the design D_2. Clearly, the number of treatments in D_2 is $v_2 = k$ and the number of blocks in D_2 is $b_2 = b - 1$. Since D is a SBIBD, there are λ treatments common between B_1 and B_i, $i = 2, 3, \ldots, b$ and these are retained in $\overset{*}{B}_i$, $i = 2, 3, \ldots, b$. Thus each block of D_2 contains $k_2 = \lambda$ treatments. Each treatment of B_1 occurs in exactly $r - 1$ blocks of the set $B_2, B_3, \ldots, B_b$. Hence each treatment in D_2 occurs exactly in $r - 1$ blocks of D_2. Thus we get $r_2 = r = 1$. Further, each pair of the treatments of B_1 occurs together in exactly $\lambda - 1$ blocks of the set $B_2, B_3, \ldots, B_b$. Hence each pair of treatments in D_2 will occur together in exactly $\lambda - 1$ blocks in D_2. Thus, $\lambda_2 = \lambda - 1$. Hence, D_2 is a BIBD with parameters $v_2 = k, b_2 = b - 1, r_2 = r - 1, k_2 = \lambda, \lambda_2 = \lambda - 1$.

156. (i) We construct a complete set of $(s - 1)$ mutally orthogonal Latin Squares of side s. Denote these Latin squares by $L_1, L_2, \ldots, L_{s-1}$. The $v = s^2$ treatments are arranged in a $s \times s$ square and call this square as L. Take rows and columns of L as blocks. We get thus 2s blocks. The blocks obtained from rows and columns thus form two replications.

Now take the Latin square $L_i, i = 1, 2, \ldots, s - 1$ and superimpose on L, and form blocks by taking treatments which correspond to the same letters (numbers) of L_i. Since L_i contains s letters (numbers), we get s blocks. Thus from each Latin square, we get s blocks and they form one replication. Hence we get in all $2s + s(s - 1) = s^2 + s$ blocks. Clearly each block has s treatments. Also $r = 2 + s - 1 = s + 1$. Consider any pair of treatments λ and β. Then one of the following situations occurs:

(i) α and β occur in the same row of L,
(ii) α and β occur in different rows of L but not in the same column of L, or
(iii) α and β occur in different rows of L and in the same column of L.

Consider the situation (i). Here, α and β occur together only in one block they cannot occur together in blocks obtained from the columns of L. Also since these two treatments cannot correspond to the same letter when $L_1, L_2, \ldots, L_{s-1}$, they cannot occur together in blocks obtained when $L_1, L_2, \ldots, L_{s-1}$ are superimposed on L. Thus α and β occur together only in one block, Hence $\lambda = 1$.

Consider the situation (ii). Clearly α and β cannot occur together in blocks obtained from rows and columns of L. Since, $L_1, L_2, \ldots, L_{s-1}$ are mutally orthogonal Latin squares, α and β will correspond to the same letter of only one Latin Square, and the α and β will occur together only in one block. Hence, $\lambda = 1$.

Consider the situation (iii). Since α and β occur in the same column, they occur together in exactly one block from among blocks obtained from columns of L and cannot occur together in blocks obtained from rows of L. Since α and β occur in the same column, they cannot correspond to the same letter, when $L_1, L_2, \ldots, L_{s-1}$ are superimposed on L. Hence α and β cannot occur together in blocks obtained from superimposing $L_1, L_2, \ldots, L_{s-1}$ on L. Thus α and β occur together in exactly one block, and therefore $\lambda = 1$. Thus we get a BIBD with parameters $v = s^2,\ b = s^2 + s,\ r = s + 1,\ k = s,\ \lambda = 1$.

(ii) Out of $s^2 + s + 1$ treatments, we take s^2 treatments and arrange them in a $s \times s$ square and construct $s^2 + 2$ blocks as in (i). To the blocks of each replication, add one treatment from the remaining $s + 1$ treatments. The remaining $s + 1$ treatments are also taken to form a block. Thus, we get a BIBD with parameters $v = b = s^2 + s + 1, r = k = s + 1$ and $\lambda = 1$.

Alternatively we can use EG (2, s) to construct the BIBD $v = s^2,\ b = s^2 + s,\ r = s + 1,\ k = s, \lambda = 1$ and PG (2, s) to construct the BIBD $v = b = s^2 + s + 1,\ r = k = s + 1,\ \lambda = 1$. See Exercises 152 and 150 respectively.

157. Let M be a module consisting of residues mod $(2s + 1)$. Clearly the number of residues mod $(2s + 1)$ is $2s + 1$. To each element of M, associate 3 treatments. Thus, the total number of treatments is $v = 3(2s + 1)$. Now consider the following set of $3s + 1$ blocks.

first s blocks

$$\{1_1,\ (2s)_1,\ 0_2\},\ \{2_1,\ (2s-1)_1,\ 0_2\}, \ldots, \{s_1, (s+1)_1, 0_2\}$$

second s blocks

$$\{1_2,\ (2s)_2,\ 0_3\},\ \{2_2,\ (2s-1)_2,\ 0_3\}, \ldots, \{s_2, (s+1)_2, 0_3\}$$

third s blocks

$$\{1_3,\ (2s)_3,\ 0_1\},\ \{2_3,\ (2s-1)_3,\ 0_1\}, \ldots, \{s_3, (s+1)_3, 0_1\}$$

last block

$$\{0_1,\ 0_2,\ 0_3\}.$$

The pure differences will arise from the first two treatments in the first 3s blocks. Consider the pure differences of type $\{1, 1\}$, which will arise from the first s blocks. There will be 2s such differences which are as follows.

$$1 - 2s = 2,\ 2 - (2s - 1) = 4,\ 3 - (2s - 2) = 6$$
$$, \ldots, s - (s + 1) = 2s,\ 2s - 1,\ 2s - 3,$$
$$2s - 5, \ldots, s + 1 - s = 1.$$

Thus, we see that each non-zero element of M occurs exactly $\lambda = 1$ time among the pure differences of the type [1, 1]. The pure differences of the types [2, 2] and [3, 3] will respectively arise from the second s and third s blocks, and we can as above verify that each non-zero element of M will occur exactly $\lambda = 1$ time among the pure differences of the types [2, 2] and [3, 3].

We now consider the mixed differences of the type [1, 2]. These will arise from the treatments of the first s blocks and the last block. These are

$$1 - 0 = 1, \quad 2 - 0 = 2, \ldots, s - 0 = s,$$
$$2s - 0 = 2s,\ 2s - 1 - 0 = 2s - 1, \ldots, s + 1 - 0 = s + 1,$$
$$\text{and } 0 - 0 = 0.$$

Thus, we see that each element of M occurs exactly $\lambda = 1$ times among the mixed differences of the type [1, 2]. The mixed differences of the type [2, 1] will arise from the treatments of the first s blocks and the last block and it can be easily verified that each element of M will occur exactly $\lambda = 1$ times among the mixed differences of the type [2, 1}.

The mixed differences of the types [1, 3] and [3, 1] will arise from the treatments from the third s blocks and the last block. The mixed differences of the type [2, 3] and [3, 2] will arise from the treatments of the second s blocks and the last block. We can easily verify that each element of M occurs exactly $\lambda = 1$ time among the mixed differences of the types [1, 3], [3, 1], [2, 3] and [3, 2].

Thus, it follows that the differences arising from the above $3s + 1$ blocks are symmetrically repeated, each occuring $\lambda = 1$ time. Further one can easily verify that of the $v = 6s + 3$ treatments of the above $3s + 1$ blocks, exactly $r = 2s + 1$ belong to each class. Hence, by the first fundamental theorem of symmetric differences of Bose (see Exercise 34) the blocks obtained by adding the elements of M in succession to the treatments of the above $3s + 1$ blocks will form a BIBD with parameters

$$v = 6s + 3,\ b = (3s + 1)(2s + 1),\ r = 2\,s + 1,\ k = 3,\ \lambda = 1.$$

158. Let elements of GF(s), where $s = 6t + 1$, represent the $6t + 1$ treatments. Let x be a primitive element of GF(s). Take the initial set of t blocks as

$$[x^0, x^{2t}, x^{4t}],\ [x, x^{2t+1}, x^{4t+1}], \ldots, [x^{t-1},\ x^{3t-1},\ x^{5t-1}].$$

Since x is a primitive element of GF(s), we have $x^{6t} = 1$ and hence $x^{3t} = -1$. Since one treatment is associate with each element of GF(s), we see that all differences among them are pure differences. Each block of the above set of t blocks will give 6 differences which are non-null elements of GF(s). The t blocks willgive 6t pure differences, which will be all the non-null elements of GF(s). Thus, each non-null element of GF(s) will occur exactly $\lambda = 1$ time among the pure differences that will arise from the above t blocks, and hence they are symmetrically repeated. Further $r = 3t$ treatments of the t blocks belong the class of $6t + 1$ treatments. Thus we apply the first fundamental theorem of symmetric differences. Hence by adding elements of GF(s) in succession to the treatments of the above tblocks, we get t (6t + 1) blocks which form a BIBD with parameters

$$v = 6t + 1,\ b = t(6t + 1),\ r = 3t,\ k = 3,\ \lambda = 1.$$

159. Consider the elements of GF(s) and associate one treatment with each element of GF(s). The elements of GF(s) are denoted by $o, x^0 = 1, x, x^2, \ldots, x^{4t+2}$, where x is a primitive element of GF(s). Consider the following block as the initial block:

$$B = [x^0,\ x^2,\ x^4, \ldots, x^{4t}].$$

The differences arising from B are all pure differences. B contains $k = 2t + 1$ elements and hence from b, we shall get $k\ (k - 1) = 2t(2t + 1)$ differences. We can arrange these differences in t sets of $2(t + 1)$ differences each as

$$\pm\,(x^{2i} - x^0), \pm\,(x^{2i+2} - x^2), \ldots, \pm\,(x^{4t+2i} - x^4),\quad i = 1, 2, \ldots, t.$$

Now, let $x^{2i} - x_0 = x^{pi}$, then the above differences become

$$x^{p_i},\ x^{p_i+2}, \ldots, x^{p_i+4t},\ x^{p_i+2t+1},\ x^{p_i+2t+3}, \ldots, x^{p_i+6t+1},$$
$$i = 1, 2, \ldots, t$$

since $x^{2t+1} = -1$. The differences in each set are the $4t + 2$ non-null elements. Thus, each non-null element of GF (s) occurs exactly $\lambda = t$ times among the differences of the block B and hence are symmetrically repeated. Hence by adding elements of GF(s) in succession to the treatments of B, we get $4t + 3$ blocks which form a BIBD with parameters

$$v = b = 4t + 3,\ r = k = 2t + 1, \lambda = t.$$

160. Consider GF(s), where $s = 4t + 1$. Associate one treatment with each element of GF(s). Let x be a primitive element of GF(s). Take the initial set of 2 blocks as

$$B_1 = [x^0,\ x^2,\ x^4, \ldots, x^{4t-2}]$$
$$B_2 = [x,\ x^3,\ x^5, \ldots, x^{4t-1}].$$

Each block contains $k = 2t$ treatments. The differences arising from B_1 and B_2 are pure differences and their number is $2k(k - 1) = 4t(2t - 1)$. The $2t(2 - 1)$ differences arising from B_1 are arranged in $(2t - 1)$ sets of 2 t each as under

$$\pm\,(x^{2i} - x^0), \pm\,(x^{2i+2} - x^2), \ldots, \pm\,(x^{2i+4t-2} - x^{4t-2}),$$
$$i = 1, 2, \ldots, 2t - 1.$$

The $2t(2t - 1)$ differences arising from B_2 are arranged in $(2t - 1)$ sets of

2t each as

$$\pm(x^{2i+1} - x), \pm(x^{2i+3} - x^3), \ldots, \pm(x^{2i+4t-1} - x^{4t-1}),$$
$$i = 1, 2, \ldots, 2t - 1.$$

Since $x^{4t} = 1$, we have $x^{2t} = -1$. Take $x^{2i} - x^o = p_i$. Hence the above differences can be written as

$$\begin{aligned}
&x^{p_i}, x^{p_i+2}, \ldots, x^{p_i+4t-2}\\
&x^{p_i+2t}, x^{p_i+2t+2}, \ldots, x^{p_i+6t-2}\\
&x^{p_i+1}, x^{p_i+3}, \ldots, x^{p_i+4t-1}\\
&x^{p_i+2t+1}, x^{p_i+2t-3}, \ldots, x^{p_i+6t-1}, \quad i = 1, 2, \ldots, 2t - 1.
\end{aligned}$$

Thus, we see that the differences in each set i are all the 4t non-null elements of GF(s). Hence, each non-null element of GS(s) occurs $\lambda = 2t - 1$ times among the differences obtained from B_1 and B_2 and hence the differences are symmetrically repeated. Further r = number of treatments in B_1 and $B_2 = 4t$. Therefore the first fundamental theorem of symmetric differences of Bose can be applied. Hence by adding the elements of GF(s) to the treatments of B_1 and B_2, we get $2(4t + 1)$ blocks which form a BIBD, with parameters

$$v = 4t + 1, \quad b = 2(4t + 1), \quad r = 4t, \quad k = 2t, \quad \lambda = 2t - 1.$$

161. Consider the elements of GF(s), where $s = 4t + 1$. To each element of GF(s), associate 3 treatments, thus we get $3s = 12t + 3$ finite treatments. We also introduce one additional infinite treatment denoted by ∞. Thus we get $v = 12t + 4$ treatments. We take the following $(3t + 1)$ blocks as the initial blocks.

$$\begin{aligned}
&[x_1^{2i}, x_1^{2t+2i}, x_2^{2i+1}, x_2^{2t+2i+1}],\\
&[x_2^{2i}, x_2^{2t+2i}, x_3^{2i+1}, x_3^{t+2i+1}],\\
&[x_3^{2i}, x_3^{2t+2i}, x_1^{2i+1}, x_1^{2t+2i+1}], \ i = 0,\ 1, 2, \ldots, t - 1\\
&\text{and } [\infty, 0_1, 0_2, 0_3]
\end{aligned}$$

where α is chosen so that $\dfrac{x^\alpha + 1}{x^\alpha - 1} = x^q$, $q \equiv 1 \pmod 2$. Each block contains $k = 4$ treatments. From each class (there are three classes) of finite treatments, 4t finite treatments occur in the first 3t blocks and 1 finite treatment occurs in the last block. Hence $r = 4t + 1$, and $\lambda = 1$. We now consider differences arising from the finite treatments. There will be 3 different types of pure differences, namely [1, 1], [2, 2] and [3, 3]. There will be 6 types of mixed differences; [1, 2], [2, 1], [1, 3], [3, 1], [2, 3] and [3, 2]. One can verify that in each type of pure differences, there are 4t differences, which are non-null elements of GF(s) and that in each type of mixed differences arising from the first 3t blocks, each non-null element of GF(s) occurs once and 0 occurs in the mixed differences arising from the last block. Thus, the differences arising from the finite treatments of the initial blocks are symmetrically repeated, each occurring $\lambda = 1$ time. Hence, the conditions of the second fundamental theorem of symmetric differences, see Exercise 34, are satisfied. And by adding the elements of GF(s) successively to the treatments of the initial blocks, we get $(3t + 1) \times (4t + 1)$ blocks which form a BIBD with parameters

$$v = 12t + 4, \ b = (3t + 1)(4t + 1), \ r = 4t + 1, \ k = 4, \ \lambda = 1.$$

162. (i) Consider EG (2, 3). The pointes of EG (2,3) are

$$p_1 = (0 \ 0), \quad p_2 = (0 \ 1), \quad p_3 = (0 \ 2),$$
$$p_4 = (1 \ 0), \quad p_5 = (1 \ 1), \quad p_6 = (1 \ 2),$$
$$p_7 = (2 \ 0), \quad p_8 = (2 \ 1), \quad p_9 = (2 \ 2).$$

We take these 9 points as 9 treatments. The 1-flats are taken as blocks. Hence,

$$\begin{aligned} b &= \text{no. of blocks} \\ &= \text{no. of 1-flats in EG(2, 3)} \\ &= 3\,(3^2 - 1)/(3 - 1) = 12. \end{aligned}$$

Also,

$$\begin{aligned} k &= \text{no. of points on a 1-flat of EG (2, 3)} = 3, \\ r &= \text{no. of 1-flats through a given point} \\ &= \phi\,(1, 0, 3) = (3^2 - 1)/(3 - 1) = 4 \\ \lambda &= \text{no. of 1-flats through a given pair of points} \\ &= \phi\,(0, -1, 3) = 1. \end{aligned}$$

We now construct 1-flats. The equation of a 1-flat in EG(2,3) is given by

$$a_1\, x_1 + a_2\, x_2 = a_3$$

where a_1, a_2, a_3 are elements of GF(s) and $(a_1 a_2 a_3) \neq (0 \ 0 \ 0)$, $(0 \ 0 \ 1), (0 \ 0 \ 2)$. Hence there will be $27 - 3 = 24$ choices for $(a_1 \ a_2 \ a_3)$, but the choices $(a_1 \ a_2 \ a_3)$ and $(2a_1 \ 2a_2 \ 2a_3)$ are identical. Hence the number of different choices are $24/2 = 12$. Thus, there will be 12 1-flats (blocks). These 12 blocks and the points (treatments) lying on them are given below:

$a_1\ a_2\ a_3$	Eq. of a 1-flat	Block
(1 0 0)	$x_1 = 0$	$p_1\ p_2\ p_3$
(1 0 1)	$x_1 = 1$	$p_4\ p_5\ p_6$
(1 0 2)	$x_1 = 2$	$p_7\ p_8\ p_9$
(1 1 0)	$x_1 + x_2 = 0$	$p_1\ p_6\ p_8$
(1 2 0)	$x_1 + 2x_2 = 0$	$p_1\ p_5\ p_9$
(1 1 1)	$x_1 + x_2 = 1$	$p_2\ p_4\ p_9$
(1 1 2)	$x_1 + x_2 = 2$	$p_3\ p_5\ p_7$
(1 2 1)	$x_1 + 2x_2 = 1$	$p_3\ p_4\ p_8$
(1 2 2)	$x_1 + 2x_2 = 2$	$p_2\ p_6\ p_7$
(0 1 0)	$x_2 = 0$	$p_1\ p_4\ p_7$
(0 1 1)	$x_2 = 1$	$p_2\ p_5\ p_8$
(0 1 2)	$x_2 = 2$	$p_3\ p_6\ p_9$

(ii) This is a particular case of orthogonal series of Yates

$$v = b = s^2 + s + 1, \quad r = k = s + 1, \quad \lambda = 1$$

Here s = 3. Then write $v = s^2 = 9$ treatments in the form of a 3×3 square as follows

$$L: \begin{array}{ccc} t_1 & t_2 & t_3 \\ t_4 & t_5 & t_6 \\ t_7 & t_8 & t_9 \end{array}$$

The two mutually orthogonal Latin Squares of side 3 are

$$L_1: \begin{array}{ccc} 0 & 1 & 2 \\ 1 & 2 & 0 \\ 2 & 0 & 1 \end{array} \qquad L_2: \begin{array}{ccc} 0 & 1 & 2 \\ 2 & 0 & 1 \\ 1 & 2 & 0 \end{array}$$

We now form blocks by taking (a) rows of L, (b) columns of L, treatments corresponding to the same numbers of L_1 and L_2, when L_1 and L_2 are superimposed on L. Thus, we get blocks as follows

From rows	From cols.	From L_1	From L_2
$t_1\ t_2\ t_3$	$t_1\ t_4\ t_7$	$t_1\ t_6\ t_8$	$t_1\ t_5\ t_9$
$t_4\ t_5\ t_6$	$t_2\ t_5\ t_8$	$t_2\ t_4\ t_9$	$t_2\ t_6\ t_7$
$t_7\ t_8\ t_9$	$t_3\ t_6\ t_9$	$t_3\ t_5\ t_7$	$t_3\ t_4\ t_8$

We now add treatment t_{10} to the blocks obtained from rows, treatments t_{11} to the blocks obtained from columns, treatment t_{12} to the blocks obtained from L_1 and treatment t_{13} to the blocks obtained from L_2, and take one additional block consisting of $t_{10}, t_{11}, t_{12}, t_{13}$.

Hence the 13 blocks are obtained as

t_1	t_2	t_3	t_{10}	t_1	t_4	t_7	t_{11}	t_1	t_6	t_8	t_{12}
t_4	t_5	t_6	t_{10}	t_2	t_5	t_8	t_{11}	t_2	t_4	t_9	t_{12}
t_7	t_8	t_9	t_{10}	t_3	t_6	t_9	t_{11}	t_3	t_5	t_7	t_{12}
t_1	t_5	t_9	t_{13}								
t_2	t_6	t_7	t_{13}					t_{10}	t_{11}	t_{12}	t_{13}
t_3	t_4	t_8	t_{13}								

(iii) We apply the construction of Exercise 159. Consider GF(11). Clearly $x = 2$ is a primitive element of GF(11). Hence, the initial block is taken as $(x^0, x^2, x^4, x^6, x^8) = (1, 3, 4, 5, 9)$. The other blocks are writen down by adding the elements of GF(11) in succession. Thus, the 11 blocks are obtained as

1	3	4	5	9	7	9	10	0	4
2	4	5	6	10	8	10	0	1	5
3	5	6	7	0	9	0	1	2	6
4	6	7	8	1	10	1	2	3	7
5	7	8	9	2	0	2	3	4	8
6	8	9	10	3					

(iv) We apply the construction given in Exercise 157 by taking $s = 2$. We take the module M consisting of $2s + 1 = 5$ elements 0, 1, 2, 3, 4 and associate 3 treatments with each element of M. Thus, the 15 treatments are

0_1	1_1	2_1	3_1	4_1
0_2	1_2	2_2	3_2	4_2
0_3	1_3	2_3	3_3	4_3

Then following Exercise 157, the initial set of 7 blocks is

$$(1_1, 4_1, 0_2),\ (2_1, 3_1, 0_2),\ (1_2, 4_2, 0_3),$$
$$(2_2, 3_2, 0_3),\ (1_3, 4_3, 0_1),\ (2_3, 3_3, 0_1),$$
$$(0_1, 0_2, 0_3).$$

We now write down the other blocks by adding the elements 1, 2, 3, 4 to the above blocks in succession. Thus we shall get in all 35 blocks. The writing of 35 blocks is left as an exercise to the reader.

(v) This is constructed by applying Exercise 160 and taking $t = 3$. The elements of GF(13) are 0, 1, 2, 3, 4, 5, 6, 7, 8, 9, 10, 11, 12 and 2 is a primitive root of GF(13). Hence the initial two blocks are

$$(1, 3, 4, 9, 10, 12) \text{ and } (2, 5, 6, 7, 8, 11).$$

By adding the elements of GF(13) in succession to the above 2 blocks, we get all the 26 blocks.

(vi) This BIBD is constructed by the application of second fundamental theorem of symmetrical differences. We consider the module M consisting of residues mod 7, i.e. of 0, 1, 2, 3, 4, 5, 6. We associate one treatment with each element of M. We also take one additional infinite treatment as ∞. Then there are 8 treatments. Consider the following 2 blocks.

$$B_1 = (0, 1, 3, 6), \qquad \overset{*}{B}_1 = (\infty, 0, 1, 3).$$

The differences among finite treatments arising from B_1 are

$$\begin{array}{llll} 0-1=6, & 0-3=4, & 0-6=1, & 1-0=1, \\ 1-3=5, & 1-6=2, & 3-0=3, & 3-1=2, \\ 3-6=4, & 6-0=6, & 6-1=5, & 6-3=3. \end{array}$$

The differences among finite treatments arising from $\overset{*}{B}_1$ are

$$\begin{array}{llll} 0-1=6, & 0-3=4. & 1-0=1, & 1-3=5 \\ 3-0=3, & 3-1=2. & & \end{array}$$

Among all the 18 differences among the finite treatments, we see that each non-zero element of M occurs exactly $\lambda = 3$ times. Hence, the differences among the finite treatments are symmetrically repeated, each occurring $\lambda = 3$ times. Further, out of 7 finite treatments, 4 occur in B_1 and $\lambda = 3$ occur in $\overset{*}{B}_1$. Thus the conditions of the second fundamental theorem of symmetric differences are satisfied by the blocks B_1 and $\overset{*}{B}_1$. Hence, by adding the elements of M in auccession to the treatments of B_1 and $\overset{*}{B}_1$, we get the 14 blocks, which are as follows

$$\begin{array}{ll} (0, 1, 3, 6) & (\infty, 0, 1, 3) \\ (1, 2, 4, 0) & (\infty, 1, 2, 4) \\ (2, 3, 5, 1) & (\infty, 2, 3, 5) \\ (3, 4, 6, 2) & (\infty, 3, 4, 6) \\ (4, 5, 0, 3) & (\infty, 4, 5, 0) \\ (5, 6, 1, 4) & (\infty, 5, 6, 1) \\ (6, 0, 2, 5) & (\infty, 6, 0, 2) \end{array}$$

(vii) This BIBD is constructed by applying Exercise 161 and taking $t = 2$. Consider the GF(9), and let x be a primitive root of GF(9). Hence $x^8 = 1$ and $x^4 = -1$. To each element of GF(9), we associate 3 treatments and we take one additional infinite treatment as ∞. Thus we get 27 finite and 1 infinite treatment. The elements are

$$0,\ 1,\ x,\quad x^2 = (2x+1),\quad x^3 = (2x+2),\quad x^4 = 2$$
$$x^5 = 2x,\quad x^6 = (x+2),\quad x^7 = (x+1).$$

Now

$$\frac{x+1}{x-1} = \frac{x^7}{x^6} = x.$$

Note that the elements $x^7 = x+1$ and $x^6 = x-1$. Thus we take $\alpha = 1$. Then the initial set of $3t+1 = 7$ blocks is

$[x_1^{2i}, x_1^{2i+4}, x_2^{2i+1}, x_2^{2i+5}]$

$[x_2^{2i}, x_2^{2i+4}, x_3^{2i+1}, x_3^{2i+5}]$

$[x_3^{2i}, x_3^{2i+4}, x_1^{2i+1}, x_1^{2i+5}], i = 0, 1,$

$[\infty, 0_1, 0_2, 0_3].$

Taking i = 0 and 1, we get in all 7 blocks as

$[x_1^0, x_1^4, x_2^1, x_2^5] = [1_1, 2_1, x_2, (2x)_2],$

$[x_2^0, x_2^4, x_3^1, x_3^5] = [1_2, 2_2, x_3, (2x)_3],$

$[x_3^0, x_3^4, x_1^1, x_1^5] = [1_3, 2_3, x_1, (2x)_1],$

$[x_1^2, x_1^6, x_2^3, x_2^7] = [(2x+1)_1, (x+2)_1, (2x+2)_2, (x+1)_2],$

$[x_2^2, x_2^6, x_3^3, x_3^7] = [(2x+1)_2, (x+1)_2, (2x+2)_3, (x+1)_3],$

$[x_3^2, x_3^6, x_1^3, x_1^7], = [(2x+1)_3, (x+2)_3, (2x+2)_1, (x+1)_1],$

$[\infty, 0_1, 0_2, 0_3].$

By adding the elements of GF(9) to the treatments of the above 7 blocks successfully, we get 63 blocks.

163. Let D be the given SBIBD with parameters $v = b$, $r = k$, and λ. The treatments of D are denoted by integers $1, 2, \ldots, v$. We write the blocks as columns. Let S = set of all v treatments and $S_1, S_2, \ldots, S_v$ be the subsets of S, representing the columns (blocks) of D. Each column contains $r = k$ different treatments. Take any h columns $S_{i_1}, S_{i_2}, \ldots, S_{i_h}$, of D . Then, these h columns contain between them $hr = hk$ treatments. Now every treatment can occur at most r times in these h columns, hence the number of distinct elements in these h columns is at least h, that is

$$|S_{i_1} \cup S_{i_2}, \cup \ldots \cup S_{i_h}| \geq h$$

for every h and $i_1, i_2, \ldots, i_h$, where $|T|$ denotes the cardinality of the set T. Hence, from Exercise 50, the columns $S_1, S_2, \ldots, S_v$ possess a SDR. This SDR is a permutation of integers $1, 2, \ldots, v$. Take this SDR as the first row. Delete this row from $S_1, S_2, \ldots, S_v$. Denote the new columns by $\overset{*}{S}_1, \overset{*}{S}_2, \ldots, \overset{*}{S}_v$. Each $\overset{*}{S}_i$, $i = 1, 2, \ldots, v$ will contain $r-1 = k-1$ treatments. Take any h columns $\overset{*}{S}_{i_1}, \overset{*}{S}_{i_2}, \ldots, \overset{*}{S}_{i_h}$. These h columns contain between them $h(r-1)$ treatments. Now every treatment can occur at most $(r-1)$ times in these h column, hence the number of distinct treatmsnts in these h columns is at least h, that is

$$| \overset{*}{S}_{i_1} \cup \overset{*}{S}_{i_2} \cup \ldots \cup \overset{*}{S}_{i_h} | \geq h$$

for every h and $i_1, i_2, \ldots, i_h$. Thus, the columns $\overset{*}{S}_1, \overset{*}{S}_2, \overset{*}{S}_v$ possess an SDR. Take this SDR as the second row. Proceeding in this way, we get r = k systems of distinct representatives for the columns $S_1, S_2, \ldots, S_v$ which are taken as r = k rows. These r = k rows have a property that each treatment occurs exactly once in each row. Hence these r = k rows form a Youden Square.

164. Let D be the given SBIBD (v = b, r = k, λ), whose blocks are denoted by columns $S_1, S_2, \ldots, S_v$. Clearly $S_1, S_2, \ldots, S_v$ are subsets of S, the set of all treatments. Each S_i, i = 1, 2, ..., v contains r = k treatments (k < v). Consider the construction of first row, i. e., the first SDR. Then, from Exercise 50, the number of ways of constructing the first row is ≥ k!. Consider the construction of the j-th row (j = 1, 2, ..., k). At this stage, each column contains (k − j + 1) treatments, hence using this result, it follows that the number of ways of constructing the jth row is (k − j + 1)!. Hence, the number of ways of constructing the k rows, i.e., the number of ways constructing the Youden squares is greater then or equal to $\sum_{j=1}^{k}(k - j + 1)!$.

165. Consider the finite projective geometry PG (m,s). Omit one point P, say, from this geometry and take the remaining points as treatments. Thus, we get $v = [(s^{m+1} - 1)/(s - 1)] - 1 = s(s^m - 1)/(s - 1)$. Take g-flats ($1 \leq g \leq m - 1$), not passing through the omitted point P as blocks. Thus, the number of blocks is obtained as

$$\begin{aligned} b &= \text{no. of g-flats not passing through P} \\ &= [\text{no. of g-flats in PG (m, s)}] \\ &\quad - [\text{no. of g-flats in PG (m, s) throught the point P}] \\ &= \phi\,(m, g, s) - \phi\,(m - 1, g - 1, s). \end{aligned}$$

Further, the number of treatments in a block is obtained as

$$\begin{aligned} k &= \text{no. of points on a g-flat} \\ &= (s^{g+1} - 1)/(s - 1). \end{aligned}$$

Also, the number of times each treatment is replicated is obtained as

$$\begin{aligned} r &= \text{no. of g-flats containing the given point but not containing} \\ &\quad \text{the omitted point P} \\ &= [\text{no. of g-flats containing the given point}] \\ &\quad - [\text{no. of g-flats containing the given point and the point P}] \\ &= \phi\,(m - 1, g - 1, s) - \phi\,(m - 2, g - 2, s). \end{aligned}$$

Two treatments will be called first associates if they occur together is the same block, if they do not occur together in the same block, then they are called second associates. Clearly, two treatments will not occur together in a block, if the line joining them passes throught the omitted point P. Hence, the second associates of a treatment are the points lying on the line (1-flat) joining that point and the omitted point P, excluding that point and the point P.

Thus, $n_2 = (s + 1) - 2 = s - 1$. Therefore, $n_1 = (v - 1) - n_2 = s^2(s^{m-1} - 1)/(s - 1)$. Now let q and f be a pair of first associates. Then,

$$\begin{aligned}\lambda_1 &= \text{no. of g-flats containing } \theta \text{ and } \phi \text{ but not the omitted} \\ &\quad \text{point P} \\ &= \text{no. of g-flats containing } \theta \text{ and } \phi \text{ but not the omitted} \\ &\quad \text{point P} \\ &= (\text{no. of g-flats containing } \theta \text{ and } \phi) \\ &\quad - (\text{no. of g-flats containing } \theta,\ \phi \text{ and P}). \\ &= \phi\,(m - 2, g - 2, s) - \phi\,(m - 3, g - 3, s).\end{aligned}$$

Clearly, $\lambda_2 = 0$.

Now, let θ and ϕ be a pair of second associates, i.e., the line joining them passes through the omitted point P. Then, clearly the number of treatments common between the second associates of θ and ϕ are the other treatments lying on the line joining them and the point P. Hence

$$p^2_{22} = (s + 1) - 3 = s - 2 = n_2 - 1.$$

Now

$$p^2_{12} + p^2_{22} = n_2 - 1$$

and therefore

$$p^2_{12} = n_2 - 1 - p^2_{22} = 0$$

and, $p^2_{21} = 0$. Also, $p^2_{11} + p^2_{12} = n_1$, hence $p^2_{11} = n_1$. Thus, we get

$$P_2 = \begin{bmatrix} n_1 & 0 \\ 0 & n_2 - 1 \end{bmatrix}.$$

Again, $n_1 p^1_{22} = n_2 p^2_{12} = 0$. Therefore, $p^1_{22} = 0$, and from $p^1_{21} + p^1_{22} = n_2$, we get $p^1_{21} = n_2 = p^1_{12}$. Further, from $p^1_{21} + p^1_{12} = n - 1$, we get $p^1_{11} = n_1 - n_2 - 1$. Hence, we get

$$P_1 = \begin{bmatrix} n_1 - n_2 - 1 & n_2 \\ n_2 & 0 \end{bmatrix}.$$

166. Consider PG (m,s). Select one point P , say, and choose t lines passing through it and take points on these lines other than P as treatments. Since, the number of points on each line, excluding the point P iss, we get $v = st$ treatments. Take the $(m - 1)$—flats not passing through P as blocks. Thus,

the number of blocks is obtained as

$$\begin{aligned} b &= \text{no. of } (m-1)\text{ - flats not passing through P} \\ &= [\text{no. of } (m-1)\text{ - flats}] \\ &\quad - [\text{no. of } (m-1)\text{ - flats passing through P}] \\ &= \phi(m, m-1, s) - \phi(m-1, m-2, s) + s^m. \end{aligned}$$

The number of times each treatment is replicated is obtained as

$$\begin{aligned} r &= \text{no. of } (m-1)\text{ - flats containing a given point but not} \\ &\quad \text{containing the point P} \\ &= (\text{no. of } (m-1)\text{ - flats containing a given point}) \\ &\quad - (\text{no. of } (m-1)\text{ - flats containing the given point and P}) \\ &= \phi(m-1, m-2, s) - \phi(m-2, m-3, s) \\ &= s^{m+1}, \end{aligned}$$

and hence $k = vr/b = t$.

Two treatments are first associates if they lie on one of the chosen lines through the point P, otherwise they are called second associates. Hence, we get $n_1 = s - 1$, $n_2 = (v-1) - n_1 = s(t-1)$. Clearly $\lambda_1 = 0$ and λ_2 can be found from the relation $n_1\lambda_1 + n_2\lambda_2 = r(k-1)$. Hence $\lambda_2 = s^{m-2}$. Let θ and ϕ be a pair of first associates. Then the treatments common between their first associates are the other treatments on the line joining them and P. Hence $p_{11}^1 = (s+1) - 3 = s - 2 = n_1 - 1$. The values of P_1 and P_2 then follow from the properties of p_{jk}^i and are found to be

$$P_1 = \begin{bmatrix} n_1 - 1 & 0 \\ 0 & n_2 \end{bmatrix}, \qquad P_2 = \begin{bmatrix} 0 & n_1 \\ n_1 & n_2 - n_1 - 1 \end{bmatrix}.$$

167. Consider a finite Euclidian Geometry EG(m, s). Omit one point P, say and all the g-flats $(1 \le g \le m-1)$ through P. Take the remaining points as treatments and the g-flats not containing P as blocks. Then we get

$$\begin{aligned} v &= s^m - 1, \\ b &= \text{no. of g-flats not containing P} \\ &= [\text{no. of g-flats in EG}(m, s)] \\ &\quad - [\text{no. of g-flats in EG}(m, s)\text{ containing P}] \\ &= [\phi(m, g, s) - \phi(m-1, g, s,)] - \phi(m-1, g-1, s). \\ r &= \text{no. of g-flats in EG}(m, s)\text{ containing a given point but not} \\ &\quad \text{the point P} \\ &= [\text{no. of g-flats in EG}(m, s)\text{ containing a given point}] \\ &\quad - [\text{no. of g-flats in EG}(m, s)\text{ containing the given point} \\ &\quad \text{and P}] \\ &= \phi(m-1, g-1, s) - \phi(m-2, g-2, s), \end{aligned}$$

and

$$k = \text{no. of points on a g-flat in EG(m, s)} = s^g.$$

Two treatments are first associates if they occur together in the same block, otherwise they are called second associates. Now. two treatments will not occur together in a block, if the line joining them passes through P. Hence $n_2 = s - 2$ and $\lambda_2 = 0$. Then $n_1 = v - 1 - n_2 = s^m - s$. Further if θ and ϕ are a pair of first associates, then

$$\begin{aligned}\lambda_1 &= \text{no. of blocks which contain } \theta \text{ and } \phi \\ &= \text{no. of g-flats containing } \theta \text{ and } \phi \text{ but not P} \\ &= [\text{no. of g-flats containing } \theta \text{ and } \phi] \\ &\quad - [\text{no. of g-flats containing } \theta \text{ and } \phi \text{ and P}] \\ &= \phi(m-2, g-2, s) - \phi(m-3, g-3, s).\end{aligned}$$

Let θ and ϕ be a pair of second associates. Hence the line joining them passes through P. Then the treatments common between their second associates are the other treatments lying on the line joining them and P. Hence we get $p^2_{22} = s - 3 = n_2 - 1$. Then using the preperties of p^i_{jk}, we obtain P_1 and P_2 as

$$P_1 = \begin{bmatrix} n_1 - n_2 - 1 & n_2 \\ n_2 & 0 \end{bmatrix}, \qquad P_2 = \begin{bmatrix} n_1 & 0 \\ 0 & n_2 - 1 \end{bmatrix}.$$

168. Arrange the $v = pq$ treatments in an array of p rows and q columns. Blocks are formed by taking a treatment and treatments in the same row and the same column as that treatment. Thus we get $b = pq$ and $k = 1 + (p-1) + (q-1) = p + q - 1$. Clearly $r = p + q - 1$.

Two treatments are called first associates if they lie in the same row, and second associates if they lie in the same column; otherwise they are called third associates. We then have $n_1 = q - 1$, $n_2 = p - 1$, $n_3 = (p-1)(q-1)$. Clearly $\lambda_1 = q$, $\lambda_2 = p$ and $\lambda_3 = 2$. Further, one can easily verify that the values of the P-matrices are given by

$$P_1 = \begin{bmatrix} q-2 & 0 & 0 \\ 0 & 0 & p-1 \\ 0 & p-1 & (p-1)(q-2) \end{bmatrix}$$

$$P_2 = \begin{bmatrix} 0 & 0 & q-1 \\ 0 & p-2 & 0 \\ q-1 & 0 & (p-2)(q-1) \end{bmatrix}$$

$$P_3 = \begin{bmatrix} 0 & 1 & q-2 \\ 1 & 0 & p-2 \\ q-2 & p-2 & (p-2)(q-2) \end{bmatrix}.$$

169. We arrange the $v = pq$ treatments in an array of p rows and q columns. Blocks are formed by taking all treatments that occur in the same row and

column as each treatment but excluding that treatment. Thus, we get $b = pq$, and $r = k = p + q - 2$.

Two treatments are first associates if they occur in the same row; and second associates if they occur in the same column, otherwise they are third associates. Thus, we have $n_1 = q - 1$, $n_2 = p - 1$, $n_3 = (p - 1)(q - 1)$. We can easily verify that $\lambda_1 = q - 2$, $\lambda_2 = p - 2$, $\lambda_3 = 2$. Also one can easily verify that the matrices P_1, P_2, P_3 are the same as given in Exercise 168.

170. The $v = p^2$ treatments are arranged in a $p \times p$ square. Blocks are formed by taking all treatments in the same row, the same column and which correspond to the same letter of a $p \times p$ Latin square as each treatment. Clearly we have $b = p^2$, and $r = k = 1 + (p - 1) + (p - 1) + (p - 1) = 3p - 2$.

Two treatments are first associates if they occur in the same row; second associates if they occur in the same column; and third associates if they correspond to the same letter of a $p \times p$ Latin square; otherwise they are fourth associates. Then we obtain

$$n_1 = n_2 = n_3 = p - 1 \qquad n_4 = (p - 1)(p - 2)$$
$$\lambda_1 = \lambda_2 = \lambda_3 = p + 2, \quad \lambda_4 = 6.$$

Further one can easily verify the values of P-matrices.

171. (i) Consider the double triangle as shown in Chapter 1, Result 36(3)(i) on page 43.

The vertices are denoted by the numbers from 1 to 10, and take them as the 10 treatments. The blocks are fromed by taking treatments on the same lines. There are 5 lines and so we get the 5 blocks as

(1 2 5 8), (8 6 9 10), (4 2 3 10),
(4 5 6 7), (1 3 9 7).

Clearly $k = 4$. Since there are two lines through a vertex, we get $r = 2$. The first associates of any treatment are the treatments lying on the lines passing through that treatment. Thus we get $n_1 = 6$. The second associates of a treatment are the treatments which do not lie on the lines passing through that treatment. Thus, $n_2 = 3$, clearly $\lambda_1 = 1$, and $\lambda_2 = 0$.

Consider treatments 2 and 3, which are first associates. The first associates of 2 are 4, 3, 10, 1, 5, 8. The first associates of 3 are 4, 2, 10, 1, 9, 7. Hence, the treatments common between the first associates of 2 and 3 are 4, 10, 1. Thus we get $p^1_{11} = 3$. The other values of the matrices P_1 and P_2can be determined similarly or from the properties of p^i_{jk} and we obtain P_1 and P_2 as

$$P_1 = \begin{bmatrix} 3 & 2 \\ 2 & 1 \end{bmatrix}, \qquad P_2 = \begin{bmatrix} 4 & 2 \\ 2 & 0 \end{bmatrix}.$$

(ii) Consider the parallelepiped as shown in Chapter 1, Result 36(3)(ii) on page 44.

The vertices of the parallelepiped are denoted by numbers from 1 to 8 and are taken as the $v = 8$ treatments. The 6 faces of the parallelepiped are taken as blocks. Thus we get $b = 6$ and clearly $k = 4$. Since 3 faces pass through a vertex, we get $r = 3$. The first associates of a treatment are defined as those treatments which lie on the same faces as that treatment but not on the same edges as that treatment, its second associates are those treatments which lie on the same edges as that treatment; and its third associates are treatments diagonally opposite to that treatment. Thus the first, second and third associates of the treatment 1 are respectively 3, 8, 6; 2, 4, 5; 7. Hence $n_1 = 3, n_2 = 3, n_3 = 1$, and $\lambda_1 = 1, \lambda_2 = 2, \lambda_3 = 0$. One can easily verify that the P-matrices are given by

$$P_1 = \begin{bmatrix} 2 & 0 & 0 \\ 0 & 2 & 1 \\ 0 & 1 & 0 \end{bmatrix}, \quad P_2 = \begin{bmatrix} 0 & 2 & 1 \\ 2 & 0 & 0 \\ 1 & 0 & 0 \end{bmatrix}, \quad P_3 = \begin{bmatrix} 0 & 3 & 0 \\ 3 & 0 & 0 \\ 0 & 0 & 0 \end{bmatrix}.$$

(iii) This PBIBD will be constructed by applying the method of Exercise 165. Consider PG (2, 3). The 13 points are

$$\begin{aligned} &p_1 = (0\,0\,1), && p_2 = (0\,1\,0), && p_3 = (0\,1\,1), \\ &p_4 = (0\,1\,2), && p_5 = (1\,0\,0), && p_6 = (1\,0\,1), \\ &p_7 = (1\,0\,2), && p_8 = (1\,1\,0), && p_9 = (1\,1\,1), \\ &p_{10} = (1\,1\,2), && p_{11} = (1\,2\,0), && p_{12} = (1\,2\,1), \\ &p_{13} = (1\,2\,2). \end{aligned}$$

We omit the point p_5 and take the remaining 12 points as the 12 treatments. Thus we get $v = 12$ treatments as $p_1, p_2, p_3, p_4, p_6, p_7, p_8, p_9, p_{10}, p_{11}, p_{12}, p_{13}$. In Exercise 151, we have listed the 1-flats of PG (2, 3). We omit the 1-flats which pass through p_5, and take the remaining 1-flats as blocks. Thus, we obtain $b = 9$ blocks as

$$\begin{aligned} &(p_1\ p_2\ p_3\ p_4) && (p_3\ p_6\ p_{10}\ p_{11}) \\ &(p_1\ p_8\ p_9\ p_{10}) && (p_3\ p_7\ p_8\ p_{13}) \\ &(p_1\ p_{11}\ p_{12}\ p_{13}) && (p_4\ p_6\ p_8\ p_{13}) \\ &(p_2\ p_6\ p_9\ p_{12}) && (p_4\ p_7\ p_9\ p_{11}) \\ &(p_2\ p_7\ p_{10}\ p_{13}) \end{aligned}$$

Clearly $k = 4$ and $r = 3$.

Two treatments are first associates, if they occur together in a block, and are second associates if they do not occur together in a block. Thus the first associates of p_1 and p_2 are respectively $p_2, p_3, p_4, p_8, p_9, p_{10}, p_{11}, p_{12}, p_{13}$, and $p_1, p_3, p_4, p_6, p_9, p_{12}, p_7, p_{10}, p_{13}$. Thus we get $n_1 = 9$ and hence $n_2 = 2$. Also the common treatments between the first associates of p_1 and p_2 are $p_3, p_4, p_9, p_{10}, p_{12}, p_{13}$. Hence $p^1_{11} = 6$. Also p_1 and p_2

occur together only in one block, hence $\lambda_1 = 1$. Then $\lambda_2 = 0$. The other values of p^i_{jk} can be obtained similarly from the properties of p^i_{jk}. Thus one can easily verify.

$$P_1 = \begin{bmatrix} 6 & 2 \\ 2 & 0 \end{bmatrix}, \qquad P_2 = \begin{bmatrix} 9 & 0 \\ 0 & 1 \end{bmatrix}.$$

(iv) This PBIBD can be constructed by applying the method of Exercise 167. Consider the EG(2,3). The points and 1-flats in EG (2, 3) are

$$\begin{aligned} \text{Points}: \; & p_1 = (0\,0), \quad p_2 = (0\,1), \quad p_3 = (0\,2), \\ & p_4 = (1\,0), \quad p_5 = (1\,1), \quad p_6 = (1\,2), \\ & p_7 = (2\,0), \quad p_8 = (2\,1), \quad p_9 = (2\,2). \end{aligned}$$

$$\begin{aligned} \text{1-flats}: \; & (p_1\; p_2\; p_3), \quad (p_4\; p_5\; p_6), \quad (p_7\; p_8\; p_9), \\ & (p_1\; p_4\; p_7), \quad (p_2\; p_5\; p_8), \quad (p_3\; p_6\; p_9), \\ & (p_1\; p_5\; p_9), \quad (p_2\; p_6\; p_9), \quad (p_3\; p_4\; p_8), \\ & (p_1\; p_6\; p_8), \quad (p_2\; p_4\; p_9), \quad (p_3\; p_5\; p_7). \end{aligned}$$

Omit the point p_1 and take the remaining points as treatments. Thus, we get $v = 8$ treatments as $p_2, p_3, p_4, p_5, p_6, p_7, p_8, p_9$. Omit the 1-flats passing through p_1 and take the remaining 1-flats as blocks. Thus, we get $b = 8$ blocks as

$$\begin{aligned} & (p_4\; p_5\; p_6), \quad (p_2\; p_5\; p_8), \quad (p_2\; p_6\; p_9), \\ & (p_2\; p_4\; p_9), \quad (p_7\; p_8\; p_9), \quad (p_3\; p_6\; p_9), \\ & (p_3\; p_4\; p_8), \quad (p_3\; p_5\; p_7). \end{aligned}$$

Clearly $k = 3$ and $r = 3$.

Two treatments are first associates if they occur together in a block; they are second associates if they do not occur together in a block. Thus, the first and second associates of p_4 are p_5, p_6, p_3, p_8, p_9 and p_7. Hence $n_1 = 6$ and $n_2 = 1$. Clearly $\lambda_1 = 1$ and $\lambda_2 = 0$. Consider p_4 and p_5 which are first associates. The first associates of p_4 and p_5 are respectively $p_5, p_6, p_3, p_8, p_2, p_9$ and $p_4, p_6, p_2, p_8, p_3, p_7$. The common treatments between the first associates of p_4 and p_5 are p_6, p_8, p_3, p_2. Hence $p^1_{11} = 4$. The other values of p^i_{jk} can be similarly determined or from the properties of p^i_{jk}. Thus one can easily verify that

$$P_1 = \begin{bmatrix} 4 & 1 \\ 1 & 0 \end{bmatrix}, \; P_2 = \begin{bmatrix} 6 & 0 \\ 0 & 0 \end{bmatrix}.$$

(v) This PBIBD can be constructed by applying the method of Exercise 168. We arrange the 9 treatments in an array of 3 rows and

3 columns as

$$\begin{matrix} t_1 & t_2 & t_3 \\ t_4 & t_5 & t_6 \\ t_7 & t_8 & t_9 \end{matrix}$$

The blocks are formed by taking each treatment and treatments in the same row, the same column as that treatment. Thus we get 9 blocks as

$$\begin{matrix} t_1 & t_2 & t_3 & t_4 & t_7 & & t_6 & t_4 & t_5 & t_3 & t_9 \\ t_2 & t_1 & t_3 & t_5 & t_8 & & t_7 & t_8 & t_9 & t_1 & t_4 \\ t_3 & t_1 & t_2 & t_6 & t_9 & & t_8 & t_7 & t_9 & t_2 & t_5 \\ t_4 & t_5 & t_6 & t_1 & t_7 & & t_9 & t_7 & t_8 & t_3 & t_6 \\ t_5 & t_4 & t_6 & t_2 & t_8. & & & & & & \end{matrix}$$

Clearly $k = 5$ and $r = 5$.

Two treatments are first associates if they are in the same row; they are second associates if they are in the same column; otherwise they are third associates. Thus, the first, the second and the third associates of t_1 are respectively

$$t_2, t_3; \qquad t_4, t_7; \qquad t_5, t_6, t_8, t_9.$$

Hence $n_1 = n_2 = 2$ and $n_3 = 4$.

Consider t_1 and t_2 which are first associates. They occur together in 3 blocks. Hence $\lambda_1 = 3$. Consider treatments t_1 and t_4 which are second associates. They occur together in 3 blocks, hence $\lambda_2 = 3$. Consider treatments t_1 and t_5 which are third associates. They occur together in 2 blocks. Hence $\lambda_3 = 2$. Now, consider

Treatment	**First Associate**	**Second Associate**	**Third Associate**
t_1	t_2, t_3	t_4, t_7	t_5, t_6, t_8, t_9
t_2	t_1, t_3	t_5, t_8	t_4, t_6, t_7, t_9
t_4	t_5, t_6	t_1, t_7	t_2, t_3, t_8, t_9
t_5	t_4, t_6	t_2, t_8	t_1, t_3, t_7, t_9

Here, t_1 and t_2, are first associates, hence by finding the common treatments between their different associates, we find

$$p_1 = \begin{bmatrix} 1 & 0 & 0 \\ 0 & 0 & 2 \\ 0 & 2 & 2 \end{bmatrix}.$$

Next, t_1 and t_4 are second associates hence, by finding the common treatments between their different associates, we obtain P_2 as

$$p_2 = \begin{bmatrix} 0 & 0 & 2 \\ 0 & 1 & 0 \\ 2 & 0 & 2 \end{bmatrix}.$$

Lastly, consider t_1 and t_5 which are third associates. Hence, by finding the common treatments between their different associates, we obtain p_3 as

$$p_3 = \begin{bmatrix} 0 & 1 & 1 \\ 1 & 0 & 1 \\ 1 & 1 & 1 \end{bmatrix}.$$

(vi) This PBIBD can be constructed by applying the method of Exericse 169. We arrange the $v = 9$ treatments in an array of 3 rows and 3 columns as under

$$\begin{matrix} t_1 & t_2 & t_3 \\ t_4 & t_5 & t_6 \\ t_7 & t_8 & t_9 \end{matrix}$$

The blocks are formed by taking treatments in the same row and column as each treatment, but excluding that treatment. Thus, we obtain 9 blocks as under

$(t_2\ t_3\ t_4\ t_7)$, $(t_1\ t_3\ t_5\ t_8)$
$(t_1\ t_2\ t_6\ t_9)$, $(t_5\ t_6\ t_1\ t_7)$
$(t_4\ t_6\ t_2\ t_8)$, $(t_4\ t_5\ t_3\ t_9)$
$(t_8\ t_9\ t_1\ t_4)$, $(t_7\ t_9\ t_2\ t_5)$
$(t_7\ t_8\ t_3\ t_6)$.

Clearly $k = 4$ and $r = 4$. Two treatments are first associates if they occur in the same row; they are second associates if they occur in the same column, otherwise they are third associates. Thus, we can easily get the following:

Treatment	**First Associate**	**Second Associate**	**Third Associate**
t_1	$t_2\ t_3$	$t_4\ t_7$	$t_5\ t_6\ t_8\ t_9$
t_2	$t_1\ t_3$	$t_5\ t_8$	$t_4\ t_6\ t_7\ t_9$
t_4	$t_5\ t_6$	$t_1\ t_7$	$t_2\ t_3\ t_8\ t_9$
t_5	$t_4\ t_6$	$t_2\ t_8$	$t_1\ t_3\ t_7\ t_9$

Therefore, we get $n_1 = n_2 = 2$ and $n_3 = 4$. Since t_1 and t_2 occur together only in one block, $\lambda_1 = 1$. Now t_1 and t_4 occur togehter only in one block, hence $\lambda_2 = 1$. Moreover t_1 and t_5 occur together in 2 blocks, and hence $\lambda_3 = 2$. The above table of treatments and their associates is the same as found in (v), and hence we find the P-matrices same as found in (v).

(vii) This PBIBD can be constructed by applying the method of Exericse 170. We arrange the $v = 9$ treatments in an array of 3 rows and 3 columns as under

t_1	t_2	t_3
t_4	t_5	t_6
t_7	t_8	t_9

We also take a Latin square of side 3 as

A	B	C
B	C	A
C	A	B

The blocks are formed by taking treatments in the same row, the same column and those which correspond to the same letter of a 3×3 Latin square as each treatment. Thus, we obtain 9 blocks as under

$(t_1\ t_2\ t_3\ t_4\ t_7\ t_8\ t_9)$, $(t_6\ t_4\ t_5\ t_3\ t_9\ t_1\ t_8)$,
$(t_2\ t_1\ t_3\ t_5\ t_8\ t_4\ t_9)$, $(t_7\ t_8\ t_9\ t_1\ t_4\ t_3\ t_5)$,
$(t_3\ t_1\ t_2\ t_6\ t_9\ t_5\ t_7)$, $(t_8\ t_7\ t_9\ t_2\ t_5\ t_1\ t_6)$,
$(t_4\ t_5\ t_6\ t_1\ t_7\ t_2\ t_9)$, $(t_9\ t_7\ t_8\ t_3\ t_6\ t_2\ t_4)$,
$(t_5\ t_4\ t_6\ t_2\ t_8\ t_3\ t_7)$.

Clearly $k = 7$ and $r = 7$. Two treatments are first associates if they occur in the same row; they are second associates if they occur in the same column; they are third associates if they correspond to the same letter of the 3×3 Latin square; otherwise they are fourth associates. Thus one can easily verify the following table.

Treatment	**First Associate**	**Second Associate**	**Third Associate**	**Fourth Associate**
t_1	$t_2\ t_3$	$t_4\ t_7$	$t_6\ t_8$	$t_5\ t_9$
t_2	$t_1\ t_3$	$t_5\ t_8$	$t_4\ t_9$	$t_6\ t_7$
t_4	$t_5\ t_6$	$t_1\ t_7$	$t_2\ t_9$	$t_3\ t_8$
t_6	$t_4\ t_5$	$t_3\ t_9$	$t_1\ t_8$	$t_2\ t_7$
t_5	$t_4\ t_6$	$t_2\ t_8$	$t_3\ t_7$	$t_1\ t_9$

By considering t_1 and t_2 which are first associates and by finding the common treatments between their different associates, we find P_1 as

$$P_1 = \begin{bmatrix} 1 & 0 & 0 & 0 \\ 0 & 0 & 1 & 1 \\ 0 & 1 & 0 & 1 \\ 0 & 1 & 1 & 0 \end{bmatrix}.$$

Consider t_1 and t_4 which are second associates. By finding the common treatments between their different associates we find P_2 as

$$P_2 = \begin{bmatrix} 0 & 0 & 1 & 1 \\ 0 & 1 & 0 & 0 \\ 1 & 0 & 0 & 1 \\ 1 & 0 & 1 & 0 \end{bmatrix}.$$

Next consider t_1 and t_6 which are third associates. By finding the common treatments between their different associates, we find P_3 as

$$P_3 = \begin{bmatrix} 0 & 1 & 0 & 1 \\ 1 & 0 & 0 & 1 \\ 0 & 0 & 1 & 0 \\ 1 & 1 & 0 & 0 \end{bmatrix}.$$

Lastly, consider t_1 and t_5 which are fourth associates. By finding the common treatments between their different associates, we find P_4 as

$$P_4 = \begin{bmatrix} 0 & 1 & 1 & 0 \\ 1 & 0 & 1 & 0 \\ 1 & 1 & 0 & 0 \\ 0 & 0 & 0 & 1 \end{bmatrix}.$$

172. The key block B_1 is constructed by taking all treatments having even number of letters in common with ABC, BCE, and ABDE. The treatments in the key block B_1 are: (1), ace, bcd, abde. The block B_i $(i = 2, 3, \ldots, 7)$ is constructed by taking treatments obtained by multiplying a treatment not included in the previous blocks $B_1, B_2, \ldots, B_{i-1}$ with the treatments of the block B_1 and replacing the square of any letter by unity. Thus the 8 blocks are obtained as

Block	Treatments	Block	Treatments
1	(1), ace, bcd, abde	5	d, acde, be, abe
2	a, ce, abcd, bde	6	e, ac, bcde, abd
3	b, abce, cd, ade	7	ab, bce, acd, de
4	c, ae, bd, abcde	8	ad, cde, abc, be

The other interactions that are confounded are:

$$(ABC)(BCE) = AE$$
$$(ABC)(ABDE) = CDE$$
$$(BCE)(ABDE) = ACD$$
$$(ABC)(BCE)(ABDE) = BD.$$

173. The pencil corresponding to the interaction AB^2C^2 is P(1 2 2). The equations of P(1 2 2) are

$$x_1 + 2x_2 + 2x_3 = 0, 1, 2.$$

We construct the blocks by taking treatments satisfying each equation. Consider the equation

$$x_1 + 2x_2 + 2x_3 = 0.$$

We can write this equation as

$$x_1 = x_2 + x_3.$$

Thus, the solutions of this equation are obtained by writing down all possible combinations of x_2 and x_3 and then taking x_1 to be their sum. Thus, the 9 treatments satisfying the above equation and forming the first block are

$$\begin{array}{llll} \text{Block 1:} & (0\ \ 0\ \ 0), & (1\ \ 0\ \ 1), & (2\ \ 0\ \ 2), \\ & (1\ \ 1\ \ 0), & (2\ \ 1\ \ 1), & (0\ \ 1\ \ 2), \\ & (2\ \ 2\ \ 0), & (0\ \ 2\ \ 1), & (1\ \ 2\ \ 2). \end{array}$$

Consider the second equation

$$x_1 + 2x_2 + 2x_3 = 1$$

which can be written as

$$x_1 = x_2 + x_3 + 1.$$

We write all combinations of x_2 and x_3 and take x_1 to be their sum plus 1. Thus, we get the second block as

$$\begin{array}{llll} \text{Block 2:} & (1\ \ 0\ \ 0), & (2\ \ 0\ \ 1), & (0\ \ 0\ \ 2), \\ & (2\ \ 1\ \ 0), & (0\ \ 1\ \ 1), & (1\ \ 1\ \ 2), \\ & (0\ \ 2\ \ 0), & (1\ \ 2\ \ 1), & (2\ \ 2\ \ 2). \end{array}$$

Solving the third equation as $x_1 = x_2 + x_3 + 2$, we obtain the third block as

$$\begin{array}{llll} \text{Block 3:} & (2\ \ 0\ \ 0), & (0\ \ 0\ \ 1), & (1\ \ 0\ \ 2), \\ & (0\ \ 1\ \ 0), & (1\ \ 1\ \ 1), & (2\ \ 1\ \ 2), \\ & (1\ \ 2\ \ 0), & (2\ \ 2\ \ 1), & (0\ \ 2\ \ 2). \end{array}$$

It may be noted that the blocks 2 and 3 are obtained from block 1 by adding 1 and 2 successively to any coordinate of the treatments of block 1. The order of blocks is immaterial.

174. The pencils corresponding to the interactions AB and BC are respectively P(1 1 0) and P(0 1 1). The blocks are obtained by solving pairs of equations one belonging to P(1 1 0) and the other to P(0 1 1). The key block is obtained by solving the equations

$$x_1 + x_2 = 0, \qquad x_2 + x_3 = 0,$$

which give $x_1 = x_3$ and $x_2 = 2x_3$. Taking $x_3 = 0, 1, 2$, we obtain the solutions as (0 0 0), (1 2 1), (2 1 2). Thus the key block is

$$\text{Block 1:} \quad (0\ \ 0\ \ 0), \quad (1\ \ 2\ \ 1), \quad (2\ \ 1\ \ 2).$$

The other blocks can be similarly obtained by solving pairs of equations

$$x_1 + x_2 = i \text{ and } x_2 + x_3 = j, \qquad i, j = 0, 1, 2 \text{ excluding } i = j = 0.$$

However we shall give a short-cut method of writing down other blocks from the key block. We select the first two coordinates in all possible ways as

(0 0), (0 1), (0 2), (1 0), (1 1), (1 2), (2 0), (2 1), (2 2).

We omit the combination (0 0). Add these combinations to the first two coordinates of the treatments of the key block in succession. We then get the other 8 blocks. The key block and the other 8 blocks so obtained are given below.

Block	Treatments	Block	Treatments
1	(0 0 0), (1 2 1), (2 1 2)	6	(1 2 0), (2 1 1), (0 0 2)
2	(0 1 0), (1 0 1), (2 2 2)	7	(2 0 0), (0 2 1), (1 1 2)
3	(0 2 0), (1 1 1), (2 0 2)	8	(2 1 0), (0 0 1), (1 2 2)
4	(1 0 0), (2 2 1), (0 1 2)	9	(2 2 0), (0 1 1), (1 0 2)
5	(1 1 0), (2 0 1), (0 2 2)		

175. The pencils belonging to the 2-factor interactions in a 3^2 design are P(1 1) and P(1 2). Then, we get two replications, confounding P(1 1) in the first replication and P(1 2) in the second replication. The key block in the first replication in which P(1 1) is confounded is obtained by solving

$$x_1 + x_2 = 0, \text{ i.e. } x_2 = 2x_1$$

and is obtained as (0 0), (1 2), (2 1). By adding 1 and 2 to the first coordinates of these treatments, we get the other blocks of the first replication.

The key block in the second replication in which P(1 2) is confounded is obtained by solving

$$x_1 + 2x_2 = 0, \text{ i.e. } x_1 = x_2$$

and is obtained as (0 0), (1 1), (2 2). By adding 1 and 2 to the first coordinates of these treatments, we obtain the other blocks of the second replication.

The replications I and II are given below.

Replication I	Replication II
(0 0), (1 2), (2 1)	(0 0), (1 1), (2 2)
(1 0), (2 2), (0 1)	(1 0), (2 1), (0 2)
(2 0), (0 2), (1 1)	(2 0), (0 1), (1 2).

Interaction P(1 1) is confounded in Replication I and is unconfounded in Replication II. Thus, there is a loss of $\frac{1}{2}$ on each d.f. belonging to P(1 1). Similar remark applies to the interaction P(1 2).

176. We select 4 vertices $(\alpha_1, \alpha_2, \alpha_3)$ where each α is a non-null element of GF(3) as

$$(1 \quad 1 \quad 1), (1 \quad 1 \quad 2), (1 \quad 2 \quad 1), (1 \quad 2 \quad 2).$$

Each replication is constructed by taking one vertex and two pencils orthogonal to that vertex. The vertices and the corresponding orthogonal pencils for different replications are given below.

Replication	Vertex	Pencils
1	(1 1 1)	P(1 2 0), P(1 0 2)
2	(1 1 2)	P(1 2 0), P(0 1 1)
3	(1 2 1)	P(1 1 0), P(0 1 1)
4	(1 2 2)	P(1 1 0), P(1 0 1)

The key block in Replication 1 is obtained by solving the equations

$$x_1 + 2x_2 = 0, \qquad x_1 + 2x_3 = 0,$$

that is, $x_1 = x_2 = x_3$. Similarly we find the key blocks in other replications. The key blocks in the 4 replications are given below.

Replication	Key Block
1	(0 0 0), (1 1 1), (2 2 2)
2	(0 0 0), (1 1 2), (2 2 1)
3	(0 0 0), (1 2 1), (2 1 2)
4	(0 0 0), (1 2 2), (2 1 1)

The other blocks in each replication are obtained from the key blocks by the short-cut method described in Exercise 174, that is, by adding

$$(0 \ 1), \quad (0 \ 2), \quad (1 \ 0), \quad (1 \ 1), \quad (1 \ 2), \quad (2 \ 0), \quad (2 \ 1), \quad (2 \ 2)$$

to the first coordinates of the treatments of the key blocks. The blocks in different replications are given below.

Replication 1	Replication 2
(0 0 0), (1 1 1), (2 2 2)	(0 0 0), (1 1 2), (2 2 1)
(0 1 0), (1 2 1), (2 0 2)	(0 1 0), (1 2 2), (2 0 1)
(0 2 0), (1 0 1), (2 1 2)	(0 2 0), (1 0 2), (2 1 1)
(1 0 0), (2 1 1), (0 2 2)	(1 0 0), (2 1 2), (0 2 1)
(1 1 0), (2 2 1), (0 0 2)	(1 1 0), (2 2 2), (0 0 1)
(1 2 0), (2 0 1), (0 1 2)	(1 2 0), (2 0 2), (0 1 1)
(2 0 0), (0 1 1), (1 2 2)	(2 0 0), (0 1 2), (1 2 1)
(2 1 0), (0 2 1), (1 0 2)	(2 1 0), (0 2 2), (1 0 1)
(2 2 0), (0 0 1), (1 1 2)	(2 2 0), (0 0 2), (1 1 1)

Replication 3

(0 0 0),	(1 2 1),	(2 1 2)
(0 1 0),	(1 0 1),	(2 2 2)
(0 2 0),	(1 1 1),	(2 0 2)
(1 0 0),	(2 2 1),	(0 1 2)
(1 1 0),	(2 0 1),	(0 2 2)
(1 2 0),	(2 1 1),	(0 0 2)
(2 0 0),	(0 2 1),	(1 1 2)
(2 1 0),	(0 0 1),	(1 2 2)
(2 2 0),	(0 1 1),	(1 0 2)

Replication 4

(0 0 0),	(1 2 2),	(2 1 1)
(0 1 0),	(1 0 2),	(2 2 1)
(0 2 0),	(1 1 2),	(2 0 1)
(1 0 0),	(2 2 2),	(0 1 1)
(1 1 0),	(2 0 2),	(0 2 1)
(1 2 0),	(2 1 2),	(0 0 1)
(2 0 0),	(0 2 2),	(1 1 1)
(2 1 0),	(0 0 2),	(1 2 1)
(2 2 0),	(0 1 2),	(1 0 1)

The interactions confounded in each replication are given below.

Replication	Confounded Interactions
1	AB^2, AC^2 ABC, BC^2
2	AB^2, BC, AC, ABC^2
3	AB, BC, AB^2C, AC^2
4	AB, AC, AB^2C^2, BC^2

The first order interactions are AB, AB^2, AC, AC^2, BC, BC^2. From the above table, we see that each first order interaction is confounded in 2 replications and is unconfounded in the other two replications. Thus, there is uniform loss of information of $\frac{1}{2}$ on each d.f. belonging to the first order interaction. The second order interactions are ABC, AB^2C, ABC^2, AB^2C^2. From the above table, we see that each second order interaction is confounded in one replication and is unconfounded in the other three replications. Hence, there is a uniform loss of information of $\frac{1}{4}$ on each d.f. belonging to the second order interaction.

177. ABC and ACDE are defining interactions. Hence, their generalized interaction BDE is also defining interaction. We write down the $\frac{1}{4}$th replicate by taking treatments that have odd number of letters in common with BDE and even number letters in common with ACDE. Thus the $\frac{1}{4}$th replicate of a 2^5 design consists of the following treatments:

b, ad, cd, ae, ce, bde, abc, abcde.

Note that if we suppress letters a and b in the above set of 8 treatments, we obtain the 8 treatements in a 2^3 design with factors C, D, E. The 7 alias sets correspond to the 7 factorial effects of this 2^3 design and are obtained as below.

$$C = AB = ADE = BCDE$$

$$D = ABCD = ACE = BE$$

$$CD = ABD = AE = BCE$$

$$E = ABCE = ACD = BD$$

$$CE = ABE = AD = BCD$$
$$DE = ABCDE = AC = B$$
$$CDE = ABDE = A = BC$$

178. The defining interaction ABCD contains even number of letters, hence the half replicate of a 2^4 design will consist of treatments which have even number of letters in common with ABCD. Thus, the half replicate consists of the following 8 treatments

(1), ab, ac, bc, ad, bd, cd, abcd.

If we suppress letter d, then the treatments become

(1), (d)a, (d)b, ab, (d)c, ac, bc, abc(d)

which are the treatments of a 2^3 design with factors A, B, C. Now we want to confound AD = BC. The first block is written down by taking treatments that have even number of letters in common with BC. The second block is written down by taking treatments obtained by multiplying a treatment not included in the first block with the treatments of the first block and replacing the square of any letter by unit. Thus the two blocks are:

Block 1 : (1), a(d), bc, abc(d)
Block 2 : (d)b, ab, (d)c, ac.

We now introduce the suppressed letter, so that the blocks now become

Block 1 : (1), ad, bc, abcd
Block 2 : bd, ab, cd, ac.

179. Here ABCDE contains odd number of letters; hence we select in the half replicate only those treatments which have odd number of letters in common with ABCDE. Further, we suppress the letter e and write down these 16 treatments as 16 treatments of a 2^4 design as under

(1)(e), a, b, ab(e), c, ac(e), bc(e), abc, d, ad(e), bd(e),
abd, cd(e), acd, bcd, abcd(e).

We wish to confound the interactions BE and CDE. Now we write these interactions so that they contain only letters from A, B, C and D. Thus, BE = ACD and CDE = AB. So we confound ACD and AB. We write the first block B_1 by writing the treatments having even number of letters in common with ACD and AB. Then, the block B_i, $i = 2, 3, 4$ is constructed by taking a treatment not included in blocks $B_1, B_2, \dots, B_{i-1}$ and multiplying it with the treatments of the block B_1 and replacing the square of any letter by unit. Thus, the 4 blocks are

Block B_1 : (1)(c), abc, abd, cd(e)
Block B_2 : a, bc(e), bd(e), acd

Block B_3 : b, ac(e), ad(e), bcd

Block B_4 : c, ab(e), abcd(e), d.

The final blocks are written down after removing the brackets enclosing the letter e.

Note that the generalized interaction of BE and CDE, namely BCD is also confounded, and their aliases are confounded. Thus, the alias sets BE = ACD, CDE = AB, BCD = AE are confounded.

180. The defining interactions ABC and ADE contain odd number of letters and hence the $(\frac{1}{4})$th replicate will consist of treatments which have odd number of letters in common with ABC and ADE. Thus, the $(\frac{1}{4})$th replicate consists of the following treatments in which we have suppressed two letters a and b, so that they become treatments of a 2^4 design with factors C, D, E and F.

(a)(1), (ab)c, (b)d, cd, (b)e, ce, (a)de, (ab)cde, (a)f,

(ab)cf, (b)df, cef, (a)def, (ab)cdef.

The generalized interaction of ABC and ADE, that is, BCDE is also the defining contrast.

The generalized interaction of ACE and ACDF, that is, DEF and their aliases will also be confounded. Thus, the following alias sets will be confounded.

ACE = BE = CD = ABD

ACDF = BDF = CEF = ABEF

DEF = ABCDEF = AF = BCF.

Now ACE and ACDF contain the suppressed letters; hence for confounding we shall take their such aliases which do not contain the suppressed letters. Thus, we select CD in place of ACE and CEF in place of ACDF. Thus, we confounded interactions CD and CEF. We construct block B_1 by writing the treatments having even number of letters in common with CD and CEF. The block B_i, i = 2, 3, 4 is constructed by taking a treatment not included in blocks $B_1, B_2, \ldots, B_{i-1}$ and multiplying it with the treatments of the block B_1. Thus, the 4 blocks are found as

Block B_1: (a)(1), (b)ef, cdf, (ab)cde

Block B_2: (ab)c, cef, (b)df, (a)de

Block B_3: (b)d, (a)def, (ab)cf, ce

Block B_4: (a)f, (b)e, cd, (ab)cdef.

The final blocks are written down after removing the brackets which enclose letters a and b.

181. The $(\frac{1}{3})$rd replicate of a 3^3 design, when the pencil P(1 1 1) is used as a defining pencil will consist of treatments which satisfy

$$x_1 + x_2 + x_3 = 0$$

and are given by

$$(0\ \ 0\ \ 0),\quad (0\ \ 1\ \ 2),\quad (0\ \ 2\ \ 1),\quad (1\ \ 0\ \ 2),$$
$$(2\ \ 0\ \ 1),\quad (1\ \ 2\ \ 0),\quad (2\ \ 1\ \ 0),\quad (1\ \ 1\ \ 1),$$
$$(2\ \ 2\ \ 2).$$

In a 3^3design, there are 13 indepedent pencils, one of which is used for constructing the $\left(\frac{1}{3}\right)$rd replicate. The remaining 12 will be divided into 4 alias sets of 3 pencils each. The alias set corresponding to a pencil $P(b_1\ b_2\ b_3)$ will consist of pencils

$$P(b_1 + a_1, b_2 + a_2, b_3 + a_3)$$

where $\lambda = 0, 1, 2$. Hence the following 4 alias sets:

$$P(1\ \ 0\ \ 0) = P(1\ \ 2\ \ 2) = P(0\ \ 1\ \ 1)$$
$$P(0\ \ 1\ \ 0) = P(1\ \ 2\ \ 1) = P(1\ \ 0\ \ 1)$$
$$P(0\ \ 0\ \ 1) = P(1\ \ 1\ \ 2) = P(1\ \ 1\ \ 0)$$
$$P(1\ \ 2\ \ 0) = P(1\ \ 0\ \ 2) = P(0\ \ 1\ \ 2).$$

182. The treatments of the $\left(\frac{1}{3}\right)$rd replicate of a 3^5 design, when $P(1\ \ 1\ \ 1\ \ 1\ \ 1)$ and $P(1\ \ 0\ \ 0\ \ 1\ \ 1)$ are used as the defining pencils are obtained by solving the equations

$$x_1 + x_2 + x_3 + x_4 + x_5 = 0$$
$$x_1 + x_4 + x_5 = 0.$$

The above equations can be reduced to

$$x_1 = 2x_4 + 2x_5$$
$$x_2 = 2x_3.$$

We allow x_3, x_4 and x_5 to take all possible values and determine x_1 and x_2 by the above two equations. Thus, we obtain the following treatments for the $\left(\frac{1}{9}\right)$th replicate.

(0 0 0 0 0)	(0 2 1 0 0)	(0 1 2 0 0)
(2 0 0 0 1)	(2 2 1 0 1)	(2 1 2 0 1)
(1 0 0 0 2)	(1 2 1 0 2)	(1 1 2 0 2)
(2 0 0 1 0)	(2 2 1 1 0)	(2 1 2 1 0)
(1 0 0 1 1)	(1 2 1 1 1)	(1 1 2 1 1)
(0 0 0 1 2)	(0 2 1 1 2)	(0 1 2 1 2)
(1 0 0 2 0)	(1 2 1 2 0)	(1 1 2 2 0)
(0 0 0 2 1)	(0 2 1 2 1)	(0 1 2 2 1)
(2 0 0 2 2)	(2 2 1 2 2)	(2 1 2 2 2)

We lose information on the pencils

$$P(\lambda_1 + \lambda_2, \lambda_1, \lambda_1, \lambda_1 + \lambda_2, \lambda_1 + \lambda_2),$$

where λ_1, λ_2 are elements of GF(3) not all zero. Taking independent combinations of (λ_1, λ_2) as (1,0), (0,1), (1,1), (1,2), we obtain the following 4 pencils on which information is lost.

$$P(1\ \ 1\ \ 1\ \ 1\ \ 1),\qquad P(1\ \ 0\ \ 0\ \ 1\ \ 1),$$
$$P(2\ \ 1\ \ 1\ \ 2\ \ 2) = P(1\ \ 2\ \ 2\ \ 1\ \ 1),$$
$$P(0\ \ 1\ \ 1\ \ 0\ \ 0).$$

The aliases of any pencil $P(b_1b_2b_3b_4b_5)$ are the pencils

$$P(b_1 + \lambda_1 + \lambda_2, b_2 + \lambda_1, b_2 + \lambda_1, b_4 + \lambda_1 + \lambda_2, b_5 + \lambda_1 + \lambda_2)$$

where λ_1 and λ_2 are elements of GF(3), not all zero. Taking $P(b_1b_2b_3b_4b_5) = P(11101)$ and $(\lambda_1, \lambda_2) = (1, 0), (1, 1), (1, 2), (2, 0), (2, 1), (2, 2), (0, 1), (0, 2)$, we obtain the following aliases of P(1 1 1 0 1):

$$P(1\ \ 1\ \ 1\ \ 2\ \ 1)\qquad P(0\ \ 1\ \ 1\ \ 1\ \ 0),$$
$$P(1\ \ 2\ \ 2\ \ 0\ \ 1)\qquad P(0\ \ 0\ \ 0\ \ 1\ \ 0),$$
$$P(1\ \ 0\ \ 0\ \ 0\ \ 1)\qquad P(1\ \ 0\ \ 0\ \ 2\ \ 1),$$
$$P(1\ \ 2\ \ 2\ \ 2\ \ 1)\qquad P(0\ \ 1\ \ 1\ \ 2\ \ 0).$$

183. The blocks are constructed by taking treatments which satisfy the equations

$$x_1 + x_2 + x_3 + 2x_4 = 0$$
$$x_1 + x_3 + x_4 = \alpha,$$

where α can be 0, 1 or 2. The above equations can be written as

$$x_1 = 2x_3 + 2x_4 + \alpha$$
$$x_2 = 2x_4 + 2\alpha.$$

We allow x_3 and x_4 to take all possible values of GF(3) and determine x_1 and x_2 by the above equations. Taking $\alpha = 0, 1, 2$, we get 3 blocks as follows.

Block 1:	(0 0 0 0),	(2 2 0 1),	(1 1 0 2),
	(2 0 1 0),	(1 2 1 1),	(0 1 1 2),
	(1 0 2 0),	(0 2 2 1),	(2 1 2 2).
Block 2:	(1 2 0 0),	(0 1 0 1),	(2 0 0 2),
	(0 2 1 0),	(2 1 1 1),	(1 0 1 2),
	(2 2 2 0),	(1 1 2 1),	(0 0 2 2).
Block 3:	(2 1 0 0),	(1 0 0 1),	(0 2 0 2),
	(1 1 1 0),	(0 0 1 1),	(2 2 1 2),
	(0 1 2 0),	(2 0 2 1),	(1 2 2 2).

The aliases of P(1 0 1 1) are also confounded. The aliases of P(1 0 1 1) are the pencils

$$P(1 + \lambda, \lambda, 1 + \lambda, 1 + 2\lambda)$$

where $\lambda = 1$ or 2. Hence, the aliases of P(1 0 1 1) are P(2 1 2 0) = P(1 2 1 0) and P(0 2 0 2) = P(0 1 0 1). Hence, the alias set

$$P(1\quad 0\quad 1\quad 1) = P(1\quad 2\quad 1\quad 0) = P(0\quad 1\quad 0\quad 1)$$

is confounded.

184 The blocks are obtained by taking treatmenets which satisfy the equations

$$x_1 + x_2 + x_3 = 0$$
$$x_1 + x_3 + x_4 = \alpha$$
$$x_1 + x_2 + x_4 = \beta$$

where α and β are 0, 1 or 2. Taking all possible combinations of α and β, we shall get 9 blocks. For any particular combination of α and β, the above equations can be reduced to

$$x_1 = x_4 + \alpha + \beta$$
$$x_2 = x_4 + 2\alpha$$
$$x_3 = x_4 + 2\beta.$$

We allow x_4 to take the values 0, 1, 2 and determine x_1, x_2 and x_3 by the above equations. Taking $(\alpha, \beta) = (0, 0), (0, 1), (0, 2), (1, 0), (1, 1), (1, 2), (2, 0), (2, 1)$ and (2,2), we get 9 blocks as

Block 1	**Block 2**	**Block 3**
(0 0 0 0)	(1 0 2 0)	(2 0 1 0)
(1 1 1 1)	(2 1 0 1)	(0 1 2 1)
(2 2 2 2)	(0 2 1 2)	(1 2 0 2)

Block 4	**Block 5**	**Block 6**
(1 2 0 0)	(2 2 2 0)	(0 2 1 0)
(2 0 1 1)	(0 0 0 1)	(1 0 2 1)
(0 1 2 2)	(1 1 1 2)	(2 1 0 2)

Block 7	**Block 8**	**Block 9**
(2 1 0 0)	(0 1 2 0)	(1 1 1 0)
(0 2 1 1)	(1 2 0 1)	(2 2 2 1)
(1 0 2 2)	(2 0 1 2)	(0 0 0 2)

The generalized pencils of P(1 0 1 1) and P(1 1 0 1) and their aliases are confounded. The generalized pencils of P(1 0 1 1) and P(1 1 0 1) are

$$P(\lambda_1 + \lambda_2, \lambda_2, \lambda_1, \lambda_1 + \lambda_2)$$

where $(\lambda_1, \lambda_2) = (1, 0), (1, 1), (1, 2), (0, 1)$. Hence the pencils P(1 0 1 1), P(1 1 0 1), P(1 2 2 1) and P(0 1 2 0) are confounded.

The alias sets of these pencils are also confounded. Thus, the following alias sets are confounded:

$$P(1\ 0\ 1\ 1) = P(1\ 2\ 1\ 2) = P(0\ 1\ 0\ 2)$$
$$P(1\ 1\ 0\ 1) = P(1\ 1\ 2\ 2) = P(0\ 0\ 1\ 2)$$
$$P(1\ 2\ 2\ 1) = P(1\ 0\ 0\ 2) = P(0\ 1\ 1\ 1)$$
$$P(0\ 1\ 2\ 0) = P(1\ 2\ 0\ 0) = P(1\ 0\ 2\ 0).$$

BIBLIOGRAPHY

Anderson, V. L. and McLean, Robert (1974). *Design of Experiments, A Realistic Approach*, Dekker, New York.

Cochran, W. G. and Cox, G. M. (1992). *Experimental Designs*, 2nd ed., Wiley, New York.

Hinkelmann, K. and Kempthorne, O. (1994). *Design and Analysis of Experiments, Vol. I*, Wiley, New York.

John, P. M. W. (1998). *Statistical Design and Analysis of Experiments*, SIAM, Philadelphia.

Montgomery, D. C. (2005). *Design and Analysis of Experiments*, 6th ed., Wiley, New York.

Pukelsheim, F. (1993). *Optimal Design of Experiments*, Wiley, New York.

Winer, B. J. (1971). *Statistical Principles in Experimental Design*, 2nd ed., McGraw-Hill, New York.

INDEX

9 780387 338927